T0140045

Studies in Computational Intelligence

Volume 1100

The series "Studies in Computational Intelligence" (SCI) publishes new developments and advances in the various areas of computational intelligence—quickly and with a high quality. The intent is to cover the theory, applications, and design methods of computational intelligence, as embedded in the fields of engineering, computer science, physics and life sciences, as well as the methodologies behind them. The series contains monographs, lecture notes and edited volumes in computational intelligence spanning the areas of neural networks, connectionist systems, genetic algorithms, evolutionary computation, artificial intelligence, cellular automata, self-organizing systems, soft computing, fuzzy systems, and hybrid intelligent systems. Of particular value to both the contributors and the readership are the short publication timeframe and the world-wide distribution, which enable both wide and rapid dissemination of research output.

Indexed by SCOPUS, DBLP, WTI Frankfurt eG, zbMATH, SCImago.

All books published in the series are submitted for consideration in Web of Science.

Witold Pedrycz · Shyi-Ming Chen
Editors

Advancements in Knowledge Distillation: Towards New Horizons of Intelligent Systems

Editors
Witold Pedrycz
Department of Electrical and Computer Engineering
University of Alberta
Edmonton, AB, Canada

Shyi-Ming Chen
Department of Computer Science and Information Engineering
National Taiwan University of Science and Technology
Taipei, Taiwan

ISSN 1860-949X ISSN 1860-9503 (electronic)
Studies in Computational Intelligence
ISBN 978-3-031-32097-2 ISBN 978-3-031-32095-8 (eBook)
https://doi.org/10.1007/978-3-031-32095-8

This Springer imprint is published by the registered company Springer Nature Switzerland AG
The registered company address is: Gewerbestrasse 11, 6330 Cham, Switzerland

Preface

The visible, important, and challenging quest in machine learning is about deploying advanced and complex deep models on devices with limited resources, e.g., mobile phones, autonomous sensors, and embedded devices. This is attributed to the high computational complexity and large storage requirements inherently associated with the increasingly complex machine learning methods.

To address these challenges, a variety of model compression and acceleration techniques have been developed. As a representative type of model compression and acceleration, knowledge distillation positioned in a general setting of transfer learning effectively learns a small and lightweight student model from a large teacher model. The problem becomes of particular relevance in commonly used convolutional neural networks (CNNs). There have been a number of knowledge distillation categories involved there, numerous training schemes, teacher-student architecture, and distillation algorithms.

The objective of this volume is to provide the reader with a comprehensive and up-to-date treatise of the area of knowledge distillation cast in a general framework of transfer learning by focusing a spectrum of methodological and algorithmic issues, discussing implementations and case studies, identifying the best design practices, assessing business models and practices of the methodology of this direction of machine learning as encountered nowadays in industry, health care, science, natural language processing, image understanding, administration, and business.

The volume brings forward a collection of timely, representative, and insightful contributions to the area of knowledge distillation covering a spectrum of timely and far reaching theoretical and application-driven implications.

In the chapter entitled "Categories of Response-Based, Feature-Based, and Relation-Based Knowledge Distillation", the authors offer a systematic overview of knowledge discovery schemes including knowledge categories, distillation schemes, and algorithms. They also elaborate on empirical studies focused on performance comparison and highlight directions of future research.

The next chapter entitled "A Geometric Perspective on Feature-Based Distillation" brings an idea of feature-based knowledge distillation, which concerns the use of the intermediate layers of a CNN in the process of knowledge distillation. A new family of efficient loss functions that aim to transfer the geometry of activations from intermediate layers of a teacher CNN to the activations of a student CNN model at matching spatial resolutions is proposed and investigated. The studies are augmented by a case study data-free knowledge distillation in the field of offline handwritten signature verification.

Recent years have witnessed the fast development of computer vision and natural language processing. They are the two profoundly visible domains of application of machine learning, in particular its subfield of deep learning. They are also very much computationally intensive, facing challenges of different type of data and various aspects of learning schemes. The authors in the "Knowledge Distillation Across Vision and Language" elaborate on the role of knowledge distillation in Vision and Language (VL) learning. They demonstrate how knowledge distillation can play a vital role in numerous disciplines of cross-modal tasks, including image/video captioning, visual-question answering, image/video retrieval, among others.

Fuzzy relational calculus and fuzzy relational equations have been studied for a long time as an interesting and useful alternative to represent knowledge. The essence of the proposed approach studied in the "Knowledge Distillation in Granular Fuzzy Models by Solving Fuzzy Relation Equations" is about the integration of the set of rules into a hierarchical distillation structure based on granular solutions to the system of fuzzy relational equations. The multi-task distillation scheme ensures transferring distributed knowledge into the granular student model.

The authors of "Ensemble Knowledge Distillation for Edge Intelligence in Medical Applications" investigate knowledge distillation approaches where an ensemble of "student" deep neural networks can be trained with regard to set of various facets of teacher family with the same architecture. The function and performance of such bucket of student-teacher models are explored and analyzed in case of medical datasets for Edge Intelligence devices with the limited computational abilities.

The problem of knowledge distillation for multi-task learning (MTL) is a focal point of study reported in the paper entitled "Self-Distillation with the New Paradigm in Multi-Task Learning". The new architecture is proposed leading to the enhancement of the soft and hard-sharing-based MTL models. In the sequel, it has been demonstrated the improved performance of such models when processing visual indoor and outdoor scenarios.

The key feature of advanced Unmanned Systems is their high level of autonomy which is critical to the success and performance required in a broad range of applications. The need to narrow a gap between the processing realized at the sensor level and decision-making commonly encountered at the level of domain knowledge is addressed by bringing the ideas of information. This point is advocated in the chapter carrying a title "Knowledge Distillation for Autonomous Intelligent Unmanned System". Information granules are instrumental in tasks of sense distilling aggregating distillation from data and semantics. A case study involving mechanisms of fuzzy systems is provided as a sound exemplification of the developed proposal.

We would like to take this opportunity and express thanks to the authors for their timely and impactful chapters making a tangible contribution to the body of knowledge of advanced Machine Learning. Professor Janusz Kacprzyk deserves our big thanks for his ongoing encouragement and support during the realization of the project. The editorial staff of Springer provided their professional experience and helped us to arrive at the fruition of this book. Last but not the least, we hope the readers will enjoy this volume when pursuing their research studies in the area of knowledge transfer and knowledge distillation.

Edmonton, Canada Witold Pedrycz
Taipei, Taiwan Shyi-Ming Chen

Contents

Categories of Response-Based, Feature-Based, and Relation-Based Knowledge Distillation

Chuanguang Yang, Xinqiang Yu, Zhulin An, and Yongjun Xu

Abstract Deep neural networks have achieved remarkable performance for artificial intelligence tasks. The success behind intelligent systems often relies on large-scale models with high computational complexity and storage costs. The over-parameterized networks are often easy to optimize and can achieve better performance. However, it is challenging to deploy them over resource-limited edge-devices. Knowledge Distillation (KD) aims to optimize a lightweight network from the perspective of over-parameterized training. The traditional offline KD transfers knowledge from a cumbersome teacher to a small and fast student network. When a sizeable pre-trained teacher network is unavailable, online KD can improve a group of models by collaborative or mutual learning. Without needing extra models, Self-KD boosts the network itself using attached auxiliary architectures. KD mainly involves knowledge extraction and distillation strategies these two aspects. Beyond KD schemes, various KD algorithms are widely used in practical applications, such as multi-teacher KD, cross-modal KD, attention-based KD, data-free KD and adversarial KD. This paper provides a comprehensive KD survey, including knowledge categories, distillation schemes and algorithms, as well as some empirical studies on performance comparison. Finally, we discuss the open challenges of existing KD works and prospect the future directions.

Keywords Knowledge distillation · Knowledge category · Distillation algorithms

---

C. Yang (✉) · X. Yu · Z. An · Y. Xu
Institute of Computing Technology, Chinese Academy of Sciences, Beijing, China
e-mail: yangchuanguang@ict.ac.cn

X. Yu
e-mail: yuxinqiang21s@ict.ac.cn

Z. An
e-mail: anzhulin@ict.ac.cn

Y. Xu
e-mail: xyj@ict.ac.cn

C. Yang · X. Yu
University of Chinese Academy of Sciences, Beijing, China

W. Pedrycz and S.-M. Chen (eds.), *Advancements in Knowledge Distillation: Towards New Horizons of Intelligent Systems*, Studies in Computational Intelligence 1100,
https://doi.org/10.1007/978-3-031-32095-8_1

1 Categories of Response-Based, Feature-Based, and Relation-Based Knowledge Distillation

The current offline knowledge distillation methods often involve knowledge type and distillation strategies. The former focuses on exploring various information types for student mimicry. The latter aims to help the student to learn teacher effectively. In this section, we investigate response-based, feature-based, and relation-based knowledge, commonly utilized types based on their pre-defined information. Response-based KD guides the teacher's final output to instruct the student's output [1–3]. This makes intuitive sense to let the student know how a powerful teacher produces the predictions. Besides the final output, intermediate features encode the process of knowledge abstract from a neural network. Feature-based KD [4–8] can teach the student to obtain more meaningful semantic information throughout the hidden layers. Response-based and feature-based KD often consider knowledge extraction from a single data sample. Instead, relation-based KD [7, 9–13] attempts to excavate cross-sample relationships across the whole dataset. In this section, we survey some representative approaches for each knowledge type and summarize their difference. The overview of schematic illustrations is shown in Fig. 1.

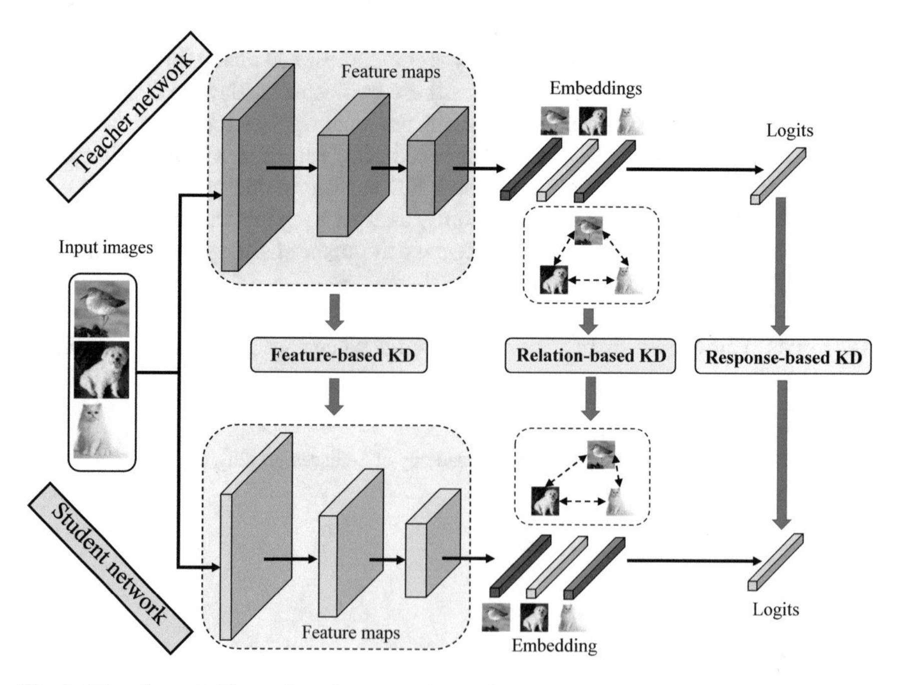

Fig. 1 The schematic illustration of response-based, feature-based, and relation-based offline KD between teacher and student networks

1.1 Response-Based Knowledge Distillation

Response-based KD focuses on learning knowledge from the last layer as the response. It aims to align the final predictions between the teacher and student. The property of response-based KD is outcome-driven learning that makes it be readily extended to various tasks. The seminal KD dates back to Hinton et al. [1]. The core idea is to distill class probability distribution via softened softmax (namely 'soft label'). For the classification task, the soft probability distribution p is formulated as Eq. 1:

$$p(z_i; T) = \frac{\exp(z_i/T)}{\sum_{j=1}^{N} \exp(z_j/T)}, \tag{1}$$

where z_i is the logit value of the i-th classes, N is the number of classes and T is a temperature parameter to adjust the smoothness of class probability distribution. Given the class probability distribution from the teacher as $p(z^T; T)$ and the student as $p(z^S; T)$, response-based KD attempts to match $p(z^S; T)$ with $p(z^T; T)$ using a distance function denoted as $\mathcal{L}_{dis}$:

$$\mathcal{L}_{response\_kd}(p(z^S; T), p(z^T; T)) = \mathcal{L}_{dis}(p(z^S; T), p(z^T; T)), \tag{2}$$

where $\mathcal{L}_{dis}$ can be formulated as Kullback-Leibler divergence loss [1], Mean Squared Error loss [14] or pearson correlation coefficient [15]. For example, the loss of the conventional KD is formulated as:

$$\mathcal{L}_{KD}(p(z^S; T), p(z^T; T)) = \sum_{n=1}^{N} p(z^T; T)[n] \log \frac{p(z^T; T)[n]}{p(z^S; T)[n]}, \tag{3}$$

where $[n]$ denotes the index of the n-th probability value.

Interpret response-based KD. The core idea of response-based KD is straightforward to understand. It guides the student to learn the final results generated from the teacher network. The efficacy of response-based KD can also be connected to label smoothing regularization [16–20]. Yuan et al. [20] resolves the equation of KL-divergence-based KD to label smoothing. Muller et al. [18] observed that label smoothing suppresses the effectiveness of KD. They think that label smoothing leads to information loss in the logits on similarities among samples from various classes, which is valuable for response-based KD. However, Shen et al. [19] provided an empirical study to illuminate that label smoothing does not suppress KD's effectiveness generally. They found that label smoothing may result in a negative impact under two scenarios: long-tailed class distribution and increased number of classes. Mobahi et al. [21] showed a theoretical perspective that distillation between two identical network architectures amplifies regularization in Hilbert space. The latest DKD [22] decouples the original KD loss [1] into target class KD (TCKD) and non-target class KD (NCKD). By only introducing two hyper-parameters to balance two terms flexibly, DKD [22] improves the effectiveness of the original KD.

Reduce performance gap with auxiliary architecture. A performance gap may exist in KD due to the capacity gap between teacher and student, leading to a performance degradation issue. To alleviate this problem, TAKD [23] introduces an intermediate-sized network as a teacher assistant and performs a sequential KD process. Along the vein, HKD [24] applies an auxiliary teacher network to transfer layer-wise information flow. DGKD [25] proposes a densely guided pattern and performs multi-step KD by all previous teachers. SFTN [26] trains a teacher along with student branches at first and then transfers more easy-to-transfer knowledge to the student. TOFD [27] and HSAKD [28] attach several auxiliary branches between teacher and student to facilitate knowledge interaction. However, these methods introduce extra architectures to the training graph and increase training costs.

Reduce performance gap with adaptive distillation. Some works attempted to investigate the sample-aware adaptive distillation to improve performance. WSL [3] proposes sample-wise weighted soft labels from the perspective of the bias-variance tradeoff. ATKD [29] uses an adaptive temperature based on standard deviation to reduce the sharpness gap between teacher and student. MKD [30] utilizes meta-learning to search a learnable temperature parameter. SCKD [31] examines the mismatch problem from the view of gradient similarity, making the student adaptively learn its beneficial knowledge. PAD [32] proposes an adaptive sample weighting mechanism with data uncertainty based on the observation that hard instances may be intractable for KD. Beyond exploring sample-dependent distillation, Song et al. [33] and Li et al. [34] proposed a hybrid forward scheme to make the student learn the teacher's knowledge implicitly via joint training. Inspried by curriculum learning, RCO [2] forces the student to mimic training trajectories of the teacher from scratch to convergence. ESKD [35] stops the teacher training early to produce more softened logits.

Discussion. We provide an empirical study of various response-based KD approaches in Table 1. Different distillation algorithms may be superior to various network architectures. Most methods are applied to convolutional neural networks, while recent MKD [30] further aims to improve vision transformers. The core idea of response-based KD is outcome-driven learning and readily applicable to existing recognition tasks. For object detection, Chen et al. [36] proposed to guide the student to mimic the teacher's object probabilities and regression bounding boxes. For semantic segmentation, the outcome-driven knowledge is pixel-wise class probability distribution [37]. Analogously, for Bert compression in natural language processing, DistilBERT [38] transfers class predictions of masked tokens. Although response-based KD has been successfully applied to many tasks, the performance gap is still an open issue. When the capacity gap between teacher and student is quite large, the student may not be able to absorb meaningful knowledge. This may lead to adverse supervisory effects. Furthermore, response-based KD ignores the intermediate information encoded in the hidden layers of a neural network, resulting in limited performance improvements.

Table 1 Top-1 accuracy(%) of response-based offline KD methods for ImageNet [39] classification. The compared works are sorted according to the published time. The networks are selected as ResNets [40], MobileNets [41], CaiT [42] and ViT [43]. All results are referred to the original papers

Method	Venue	Algorithm	Teacher (baseline)	Student (baseline)	After KD
KD [1]	ArXiv-2015	Vanilla KD	ResNet-34(73.3)	ResNet-18(69.8)	70.7
RCO [2]	ICCV-2019	Curriculum learning	ResNet-50(75.5)	MobileNet V2(64.2)	68.2
TAKD [23]	AAAI-2020	Teacher assistant	ResNet-50(76.1)	ResNet-14(65.2)	67.4
PAD [32]	ECCV-2020	Uncertainty learning	ResNet-34(73.3)	ResNet-18(69.8)	71.7
TOFD [27]	NeurIPS-2020	Task-oriented	ResNet-152(78.3)	ResNet-18(69.8)	70.9
DGKD [25]	ICCV-2021	Densely guidance	ResNet-34(73.3)	ResNet-18(69.8)	71.7
SFTN [26]	NeurIPS-2021	Prior training	ResNet-50(77.4)	ResNet-34(73.8)	75.5
HSAKD [28]	IJCAI-2021	Self-supervision	ResNet-34(73.3)	ResNet-18(69.8)	72.4
SCKD [31]	ICCV-2021	Gradient similarity	ResNet-101(77.4)	ResNet-18(70.3)	71.3
WSL [3]	ICLR-2021	Bias-variance tradeoff	ResNet-34(73.3)	ResNet-18(69.8)	72.0
ATKD [29]	Openreview-2021	Sharpness gap	ResNet-34(73.3)	ResNet-18(69.8)	72.8
DKD [22]	CVPR-2022	Balancing losses	ResNet-34(73.3)	ResNet-18(69.8)	71.7
MKD [30]	Arxiv-2022	Meta-learning	CaiT-S24(82.4)	ViT-T(72.2)	76.4

1.2 Feature-Based Knowledge Distillation

As we discussed above, response-based KD neglects intermediate-level supervision for complete guidance. To address this defect, feature-based KD focuses on exploring intermediate feature information to provide comprehensive supervisory, such as feature maps and their refined information. The common feature-based distillation loss can be formulated as Eq. 4:

$$\mathcal{L}_{feature\_kd}(F^S, F^T) = \mathcal{L}_{dis}(\phi^S(F^S), \phi^T(F^T)), \tag{4}$$

where F^S and F^T represent intermediate feature maps from student and teacher. ϕ^S and ϕ^T are meaningful transformation functions to produce refined information, such as attention mechanism [44], activation boundary [4], neuron selectivity [5] and probability distribution [45], etc. $\mathcal{L}_{dis}$ is a distance function that measures the similarity of matched feature information, for example, Mean Squared Error loss [4, 6, 44] and Kullback-Leibler divergence loss [45]. For example, the seminal FitNet [6] is formulated as:

$$\mathcal{L}_{FitNet}(F^S, F^T) = \frac{1}{H \times W \times C} \sum_{h=1}^{H} \sum_{w=1}^{W} \sum_{c=1}^{C} (F^S[h, w, c] - F^T[h, w, c])^2. \quad (5)$$

Here, we assume $F^S \in \mathbb{R}^{H \times W \times C}$ and $F^T \in \mathbb{R}^{H \times W \times C}$, where H, W, C denote the feature map's height, weight and channel number, respectively.

Knowledge exploration: transform intermediate feature maps to meaningful knowledge. The seminal FitNet [6] is the first feature-based KD method. Its core idea is to align the intermediate feature maps generated from the hidden layers in a layer-by-layer manner between teacher and student. This simple and intuitive work may not use high-level knowledge. Subsequent approaches attempted to explore more meaningful information encoded in the raw feature maps that is more suitable for feature-based KD.

AT [44] transforms the feature map to a spatial attention map as valuable information. NST [5] extracts activation heatmaps as neuron selectivity for transfer. Srinivas et al. [46] applied Jacobian matching between feature maps. PKT [45] formulates the feature map as a probability distribution and mimicked by KL-divergence. FSP [7] introduces Gramian matrix [47] to measure the flow of solution procedure across feature maps from various layers. Seung et al. [48] used singular value decomposition to resolve the feature knowledge. FT [49] introduces an auto-encoder to parse the teacher's feature map as "factors" in an unsupervised manner and a translator to transform "factors" to easily understandable knowledge. AB [4] considers activation boundaries in the hidden feature space and forces the student to learn consistent boundaries with the teacher. Overhaul [50] rethinks the distillation feature position with a newly designed margin ReLU and a partial L2 distance function to filter redundant information.

More recently, TOFD [27] and HSAKD [28] attach auxiliary classifiers to intermediate feature maps supervised by extra tasks to produce informative probability distributions. The former leverages the original supervised task, while the latter introduces a meaningful self-supervised augmented task. MGD [51] performs fine-grained feature distillation with an adaptive channel assignment algorithm. ICKD [52] excavates inter-channel correlation from feature maps containing the feature space's diversity and homology. Beyond the channel dimension, TTKD [53] conducts spatial-level feature matching using self-attention mechanism [54]. In summary, previous methods often resort to extracting richer feature information for KD, leading to better performance than vanilla FitNet [6].

Knowledge transfer: good mimicry algorithm to let the student learn better. Beyond knowledge exploration, another valuable problem is how to transfer knowledge effectively. Most feature-based KD approaches use the simple Mean Squared Error loss for knowledge alignment. Besides this vein, VID [55] refers to the information-theoretic framework and considers KD as maximizing the mutual information between teacher and student. Wang et al. [56] regarded the student as a generator and applied an extra discriminator to distinguish features produced from student or teacher. This adversarial process guides the student to learn the similar feature distribution to the teacher. Xu et al. [57] proposed to normalize feature embeddings in penultimate layer to suppress the negative impact of noise. Beyond examining mimicry metric loss, using a shared classifier between teacher and student can also help the student to align the teacher's features implicitly [58, 59].

Distillation for vision transformer. Vision transformer (ViT) [43] has shown predominant performance for image recognition. However, ViT-based networks need high demand of computational costs. KD provides an excellent solution to train a small ViT with desirable performance. Over the relation level, Manifold Distillation [63] explores patch-wise relationships as the knowledge type for ViT KD. Over the feature level, AttnDistill [64] transfers attention maps from teacher to student. ViTKD [65] provides practical guidelines for ViT feature distillation. Besides feature mimicry between homogeneous ViTs, some works [66, 67] also attempt to distill inductive biases from CNN to ViT. Some promising knowledge types are still worth further mining, such as intermediate features, attentive relationships and distillation positions.

Discussion. We provide an empirical study of various feature-based KD approaches in Table 2. Generally, feature-based KD is a comprehensive supplement to response-based KD that provides intermediate features encapsulating the learning process. However, simply aligning the same-staged feature information between teacher and student may result in negative supervisory, especially when the capacity gap or architectural difference is large. A more valuable direction may lie in the student-friendly feature-based KD that provides semantic-consistent supervision.

1.3 Relation-Based Knowledge Distillation

Response-based and feature-based KD often consider distilling knowledge from individual samples. Instead, relation-based KD explores **cross-sample** or **cross-layer** relationships as meaningful knowledge.

1.3.1 Relation-Based Cross-Sample Knowledge Distillation

A general relation-based cross-sample distillation loss is formulated as Eq. 6:

$$\mathcal{L}_{relation\_kd}(F^S, F^T) = \sum_{i,j} \mathcal{L}_{dis}(\psi^S(v_i^S, v_j^S), \psi^T(v_i^T, v_j^T)), \tag{6}$$

Table 2 Top-1 accuracy(%) of various feature-based offline KD methods for CIFAR-100 [60] classification. The compared works are sorted according to the published time. The networks are selected as ResNets [40] and WRNs [61]. All results are referred to the original papers

Method	Venue	Knowledge	Teacher (baseline)	Student (baseline)	After KD
FitNet [6]	ICLR-2015	Feature maps	ResNet-56(72.34)	ResNet-20(69.06)	69.21
AT [44]	ICLR-2017	Attention maps	ResNet-56(72.34)	ResNet-20(69.06)	70.55
FSP [7]	CVPR-2017	Solution flow	ResNet-56(72.34)	ResNet-20(69.06)	69.95
NST [5]	arXiv-2017	Neuron selectivity	ResNet-56(72.34)	ResNet-20(69.06)	69.60
Jacobian [46]	ICML-2018	Gradient	WRN-28-4(78.91)	WRN-16-4(77.28)	77.82
FT [49]	NeurIPS-2018	Paraphrased factor	ResNet-56(72.34)	ResNet-20(69.06)	69.84
PKT [45]	ECCV-2018	Probability distribution	ResNet-56(72.34)	ResNet-20(69.06)	70.34
AB [4]	AAAI-2019	Activation boundaries	ResNet-56(72.34)	ResNet-20(69.06)	69.47
VID [55]	CVPR-2019	Mutual information	ResNet-56(72.34)	ResNet-20(69.06)	70.38
Overhaul [50]	ICCV-2019	Feature position	WRN-28-4(78.91)	WRN-16-4(77.28)	79.11
FKD [57]	ECCV-2020	Normalization	ResNet-56(81.73)	ResNet-20(78.30)	81.19
DFA [62]	ECCV-2020	Differentiable search	WRN-28-4(79.17)	WRN-16-4(77.24)	79.74
MGD [51]	ECCV-2020	Channel assignment	WRN-28-4(78.91)	WRN-16-4(77.28)	78.88
TOFD [27]	NeurIPS-2020	Task-oriented	ResNet-56(73.44)	ResNet-20(69.6)	72.02
HSAKD [28]	IJCAI-2021	Self-supervision	ResNet-56(73.44)	ResNet-20(69.6)	72.60
ICKD [52]	ICCV-2021	Inter-channel correction	ResNet-56(72.34)	ResNet-20(69.06)	71.76
SRRL [59]	ICLR-2021	Softmax regression	WRN-40-2(76.31)	WRN-40-1(71.92)	74.64
SimKD [58]	CVPR-2022	Reused classifier	WRN-40-2(76.31)	WRN-40-1(71.92)	75.56
TTKD [53]	CVPR-2022	Spatial attention	ResNet-56(72.34)	ResNet-20(69.06)	71.59

where F^S and F^T denote the feature sets of teacher and student, respectively. v_i and v_j are feature embeddings of the i-th and j-th samples, and $(v_i^S, v_j^S) \in F^S$, $(v_i^T, v_j^T) \in F^T$. ψ^S and ψ^T are similarity metric functions of (v_i^S, v_j^S) and (v_i^T, v_j^T). $\mathcal{L}_{dis}$ is a distance function that measures the similarity of an instance graph, for example, Mean Squared Error loss [11, 68] and Kullback-Leibler divergence loss [13, 69]. For example, the loss of the representative RKD [11] is formulated as:

$$\mathcal{L}_{RKD}(F^S, F^T) = \sum_{i,j,k}(cos\angle v_i^S v_j^S v_k^S - cos\angle v_i^T v_j^T v_k^T)^2, \tag{7}$$

$$cos\angle v_i v_j v_k = \langle \mathbf{e}^{ij}, \mathbf{e}^{kj} \rangle = \left\langle \frac{v_i - v_j}{\|v_i - v_j\|_2}, \frac{v_k - v_j}{\|v_k - v_j\|_2} \right\rangle. \tag{8}$$

Constructing relational graph with various edge weights. The knowledge of relation-based KD can be seen as an instance graph, where the nodes denote the feature embeddings of samples. Most relation-based KD examines various similarity metric functions to compute edge weights. DarkRank [70] is the first approach to examine cross-sample similarities based on Euclidean distance of embeddings for deep metric learning. MHGD [9] processes graph-based representations using a multi-head attention network. RKD [11] uses distance-wise and angle-wise similarities of mutual relations as structured knowledge. CCKD [68] captures the correlation between instances using kernel-based Gaussian RBF. SP [12] constructs pairwise similarity matrices given the mini-batch. IRG [10] models an instance relationship graph with vertex and edge transformation. REFILLED [71] forces the teacher to reweight the hard triplets forwarded by the student for relationship matching. All methods focus on modeling relation graphs over sample-level feature embeddings but differ in various edge-weight generation strategies.

Constructing relational graph with meaningful transformation. Directly modelling edge-weights using simple metric functions may not capture correlations or higher-order dependencies meaningfully. CRD [72] introduces supervised contrastive learning among samples based on an InfoNCE [73]-inspired loss to align the teacher's representations. Over CRD, CRCD [74] proposes complementary relation contrastive distillation according to the feature and its gradient. To extract richer knowledge upon the original supervised learning, SSKD [69] follows the SimCLR [75] framework and utilizes self-supervised contrastive distillation from image rotations. To take advantage of category-level information from labels, CSKD [76] builds intra-category and inter-category structured relations. Previous methods often focus on instance-level features and their relationships but ignore local features and details. Therefore, LKD [77] utilizes a class-aware attention module to capture important regions and then models the local relational matrices using the localized patches. GLD [78] constructs a relational graph with local features extracted by a local spatial pooling layer.

1.3.2 Relation-Based Cross-Layer Knowledge Distillation

Beyond build relationships over data samples, the cross-layer interactive information encoding inside the models is also a valuable knwoledge form. A general relation-based cross-layer distillation loss is formulated as Eq. 9:

$$\mathcal{L}_{relation\_kd}(f^S, f^T) = \mathcal{L}_{dis}(g^S(f_i^S, f_j^S), g^T(f_i^T, f_j^T)), \tag{9}$$

where f^S and f^T denote the feature sets extracted from different layers of teacher and student, respectively. f_i and f_j are feature embeddings from the i-th and j-th layer, and $(f_i^S, f_j^S) \in f^S$, $(f_i^T, f_j^T) \in f^T$. g^S and g^T are layer aggregation functions of (f_i^S, f_j^S) and (f_i^T, f_j^T). $\mathcal{L}_{dis}$ is a distance function that measures the similarity of the cross-layer aggregated feature maps, for example, Mean Squared Error loss [7, 79–81]. For example, the loss of FSP [7] is formulated as:

$$\mathcal{L}_{FSP}(f^S, f^T) = \left\|G^S(f_i^S, f_j^S) - G^T(f_i^T, f_j^T)\right\|_2^2, \tag{10}$$

$$G(f_i, f_j)[a, b] = \sum_{h=1}^{H}\sum_{w=1}^{W} \frac{f_i[h, w, a] \times f_j[h, w, b]}{H \times W}, G(f_i, f_j) \in \mathbb{R}^{X \times Y}. \tag{11}$$

Here, we assume $f_i \in \mathbb{R}^{H \times W \times X}$ and $f_j \in \mathbb{R}^{H \times W \times Y}$, where H and W denote the height and width, and X and Y represent the number of channels.

FSP [7] is the seminal method to capture relationships among cross-layer feature maps for KD. It introduces Gramian matrix [47] to represent the flow of solution procedure as knowledge. Passalis et al. [24] pointed out that the same-staged intermediate layers between teacher and student networks with different capacities may show semantic abstraction gaps. Previous methods [4, 6, 44, 50] often rely on a hand-crafted layer assignment strategy in a one-to-one manner. However, the naive alignment could result in the semantic mismatch problem between the pair-wise teacher-student layers. Many subsequent works consider modeling meaningful knowledge across multiple feature layers. Jang et al. [84] introduced meta-networks for weighted layer-level feature matching. Motivated by self-attention mechanism [54], some works [79–81] utilize the attention-based weights for adaptive layer assignment. Apart from examining the layer matching issue, some works [82, 83] attempted to aggregate all-staged feature maps to construct informative features as supervisory signals. ReviewKD [82] leverages multi-level features from the teacher to guide each layer of the student according to various feature fusion modules. LONDON [83] summarizes multi-level feature maps to model Lipschitz continuity. Besides manual strategies, DFA [62] applies a search method to find appropriate feature aggregations automatically.

Discussion. We provide an empirical study of various relation-based KD approaches in Table 3. Independent from feature-based and response-based KD, relation-based methods aim to capture high-order relationships among various samples or different layers. The relational graph captures structured dependencies across

Table 3 Top-1 accuracy(%) of various relation-based offline KD methods for CIFAR-100 [60] classification. The compared works are sorted according to the published time. The networks are selected as ResNets [40] and WRNs [61]. All results are referred to the original papers

Method	Venue	Knowledge	Teacher (baseline)	Student (baseline)	After KD
MHGD [9]	BMVC-2019	Graph attention	ResNet-56(72.34)	ResNet-20(69.06)	69.42
RKD [11]	CVPR-2019	Relational graph	ResNet-56(72.34)	ResNet-20(69.06)	69.61
IRG [10]	CVPR-2019	Instance graph	ResNet-110(72.53)	ResNet-20(68.75)	69.87
CCKD [68]	ICCV-2019	Correlation congruence	ResNet-56(72.34)	ResNet-20(69.06)	69.63
SP [12]	ICCV-2019	Similarity-preserving	ResNet-56(72.34)	ResNet-20(69.06)	69.67
REFILLED [71]	CVPR-2020	Relationship matching	WRN-40-2(74.44)	WRN-16-2(70.15)	74.01
CRD [72]	ICLR-2020	Contrastive learning	ResNet-56(72.34)	ResNet-20(69.06)	71.16
SSKD [69]	ECCV-2020	Self-supervision	ResNet-56(73.44)	ResNet-20(69.63)	71.49
LKD [69]	ECCV-2020	Local correlation	ResNet-110(75.76)	ResNet-20(69.47)	72.63
SemCKD [79]	AAAI-2021	Attention matching	ResNet-32x4(79.42)	ResNet-8x4(73.09)	76.23
AFD [80]	AAAI-2021	Attention matching	ResNet-56(72.54)	ResNet-20(69.40)	71.53
ReviewKD [82]	CVPR-2021	Feature aggregation	ResNet-56(72.54)	ResNet-20(69.40)	71.89
LONDON [83]	CVPR-2021	Lipschitz continuity	WRN-28-4(78.91)	WRN-16-4(77.28)	79.67
CRCD [74]	CVPR-2021	Complementary relation	ResNet-56(72.34)	ResNet-20(69.06)	73.21
GLD [78]	ICCV-2021	Local relationships	ResNet-110(72.53)	ResNet-20(68.75)	71.37

the whole dataset. The cross-layer feature relationships encode the information of semantic process. How to model a better relationships using more meaningful node transformation and metric functions or aggregating appropriate layer information are still core problems to be further researched (Table 4).

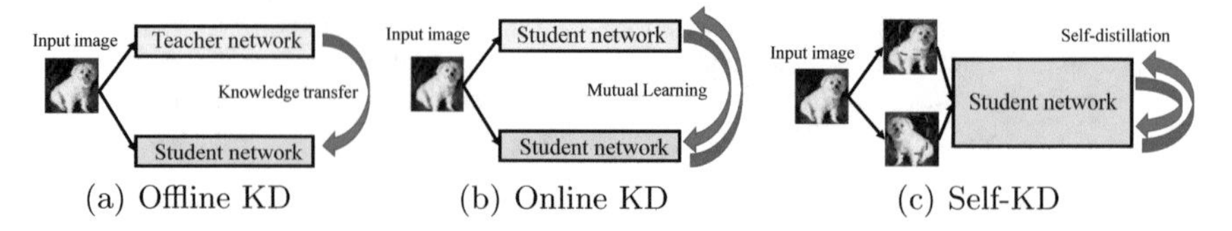

Fig. 2 The schematic illustrations of three KD schemes. a Offline KD performs unidirectional knowledge transfer from a teacher network to a student network. **b** Online KD conducts mutual learning between two peer student networks. **c** Self-KD creates two input views and regularizes similar outputs over a single student network

2 Distillation Schemes

We discuss distillation schemes of student learning, including offline KD, online KD and Self-KD. The schematic illustrations of three KD schemes are shown in Fig. 2.

2.1 Offline Knowledge Distillation

The offline KD is so-called teacher-student-based learning [1, 6], which previous works have broadly examined. The core idea of offline KD is to transfer knowledge from a large pre-trained teacher network with high performance to a small and fast student network. In practice, offline KD often conducts a two-stage training pipeline: (1) the teacher network is pre-trained over the task to achieve excellent performance; and (2) the student is guided to mimic the teacher's information during the training phase. When offline KD uses publicly available pre-trained models at hand for training a student network, offline KD can also be regarded as a one-stage pipeline. Because the teacher network is pre-trained and frozen, we call the teacher-student-based learning as offline KD, which is discussed in Sect. 1 in detail, according to different transferred knowledge types.

Trade-off between performance and distillation time. We comprehensively compare representative offline KD methods toward accuracy and distillation time. We can observe that various KD approaches have different properties. The conventional KD [1] has the lowest distillation time but only leads to a moderate gain. In contrast, HSAKD [28] achieves the best distillation performance, even matching the teacher accuracy, but the time is $3\times$ than the vanilla KD. DKD [22], a modified version of the conventional KD [1], has a desirable balance between accuracy and distillation time. From the perspective of model compression, the best-distilled ResNet-20 has $3\times$ fewer parameters and FLOPs but only results in a 0.1% performance drop compared to the teacher ResNet-56. In practice, we can choose a suitable KD algorithm according to your actual requirements and computing resources.

Table 4 Comprehensive comparison of representative offline KD methods toward accuracy and distillation time on CIFAR-100 classification when a pre-trained teacher network is available. The distillation time is evaluated on a single NVIDIA Tesla V100 GPU, and is measured from the actual time per epoch. FLOPs denote the number of floating-point operations, measuring the computational complexity of networks

Method	Venue	Teacher:ResNet-56			Student:ResNet-20			After KD	Time(s)
		Params	FLOPs	Baseline	Params	FLOPs	Baseline		
KD [1]	ArXiv-2015	0.86M	125.8M	72.34	0.28M	40.8M	69.06	70.66	20.1
FitNet [6]	ICLR-2015							69.21	23.5
AT [44]	ICLR-2017							70.55	23.2
PKT [45]	ECCV-2018							70.34	21.4
VID [55]	CVPR-2019							70.38	27.0
RKD [11]	CVPR-2019							69.61	22.7
CCKD [68]	ICCV-2019							69.63	23.4
CRD [72]	ICLR-2020							71.16	32.5
SSKD [69]	ECCV-2020							71.49	33.0
SemCKD [79]	AAAI-2021							71.54	54.0
HSAKD [28]	IJCAI-2021							72.25	75.2
DKD [22]	CVPR-2022							71.97	27.0

2.2 *Online Knowledge Distillation*

Online KD aims to train a group of student networks simultaneously from scratch and transfer knowledge from each other during the training phase. Unlike offline KD, online KD is an end-to-end optimization process and does not need an explicit pre-trained teacher network in advance. According to the knowledge type, the current online KD is mainly divided into response-based, feature-based, and relation-based approaches, as illustrated in Fig. 3. We provide an empirical study of various online KD approaches in Table 5.

Response-based Online KD. The online KD dates back to Deep mutual learning (DML) [85]. DML reveals that aligning each student's class posterior with that of other students learns better than training alone in the traditional learning scheme.

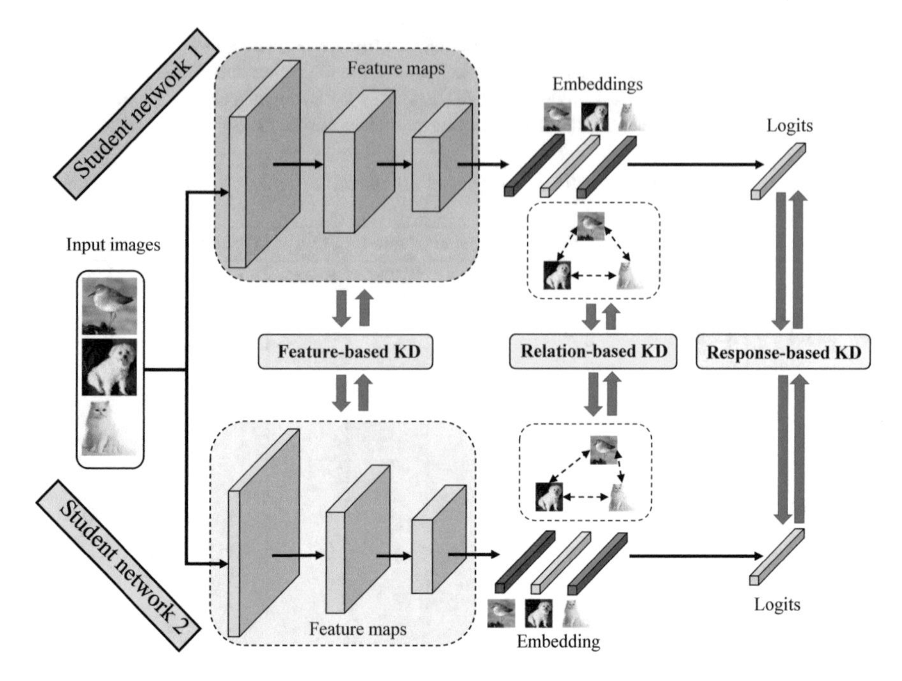

Fig. 3 The schematic illustration of online KD. Compared with offline KD in Fig. 1, online KD conducts bidirectional knowledge transfer among two student networks

This idea is further extended to a hierarchical architecture with shared low-level layers and separated high-level branches by Song et al. [86]. Anil et al. [87] applied mutual distillation to a large-scale distributed neural network. DCML [88] augments mutual learning with well-designed auxiliary classifiers added into hidden layers. MutualNet [89] performs mutual learning on sub-networks equipped with different widths using various input resolutions to explore multi-scale features. MMT [90] and PCL [91] introduce a temporal mean teacher for each peer to generate better pseudo-labels for mutual learning. Beyond the peer-teaching manner, ONE [92] assembles class probabilities to construct a virtual teacher role for providing soft labels. OKDDip [93] utilizes self-attention mechanism [54] to boost peer diversity and then transfers the ensemble knowledge of auxiliary peers to the group leader. KDCL [94] investigates generating soft ensemble targets using various aggregation strategies from two data-augmentation views. Several works [91, 95] consider using feature fusion with an extra classifier to output meaningful labels.

Feature-based Online KD. Previous online KD approaches often focus on learning class probabilities and mainly differ in various strategies or architectures, but neglect feature-level information for online learning. Walawalkar et al. [96] performed online mimicry of intermediate feature maps for model compression. Zhang et al. [85] showed that aligning feature maps directly may diminish group diversity and harm online KD. Many works [97, 98] proposed online adversarial feature dis-

Table 5 Top-1 accuracy(%) of various online KD methods for CIFAR-100 [60] classification. The compared works are sorted according to the published time. The network is selected as ResNet-32 [40]. All results are referred to the original papers

Method	Venue	Algorithm	Student (baseline)	After KD
DML [85]	CVPR-2018	Mutual learning	ResNet-32(71.28)	73.68
CL [86]	NeurIPS-2018	Hierarchical sharing	ResNet-32(71.28)	72.33
ONE [92]	NeurIPS-2018	Naive ensemble learning	ResNet-32(71.28)	73.79
OKDDip [93]	AAAI-2020	Ensemble learning with self-attention	ResNet-32(71.28)	73.25
KDCL [94]	CVPR-2020	Ensemble learning with augmentation	ResNet-32(71.28)	73.76
AFD [97]	ICML-2020	Adversarial feature distillation	ResNet-32(69.38)	74.03
AMLN [98]	ECCV-2020	Adversarial feature distillation	ResNet-32(69.71)	74.69
PCL [91]	AAAI-2021	Mean teacher	ResNet-32(71.28)	74.14
FFL [95]	ICPR-2021	Feature fusion	ResNet-32(71.28)	72.18
MCL [99]	AAAI-2022	Mutual contrastive learning	ResNet-32(70.91)	74.04
HSSAKD [100]	TNNLS-2022	Self-supervision augmentation	ResNet-32(70.91)	74.17

tillation to mutually learn feature distributions. The idea of adversarial online KD is to add a discriminator for each network that can classify the feature map from its own as fake or the other network as real.

Relation-based Online KD. MCL [99] regards each network as an individual view and introduces mutual relation-based distillation from the perspective of contrastive representation learning. Compared with previous works, MCL [99] helps each network to learn better visual feature representations. A common characteristic of previous online KD methods is that distilled knowledge types are extracted from a single original task. HSSAKD [100] attachs classifiers after feature maps to learn an extra self-supervision augmented task and guides networks to distill self-supervised distributions mutually.

2.3 Self-knowledge Distillation

Self-KD aims to distill knowledge explored from the network to teach itself. Unlike offline and online KD, Self-KD does not have additional teachers or peer networks for knowledge communication. Therefore, existing Self-KD works often utilize *auxiliary architecture* [101–105] , *data augmentation* [106–108] or *sequential snapshot distillation* [109–112] to explore external knowledge for self-boosting. Moreover, by manually designing regularization distributions to replace the teacher [20], Self-KD can also be connected to label smoothing [113]. We provide an empirical study of various Self-KD approaches in Table 6. Beyond applying Self-KD over conventional supervised learning, recent works also attempt to borrow the idea of Self-KD for self-supervised learning.

Self-KD with auxiliary architecture. The idea of this approach is to attach auxiliary architectures to capture extra knowledge to complement the primary network.

Table 6 Top-1 accuracy(%) of various Self-KD methods for CIFAR-100 [60] classification. The compared works are sorted according to the published time. The networks are selected as ResNets [40]. All results are referred to the original papers

Method	Venue	Algorithm	Student (baseline)	After KD
BAN [109]	ICML-2018	Born again	ResNet-32(68.39)	69.84
DDGSD [106]	AAAI-2019	Data-distortion invariance	ResNet-18(76.24)	76.61
DKS [102]	CVPR-2019	Pairwise knowledge transfer	ResNet-18(76.24)	78.64
SD [112]	CVPR-2019	Previous snapshot	ResNet-32(68.99)	71.78
BYOT [103]	ICCV-2019	Deep-to-shallow classifier	ResNet-18(76.24)	77.88
SAD [114]	ICCV-2019	Layer-wise attention	ResNet-18(76.24)	76.40
CS-KD [108]	CVPR-2020	Class-wise regularization	ResNet-18(76.24)	78.01
Tf-KD [20]	CVPR-2020	Manual distribution	ResNet-18(76.24)	76.61
MetaDistiller [115]	ECCV-2020	Meta-learning	ResNet-18(77.31)	79.05
BAKE [116]	ArXiv-2021	Knowledge ensembling	ResNet-18(76.24)	78.72
PS-KD [110]	ICCV-2021	Progressive refinement	ResNet-18(75.82)	79.18
FRSKD [101]	CVPR-2021	Feature refinement	ResNet-18(76.24)	77.71
DLB [111]	CVPR-2022	Last mini-batch regularization	ResNet-18(73.63)	76.12
MixSKD [107]	ECCV-2022	Mixup regularization	ResNet-18(76.24)	80.32

DKS [102] inserts several auxiliary branches and performs pairwise knowledge transfer among these branches and the primary backbone. BYOT [103] transfers probability and feature information from the deeper portion of the network to shallow ones. SAD [114] uses attention maps from the deeper layer to supervise the shallow layer's ones in a layer-wise manner. Besides peer-to-peer transfer, MetaDistiller [115] constructs a label generator by fusing feature maps in a top-down manner and optimizes it with meta-learning. FRSKD [101] aggregates feature maps in a BiFPN-like way to build a self-teacher network for providing refined feature maps and soft labels. A issue is that the auxiliary-architecture-based method highly depends on the human-designed network, and its expansibility is poor.

Self-KD with data augmentation. The data-augmentation-based methods often force similar predictions generated from two different augmented views. Along this vein, DDGSD [106] applies two different augmentation operators over the same image. CS-KD [108] randomly samples two different instances from the same category. MixSKD [107] regards the Mixup image as a view and the linearly interpolated image as the other view in the feature and probability space. To excavate cross-image knowledge, BAKE [116] attempts to absorb the other samples' knowledge via weighted aggregation to form a soft target. In general, data-augmentation-based Self-KD needs multiple forward processes compared with the baseline and improves the training costs.

Self-KD with sequential snapshot distillation. This vein considers making use of the network's counterparts along the training trajectory to provide supervisory signals. BAN [109] gradually improves the network under the supervision of previously trained counterparts in a sequential manner. SD [112] takes the network snapshots from earlier epochs to teach its later epochs. PS-KD [110] proposes progressively refining soft targets by summarizing the ground-truth and past predictions. DLB [111] performs consistency regularization between the last and current mini-batch. In general, snapshot-based Self-KD requires saving multiple copies of the training model and increases memory costs.

Self-supervised learning with Self-KD. Self-supervised learning focuses on good feature representations given unannotated data. There are some interesting connections between self-KD and self-supervised learning. In the self-supervised scenario, the framework often constructs two roles: *online* and *target* networks. The former is the training network, and the latter is a mean teacher [117] with moving-averaged weights from the online network. The target network has an identical architecture to the online network but has different weights. The target network is often used to provide supervisory signals to train the online network. MoCo [118] utilizes the target network to generate consistent positive and negative contrastive samples. Some self-supervised works regard the target network as a self-teacher to provide the regression targets, such as BYOL [119], DINO [120] and SimSiam [121]. Inspired by BYOT [103], SDSSL [122] guides the intermediate feature embeddings to contrast the features from the final layer. Although previous methods achieve desirable performance on self-supervised representation learning, there may still exist two directions to be worth exploring. First, the target network is constructed from the online network in a moving-averaged manner. Do we have a more meaningful way

Table 7 Comprehensive comparison of three knowledge distillation schemes. '*' denotes that offline KD has publicly available pre-trained teacher models at hand

Aspect	Offline KD	Offline KD (*)	Online KD	Self-KD
End-to-end training	No	Yes	Yes	Yes
Extra models	Yes	Yes	Yes	No
Computational complexity	Large	Low	Medium	Low

to build the target network? Second, the loss is often performed to align the final feature embeddings. Some intermediate features or contrastive relationships between the online and target networks could be further mined.

2.4 Comprehensive Comparison

As shown in Table 7, we comprehensively compare three KD schemes in various aspects. Offline KD needs an extra teacher model to train a student, while online KD or Self-KD trains a cohort of models or a single model with an end-to-end optimization. When publicly available pre-trained teacher models are unavailable, pre-training a teacher network for offline KD often has high capacity and thus is time-consuming for training. Self-KD utilizes a single model for self-boosting and often has low computational complexity. It is noteworthy that offline KD with publicly available pre-trained teacher models also has low complexity because the inference time for extra frozen teacher models without gradient-propagation does not introduce much cost.

3 Distillation Algorithms

3.1 Multi-teacher Distillation

In the traditional KD, knowledge is transferred from a high-capacity teacher to a compact student. But in this setting, the knowledge diversity and capacity are limited. Different teachers can provide their unique and useful knowledge to student. In this way, the student can learn various knowledge representations from multiple teacher networks. Following the vanilla KD, the knowledge in the form of logits or the intermediate features can be used as a supervision signal. The schematic illustration of multi-teacher KD is shown in Fig. 4.

KD from ensemble logits. Logits from the model ensemble are one of the direct ways in multi-teacher knowledge distillation. Based on this idea, the student is guided

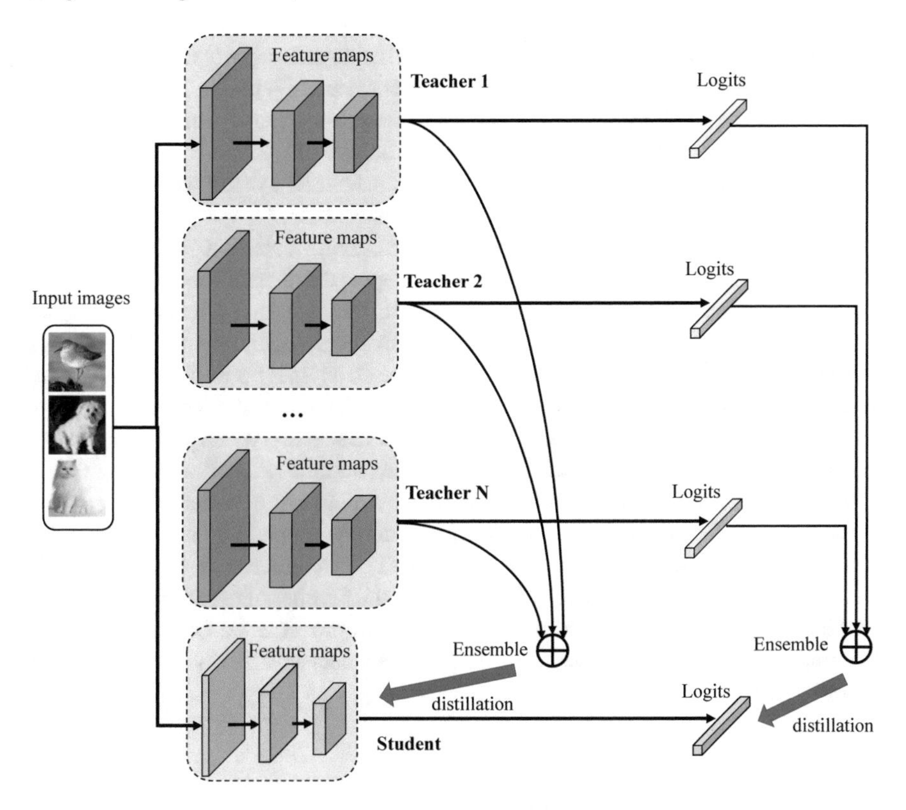

Fig. 4 The schematic illustration of multi-teacher distillation. This framework assembles multiple teachers' feature maps and logits to distill a single student network

to learn the soft output of teachers' ensemble logits in [23, 123]. However, the simple average of individual predictions may ignore the importance of variety and diversity among the teacher group. Therefore, some works [124, 125] proposed learning the student model by adaptively imitating teachers' outputs with various aggregation weights.

KD from ensemble feature representations. Besides distilling from logits, the ensemble of intermediate representations [10, 126–128] can provide more semantic information to student. However, it is more challenging for distillation from feature representations, since each member in ensemble teachers has various feature representations in specific layers. To address this issue, Park et al. [127] applied non-linear transformations to multiple teacher networks at the feature-map level. Wu et al. [128] proposed to distill the knowledge by minimizing the distance between the similarity matrices of teachers and a student. Liu et al. [10] proposed to let the student network learn the teacher models' learnable transformation matrices. To take advantage of both logits and intermediate features, Chen et al. [126] introduced double teacher networks to provide response-level and feature-level knowledge, respectively.

Computation-efficient multi-teacher KD from sub-networks. Using multi-teacher introduces extra training computation costs and decreases the training process. Thus some methods [123, 129] create some sub-teachers from a single teacher network. Nguyen et al. [129] utilized stochastic blocks and skipped connections over a teacher network to produce several teacher roles. Several works [86, 130] designed multi-headed architectures to produce many teacher roles.

Multi-task multi-teacher KD. In most cases, multi-teacher KD is based on the same task. Knowledge amalgamation [131] is proposed to learn a versatile student via learning knowledge from all teachers trained by different tasks. Luo et al. [132] aimed to learn a multi-talented student network that can absorb comprehensive knowledge from heterogeneous teachers. Ye et al. [133] focused a target network for customized tasks guided by multiple teachers pre-trained from different tasks. The student inherits desirable capabilities from heterogeneous teachers so that it can perform multiple tasks simultaneously. Rusu et al. [134] introduced a multi-teacher policy distillation approach to transfer agents' multiple policies to a single student network.

Discussion. In summary, a versatile student can be trained via multi-teacher distillation since different teachers provide diverse knowledge. However, several problems still deserve to be solved. On the one hand, the number of teachers is a trade-off problem between training costs and performance improvements. On the other hand, integrating various knowledge from multiple teachers effectively is still an open issue.

3.2 Cross-Modal Distillation

The teacher and student in common KD methods often have the same modality. However, the training data or labels for another modalities may be unavailable. Transferring knowledge between different modalities is a valuable field in practice. The core idea of cross-modal KD is to transfer knowledge from the teacher trained by a data modality to a student network from another data modality. The schematic illustration of cross-modal KD is shown in Fig. 5.

Given a teacher model pre-trained on one modality with well-labeled samples, Gupta et al. [135] transfered information between annotated RGB images and unannotated optical flow images leveraging unsupervised paired samples. The paradigm via label-guided pair-wise samples has been widely applied for cross-modal KD. Thoker et al. [136] transferred knowledge from RGB videos to a 3D human action recognition model using paired samples. Roheda et al. [137] proposed cross-modality distillation from an available modality to a missing modality using GANs. Do et al. [138] explored a KD-based visual question answering method and relied on supervised learning for cross-modal transfer using the ground truth labels. Passalis et al. [45] proposed probabilistic KD to transfer knowledge from textual modality into visual modality.

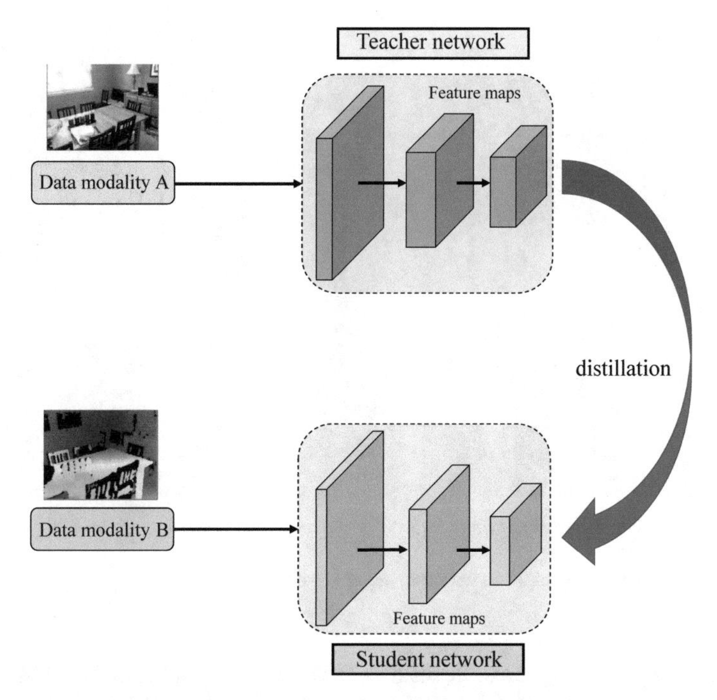

Fig. 5 **The schematic illustration of cross-modal distillation**. It transfers knowledge from the teacher trained by a data modality to a student network from another data modality

Discussion. Generally, KD performs well on cross-modal scenarios. However, cross-modal KD is difficult to model knowledge interaction when a significant modality gap exists.

3.3 Attention-Based Distillation

Attention-based distillation takes advantage of attention information for effective knowledge transfer. Current works follow two veins: (1) distilling attention maps refined from feature maps, and (2) weighted distillation based on self-attention mechanism [54], as illustrated in Fig. 6.

Distilling attention maps. Attention maps often reflect valuable semantic information and suppress unimportant parts. The seminal AT [8] constructs a spatial attention map by calculating statistics from a feature map across the channel dimension and performs alignment of attention maps between the teacher and student networks. The spatial attention map contains class-aware semantic regions that help the student to capture discriminative features. CD [139] employs a squeeze-and-excitation module [140] to generate channel attention maps and lets the student learn the teacher's channel attention weights. CWD [141] distills a spatial attention map

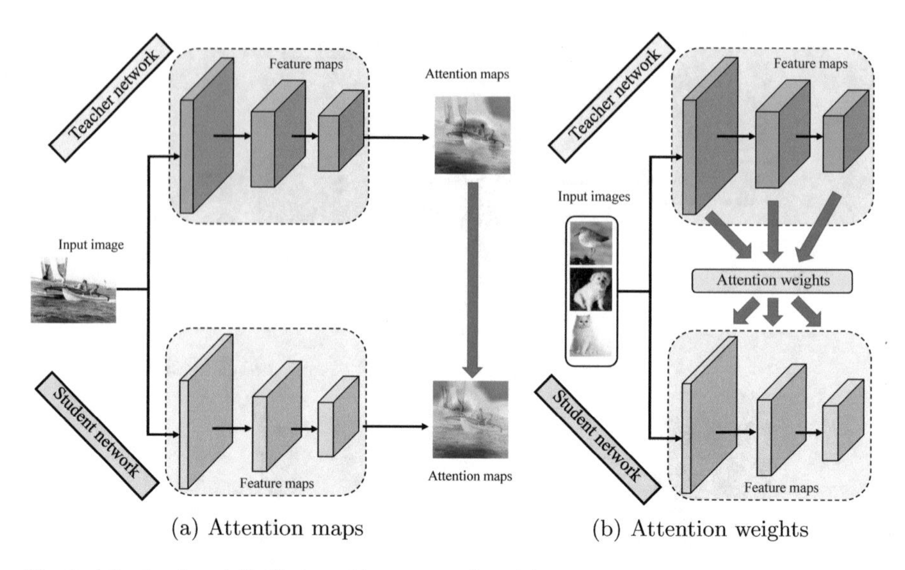

Fig. 6 Attention-based distillation with **a** maps or **b** weights

per channel representing semantic masks for dense predictions. TinyBert [142] transfers self-attention matrices for transformer-layer distillation. LKD [77] introduces a class-aware attention module to capture class-relevant regions for constructing a local correlation matrix.

Self-attention-based weighted distillation. The self-attention technique is a desirable mechanism to capture similarity relationships [54] among features. Several works [79–81] applied attention-based weights for adaptive layer-to-layer semantic matching. SemCKD [79] automatically assigns targets aggregated from suitable teacher layers with attention-based similarities for each student layer. AFD [80] proposes an attention-based meta-network to model relative similarities between teacher and student features. ALP-KD [81] fuses teacher-side information with attention-based layer projection for Bert [143] distillation. Orthogonal to the layer-wise assignment, TTKD [53] applies self-attention mechanism [54] for spatial-level feature matching.

Discussion. The attention-based map captures saliency regions and filters redundant information, helping the student learn the most critical features. However, the condensed attention map compresses the feature map's dimension and may lose meaningful knowledge. Moreover, the attention map may sometimes not focus on the correct region, resulting in an adverse supervisory impact.

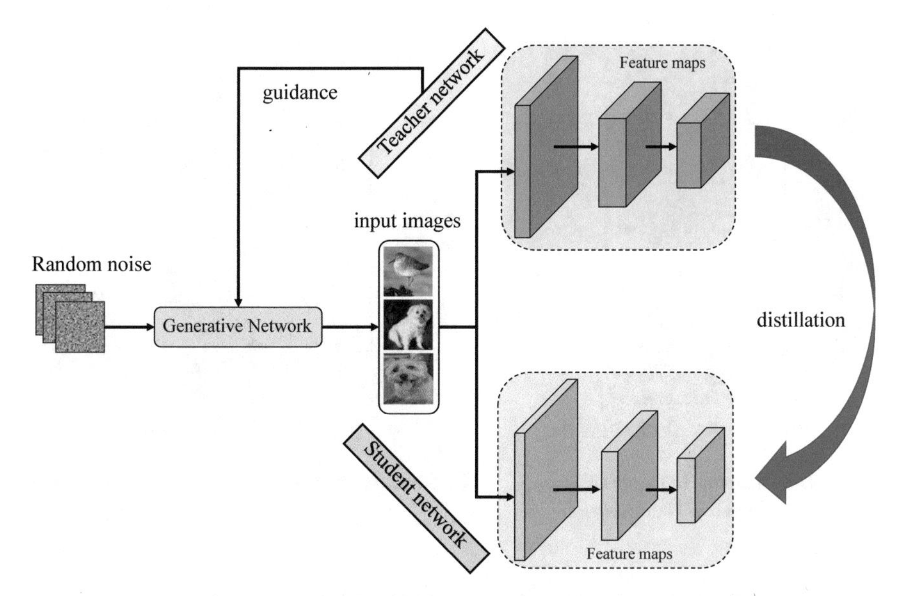

Fig. 7 **The schematic illustration of data-free distillation**. It relies on a generative network to produce images from random noise as input for distillation

3.4 Data-Free Distillation

The traditional KD methods often need large training samples. However, the training dataset may sometimes be unavailable due to privacy or safety concerns. Some methods have been proposed to handle this problem, mainly divided into data-free KD and dataset KD. The schematic illustration of data-free KD is shown in Fig. 7.

Data-free KD. Training samples are often newly or synthetically produced using generative adversarial networks (GAN) [144]. The teacher network supervises the student network with generated samples as input. Lopes et al. [145] used different types of activation records to reconstruct the original data. DeepInversion [146] explores information stored in batch normalization layers to generate synthesized samples for data-free KD. Nayak et al. [147] introduced a sample extraction mechanism by modeling the softmax space as a Dirichlet distribution from the teacher's parameters. Beyond the final output, the target data can be generated using the information from the teacher's feature representations [147, 148]. Paul et al. [148] optimized an adversarial generator to search for difficult images and then used these images to train the student. CMI [149] introduces contrastive learning to make the synthesizing instances distinguishable compared to already synthesized ones. FastDFKD [150] optimizes a meta-synthesizer to reuse the shared common features for faster data-free KD. Similar to zero-shot learning, some KD methods with few-shot learning were proposed to distill data-free knowledge from a teacher model into a student network [151, 152], where the teacher network often uses little labeled data.

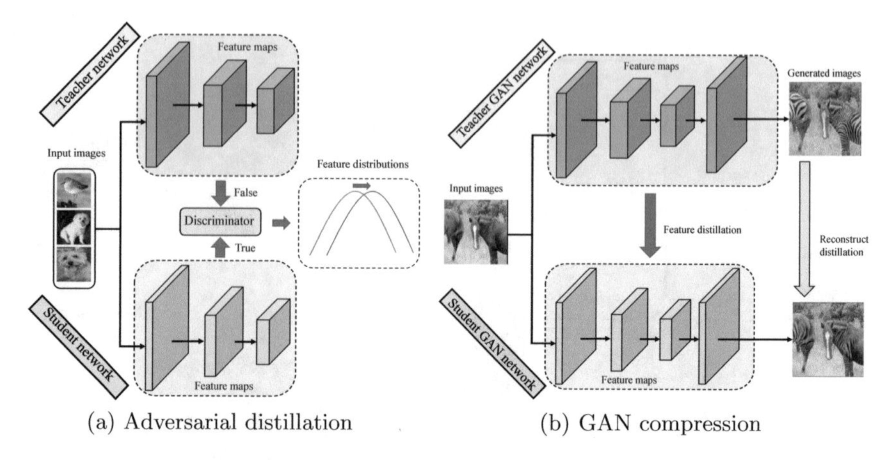

Fig. 8 The schematic illustrations of (1) KD with adversarial mechanism and (2) Generative Adversarial Networks (GAN) compression

Dataset KD. Besides data-free KD, dataset distillation [153, 154] is an essential direction for synthesizing a small dataset to represent the original full dataset without much accuracy degradation. To exploit the omni-supervised setting, Radosavovic et al. [154] assembles predictions from multiple transformations of unlabeled data by a single model to produce new training annotations. DDFlow [153] proposed to learn optical flow estimation and distill predictions from a teacher model to supervise a student network for optical flow learning. Unlabeled data may hinder Graph Convolutional Network(GCN) from learning graph-based data. In general, pseudo labels of the unlabeled data can provide extra supervision to train GCN. RDD [155] proposed a reliable data-driven semi-supervised GCN training method. It can better use high-quality data and improve graph representation learning by defining node and edge reliability. Cazenavette et al. [156] conducted long-range parameter matching along training trajectories between distilled synthetic and real data.

Discussion. In most data-free KD methods, the synthesis data is usually generated from the pre-trained teacher network's feature representations. Although current data-free KD works have shown remarkable performance in handling the data-unavailable issue, generating more high-quality and diverse training samples is still a challenge to be studied.

3.5 Adversarial Distillation

Adversarial distillation is to use the basic idea of generative adversarial networks to improve KD, mainly divided into three veins: (1) using GAN to generate extra data samples, (2) adversarial mechanism to assist general KD, and (3) compressing GAN for efficient image generation. The schematic illustrations of adversarial KD are shown in Fig. 8.

Using GAN to generate extra data samples. Almost existing KD algorithms are data-driven, i.e., relying on original or alternative data, which may be unavailable in real-world scenarios. Generative adversarial networks can be applied to learn the true data distribution and solve this problem. DFAD [157] lets the teacher and student networks play the role of discriminator jointly to reduce the discrepancy. Meanwhile, it adds an extra generator to produce hard samples to enlarge it adversarially. Several works [137, 158, 159] introduce Conditional GAN (CGAN) to generate data. Roheda et al. [137] used CGAN to distill knowledge from missing modalities given other available modalities. Lifelong-GAN [160] transfers learned knowledge from previous networks to the new network for continual conditional image generation.

Adversarial mechanism to assist general KD. The conventional KD often reduces the gap between teacher and student by aligning knowledge distributions. The adversarial mechanism can be regarded as an auxiliary method to improve the mimicry difficulty. In general, the core idea is to introduce an extra discriminator to classify feature representations from teacher or student network [56, 161–163]. Wang et al. [56] leveraged a discriminator as the teaching assistant to make the student learn similar feature distributions with the teacher for image classification. Wang et al. [163] employed adversarial KD for one-stage object detection. Liu et al. [37] transferred pixel-wise class probabilities adversarially for semantic segmentation. Beyond teacher-student-based adversarial learning, several works [97, 98] apply online adversarial KD for distilling feature maps mutually among multiple student networks.

Compressing GAN for efficient image generation. Aguinaldo et al. [164] guided a smaller "student" GAN to align a larger "teacher" GAN with mean squared loss. Chen et al. [165] let a student generator learn low- and high-level knowledge from the corresponding teacher. Moreover, the student discriminator is supervised by the teacher network via triplet loss. Li et al. [166] proposed a general compression framework for conditional GANs by transferring knowledge from intermediate representations and exploring efficient architectures via neural architecture search. Zhang et al. [167] pointed out that small GANs are often difficult to generate desirable high-frequency information. WaveletKD [167] resolves images into various frequency bands via discrete wavelet transformations and then only transfers the valuable high-frequency bands.

Discussion. Although adversarial-based KD facilitates knowledge mimicry, it may be hard to ensure the convergence of GAN-based networks in practice. For GAN compression, what information extracted from features suitable for distilling GAN is still an open issue.

4 Conclusion

In this chapter, we first survey the conventional offline KD works according to their extracted knowledge types, i.e. response, feature and relation. Benefiting the comprehensive supervision from teacher, the student could generalize better over the target task. The teacher-student-based KD has some limitations, for example, high costs for pre-training a large teacher network. Therefore, two Online KD and Self-KD schemes are proposed to improve the student without a pre-trained teacher. In practice, KD applications often face various scenarios, for example, cross-modal KD and data-free KD. Moreover, we also show some popular mechanisms to help distillation perform better, such as multi-teacher KD, attention-based KD and adversarial KD. We survey the representative works for each KD setup and summarize their main ideas and contributions. Finally, we prospect the future challenges of existing KD applications. Compared with previously seminal KD survey papers [168, 169], our paper includes some newer works published at 2022 and introduces some advanced KD directions, for example, KD for vision transformer and self-supervised learning. We hope our survey can inspire future research to develop more advanced KD algorithms for improving the student performance.

References

1. Hinton, G., et al.: Distilling the knowledge in a neural network. arXiv preprint arXiv: 1503.02531 (2015)
2. Jin, X., et al.: Knowledge distillation via route constrained optimization. In: ICCV, pp. 1345–1354 (2019)
3. Zhou, H., et al.: Rethinking soft labels for knowledge distillation: a bias-variance tradeoff perspective. ICLR (2021)
4. Heo, B., et al.: Knowledge transfer via distillation of activation boundaries formed by hidden neurons. In: AAAI, vol. 33, pp. 3779–3787 (2019)
5. Huang, Z., Wang, N.: Like what you like: knowledge distill via neuron selectivity transfer. arXiv preprint arXiv:1707.01219 (2017)
6. Romero, A., Ballas, N., Kahou, S.E., Chassang, A., Gatta, C., Bengio, Y.: Fitnets: Hints for thin deep nets. ICLR (2015)
7. Yim, J., et al.: A gift from knowledge distillation: Fast optimization, network minimization and transfer learning. In: CVPR, pp. 4133–4141 (2017)
8. Zagoruyko, S., Komodakis, N.: Paying more attention to attention: improving the performance of convolutional neural networks via attention transfer. ICLR (2017)
9. Lee, S., Song, B.C.: Graph-based knowledge distillation by multi-head attention network. BMVC (2019)
10. Liu, Y., et al.: Knowledge distillation via instance relationship graph. In: CVPR, pp. 7096–7104 (2019)
11. Park, W., et al.: Relational knowledge distillation. In: CVPR, pp. 3967–3976 (2019)
12. Tung, F., Mori, G.: Similarity-preserving knowledge distillation. In: ICCV, pp. 1365–1374 (2019)
13. Yang, C., et al.: Cross-image relational knowledge distillation for semantic segmentation. In: CVPR, pp. 12319–12328 (2022)
14. Kim, T., et al.: Comparing Kullback-Leibler divergence and mean squared error loss in knowledge distillation. arXiv preprint arXiv:2105.08919 (2021)
15. Huang, T., et al.: Knowledge distillation from a stronger teacher. arXiv preprint arXiv:2205.10536 (2022)
16. Ding, Q., et al.: Adaptive regularization of labels. arXiv preprint arXiv:1908.05474 (2019)
17. Kim, S.W., Kim, H.E.: Transferring knowledge to smaller network with class-distance loss (2017)

18. Muller, R., et al.: When does label smoothing help? NeurIPS **32** (2019)
19. Shen, Z., et al.: Is label smoothing truly incompatible with knowledge distillation: an empirical study. ICLR (2021)
20. Yuan, L., et al.: Revisiting knowledge distillation via label smoothing regularization. In: CVPR, pp. 3903–3911 (2020)
21. Mobahi, H., et al.: Self-distillation amplifies regularization in Hilbert space. NeurIPS **33**, 3351–3361 (2020)
22. Zhao, B., et al.: Decoupled knowledge distillation. In: CVPR, pp. 11953–11962 (2022)
23. Mirzadeh, S.I., et al.: Improved knowledge distillation via teacher assistant. In: AAAI, vol. 34, pp. 5191–5198 (2020)
24. Passalis, N., et al.: Heterogeneous knowledge distillation using information flow modeling. In: CVPR, pp. 2339–2348 (2020)
25. Son, W., et al.: Densely guided knowledge distillation using multiple teacher assistants. In: ICCV, pp. 9395–9404 (2021)
26. Park, D.Y., et al.: Learning student-friendly teacher networks for knowledge distillation. NeurIPS **34**, 13292–13303 (2021)
27. Zhang, L., et al.: Task-oriented feature distillation. NeurIPS **33**, 14759–14771 (2020)
28. Yang, C., et al.: Hierarchical self-supervised augmented knowledge distillation. In: IJCAI, pp. 1217–1223 (2021)
29. Guo, J.: Reducing the teacher-student gap via adaptive temperatures. Openreview (2021)
30. Liu, J., et al.: Meta knowledge distillation. arXiv preprint arXiv:2202.07940 (2022)
31. Zhu, Y., Wang, Y.: Student customized knowledge distillation: bridging the gap between student and teacher. In: ICCV, pp. 5057–5066 (2021)
32. Zhang, Y., et al.: Prime-aware adaptive distillation. In: ECCV, pp. 658–674. Springer (2020)
33. Song, L., et al.: Robust knowledge transfer via hybrid forward on the teacher-student model. In: AAAI, pp. 2558–2566 (2021)
34. Li, G., et al.: Residual distillation: towards portable deep neural networks without shortcuts. NeurIPS **33**, 8935–8946 (2020)
35. Cho, J.H., Hariharan, B.: On the efficacy of knowledge distillation. In: ICCV, pp. 4794–4802 (2019)
36. Chen, G., et al.: Learning efficient object detection models with knowledge distillation. NeurIPS **30** (2017)
37. Liu, Y., et al.: Structured knowledge distillation for semantic segmentation. In: CVPR, pp. 2604–2613 (2019)
38. Sanh, V., et al.: Distilbert, a distilled version of bert: smaller, faster, cheaper and lighter. arXiv preprint arXiv:1910.01108 (2019)
39. Deng, J., et al.: Imagenet: a large-scale hierarchical image database. In: CVPR, pp. 248–255. IEEE (2009)
40. He, K., et al.: Deep residual learning for image recognition. In: CVPR, pp. 770–778 (2016)
41. Sandler, M., et al.: Mobilenetv2: inverted residuals and linear bottlenecks. In: CVPR, pp. 4510–4520 (2018)
42. Touvron, H., et al.: Going deeper with image transformers. In: ICCV, pp. 32–42 (2021)
43. Dosovitskiy, A., et al.: An image is worth 16x16 words: transformers for image recognition at scale. arXiv preprint arXiv:2010.11929 (2020)
44. Komodakis, N., et al.: Paying more attention to attention: improving the performance of convolutional neural networks via attention transfer. In: ICLR (2017)
45. Passalis, N., Tefas, A.: Learning deep representations with probabilistic knowledge transfer. In: ECCV, pp. 268–284 (2018)
46. Srinivas, S., Fleuret, F.: Knowledge transfer with Jacobian matching. In: ICML, pp. 4723–4731. PMLR (2018)
47. Gatys, L.A., et al.: A neural algorithm of artistic style. arXiv preprint arXiv:1508.06576 (2015)
48. Lee, S.H., et al.: Self-supervised knowledge distillation using singular value decomposition. In: ECCV, pp. 335–350 (2018)

49. Kim, J., et al.: Paraphrasing complex network: network compression via factor transfer. NeurIPS **31** (2018)
50. Heo, B., et al.: A comprehensive overhaul of feature distillation. In: ICCV, pp. 1921–1930 (2019)
51. Yue, K., et al.: Matching guided distillation. In: ECCV, pp. 312–328 (2020)
52. Liu, L., et al.: Exploring inter-channel correlation for diversity-preserved knowledge distillation. In: ICCV, pp. 8271–8280 (2021)
53. Lin, S., et al.: Knowledge distillation via the target-aware transformer. In: CVPR, pp. 10915–10924 (2022)
54. Vaswani, A., et al.: Attention is all you need. In: NeurIPS, pp. 5998–6008 (2017)
55. Ahn, S., et al.: Variational information distillation for knowledge transfer. In: CVPR, pp. 9163–9171 (2019)
56. Wang, Y., et al.: Adversarial learning of portable student networks. In: AAAI, vol. 32 (2018)
57. Xu, K., et al.: Feature normalized knowledge distillation for image classification. In: ECCV, pp. 664–680. Springer (2020)
58. Chen, D., et al.: Knowledge distillation with the reused teacher classifier. In: CVPR, pp. 11933–11942 (2022)
59. Yang, J., et al.: Knowledge distillation via softmax regression representation learning. ICLR (2021)
60. Krizhevsky, A., Hinton, G., et al.: Learning multiple layers of features from tiny images. Technical Report (2009)
61. Zagoruyko, S., Komodakis, N.: Wide residual networks. arXiv preprint arXiv:1605.07146 (2016)
62. Guan, Y., et al.: Differentiable feature aggregation search for knowledge distillation. In: ECCV, pp. 469–484. Springer (2020)
63. Hao, Z., Guo, J., Jia, D., Han, K., Tang, Y., Zhang, C., et al.: Efficient vision transformers via fine-grained manifold distillation. arXiv preprint arXiv:2107.01378 (2021)
64. Wang, K., Yang, F., van de Weijer, J.: Attention distillation: self-supervised vision transformer students need more guidance. arXiv preprint arXiv:2210.00944 (2022)
65. Yang, Z., Li, Z., Zeng, A., Li, Z., Yuan, C., Li, Y.: ViTKD: Practical guidelines for ViT feature knowledge distillation. arXiv preprint arXiv:2209.02432 (2022)
66. Chen, X., Cao, Q., Zhong, Y., Zhang, J., Gao, S., Tao, D.: Dearkd: data-efficient early knowledge distillation for vision transformers. In: CVPR, pp. 12052–12062 (2022)
67. Zhang, H., Duan, J., Xue, M., Song, J., Sun, L., Song, M.: Bootstrapping ViTs: Towards liberating vision transformers from pre-training. In: CVPR, pp. 8944–8953 (2022)
68. Peng, B., et al.: Correlation congruence for knowledge distillation. In: ICCV, pp. 5007–5016 (2019)
69. Xu, G., et al.: Knowledge distillation meets self-supervision. In: ECCV, pp. 588–604. Springer (2020)
70. Chen, Y., et al.: Darkrank: accelerating deep metric learning via cross sample similarities transfer. In: AAAI, vol. 32 (2018)
71. Ye, H.J., et al.: Distilling cross-task knowledge via relationship matching. In: CVPR, pp. 12396–12405 (2020)
72. Tian, Y., Krishnan, D., Isola, P.: Contrastive representation distillation. ICLR (2020)
73. Oord, A.V.D., et al.: Representation learning with contrastive predictive coding. arXiv preprint arXiv:1807.03748 (2018)
74. Zhu, J., et al.: Complementary relation contrastive distillation. In: CVPR, pp. 9260–9269 (2021)
75. Chen, T., et al.: A simple framework for contrastive learning of visual representations. In: ICML, pp. 1597–1607. PMLR (2020)
76. Chen, Z., et al.: Improving knowledge distillation via category structure. In: ECCV, pp. 205–219. Springer (2020)
77. Li, X., et al.: Local correlation consistency for knowledge distillation. In: ECCV, pp. 18–33. Springer (2020)

78. Kim, Y., et al.: Distilling global and local logits with densely connected relations. In: ICCV, pp. 6290–6300 (2021)
79. Chen, D., et al.: Cross-layer distillation with semantic calibration. In: AAAI, vol. 35, pp. 7028–7036 (2021)
80. Ji, M., et al.: Show, attend and distill: knowledge distillation via attention-based feature matching. In: AAAI, vol. 35, pp. 7945–7952 (2021)
81. Passban, P., et al.: Alp-kd: attention-based layer projection for knowledge distillation. In: AAAI, vol. 35, pp. 13657–13665 (2021)
82. Chen, P., et al.: Distilling knowledge via knowledge review. In: CVPR, pp. 5008–5017 (2021)
83. Shang, Y., et al.: Lipschitz continuity guided knowledge distillation. In: ICCV, pp. 10675–10684 (2021)
84. Jang, Y., Lee, H., Hwang, S.J., Shin, J.: Learning what and where to transfer. In: ICML, pp. 3030–3039. PMLR (2019)
85. Zhang, Y., et al.: Deep mutual learning. In: CVPR, pp. 4320–4328 (2018)
86. Song, G., Chai, W.: Collaborative learning for deep neural networks. In: NeurIPS, pp. 1832–1841 (2018)
87. Anil, R., et al.: Large scale distributed neural network training through online distillation. ICLR (2018)
88. Yao, A., Sun, D.: Knowledge transfer via dense cross-layer mutual-distillation. In: ECCV, pp. 294–311. Springer (2020)
89. Yang, T., et al.: Mutualnet: Adaptive convnet via mutual learning from network width and resolution. In: ECCV, pp. 299–315. Springer (2020)
90. Ge, Y., et al.: Mutual mean-teaching: Pseudo label refinery for unsupervised domain adaptation on person re-identification. ICLR (2020)
91. Wu, G., Gong, S.: Peer collaborative learning for online knowledge distillation. In: AAAI, vol. 35, pp. 10302–10310 (2021)
92. Zhu, X., et al.: Knowledge distillation by on-the-fly native ensemble. In: NeurIPS, pp. 7517–7527 (2018)
93. Chen, D., et al.: Online knowledge distillation with diverse peers. In: AAAI, vol. 34, pp. 3430–3437 (2020)
94. Guo, Q., et al.: Online knowledge distillation via collaborative learning. In: CVPR, pp. 11020–11029 (2020)
95. Kim, J., et al.: Feature fusion for online mutual knowledge distillation. In: ICPR, pp. 4619–4625. IEEE (2021)
96. Walawalkar, D., Shen, Z., Savvides, M.: Online ensemble model compression using knowledge distillation. In: ECCV, pp. 18–35. Springer (2020)
97. Chung, I., et al.: Feature-map-level online adversarial knowledge distillation. In: ICML, pp. 2006–2015. PMLR (2020)
98. Zhang, X., et al.: Amln: adversarial-based mutual learning network for online knowledge distillation. In: ECCV, pp. 158–173. Springer (2020)
99. Yang, C., et al.: Mutual contrastive learning for visual representation learning. In: AAAI, vol. 36, pp. 3045–3053 (2022)
100. Yang, C., et al.: Knowledge distillation using hierarchical self-supervision augmented distribution. TNNLS (2022)
101. Ji, M., et al.: Refine myself by teaching myself: feature refinement via self-knowledge distillation. In: CVPR, pp. 10664–10673 (2021)
102. Sun, D., et al.: Deeply-supervised knowledge synergy. In: CVPR, pp. 6997–7006 (2019)
103. Zhang, L., et al.: Be your own teacher: Improve the performance of convolutional neural networks via self distillation. In: ICCV, pp. 3713–3722 (2019)
104. Zhang, L., et al.: Auxiliary training: Towards accurate and robust models. In: CVPR, pp. 372–381 (2020)
105. Zhang, L., et al.: Self-distillation: towards efficient and compact neural networks. TPAMI (2021)

106. Xu, T.B., Liu, C.L.: Data-distortion guided self-distillation for deep neural networks. In: AAAI, vol. 33, pp. 5565–5572 (2019)
107. Yang, C., et al.: Mixskd: self-knowledge distillation from mixup for image recognition. In: ECCV (2022)
108. Yun, S., et al.: Regularizing class-wise predictions via self-knowledge distillation. In: CVPR, pp. 13876–13885 (2020)
109. Furlanello, T., et al.: Born again neural networks. In: ICML, pp. 1607–1616. PMLR (2018)
110. Kim, K., et al.: Self-knowledge distillation with progressive refinement of targets. In: ICCV, pp. 6567–6576 (2021)
111. Shen, Y., et al.: Self-distillation from the last mini-batch for consistency regularization. In: CVPR, pp. 11943–11952 (2022)
112. Yang, C., et al.: Snapshot distillation: teacher-student optimization in one generation. In: CVPR, pp. 2859–2868 (2019)
113. Szegedy, C., et al.: Rethinking the inception architecture for computer vision. In: CVPR, pp. 2818–2826 (2016)
114. Hou, Y., et al.: Learning lightweight lane detection CNNs by self attention distillation. In: ICCV, pp. 1013–1021 (2019)
115. Liu, B., et al.: Metadistiller: network self-boosting via meta-learned top-down distillation. In: ECCV, pp. 694–709. Springer (2020)
116. Ge, Y., et al.: Self-distillation with batch knowledge ensembling improves imagenet classification. arXiv preprint arXiv:2104.13298 (2021)
117. Tarvainen, A., Valpola, H.: Mean teachers are better role models: weight-averaged consistency targets improve semi-supervised deep learning results. NeurIPS **30** (2017)
118. He, K., Fan, H., Wu, Y., Xie, S., Girshick, R.: Momentum contrast for unsupervised visual representation learning. In: CVPR, pp. 9729–9738 (2020)
119. Grill, J. B., Strub, F., Altché, F., Tallec, C., Richemond, P., Buchatskaya, E., et al.: Bootstrap your own latent-a new approach to self-supervised learning. NeurIPS **33**, 21271–21284 (2020)
120. Caron, M., Touvron, H., Misra, I., Jégou, H., Mairal, J., Bojanowski, P., Joulin, A.: Emerging properties in self-supervised vision transformers. In: ICCV, pp. 9650–9660 (2021)
121. Chen, X., He, K.: Exploring simple Siamese representation learning. In: CVPR, pp. 15750–15758 (2021)
122. Jang, J., Kim, S., Yoo, K., Kong, C., Kim, J., Kwak, N.: Self-distilled self-supervised representation learning. arXiv preprint arXiv:2111.12958 (2021)
123. You, S., et al.: Learning from multiple teacher networks. In: SIGKDD, pp. 1285–1294 (2017)
124. Fukuda, T., et al.: Efficient knowledge distillation from an ensemble of teachers. In: Interspeech, pp. 3697–3701 (2017)
125. Xiang, L., et al.: Learning from multiple experts: self-paced knowledge distillation for long-tailed classification. In: ECCV, pp. 247–263. Springer (2020)
126. Chen, X., et al.: A two-teacher framework for knowledge distillation. In: International symposium on neural networks, pp. 58–66. Springer (2019)
127. Park, S., Kwak, N.: Feed: feature-level ensemble for knowledge distillation. arXiv preprint arXiv:1909.10754 (2019)
128. Wu, A., et al.: Distilled person re-identification: towards a more scalable system. In: CVPR, pp. 1187–1196 (2019)
129. Nguyen, L.T., et al.: Stochasticity and skip connection improve knowledge transfer. In: EUSIPCO, pp. 1537–1541. IEEE (2021)
130. He, X., et al.: Multi-task zipping via layer-wise neuron sharing. NeurIPS **31** (2018)
131. Shen, C., et al.: Customizing student networks from heterogeneous teachers via adaptive knowledge amalgamation. In: ICCV, pp. 3504–3513 (2019)
132. Luo, S., et al.: Knowledge amalgamation from heterogeneous networks by common feature learning. arXiv preprint arXiv:1906.10546 (2019)
133. Ye, J., et al.: Amalgamating filtered knowledge: Learning task-customized student from multi-task teachers. arXiv preprint arXiv:1905.11569 (2019)
134. Rusu, A.A., et al.: Policy distillation. ICLR (2016)

135. Gupta, S., et al.: Cross modal distillation for supervision transfer. In: CVPR, pp. 2827–2836 (2016)
136. Thoker, F.M., Gall, J.: Cross-modal knowledge distillation for action recognition. In: ICIP, pp. 6–10. IEEE (2019)
137. Roheda, S., et al.: Cross-modality distillation: a case for conditional generative adversarial networks. In: ICASSP, pp. 2926–2930. IEEE (2018)
138. Do, T., et al.: Compact trilinear interaction for visual question answering. In: ICCV, pp. 392–401 (2019)
139. Zhou, Z., et al.: Channel distillation: channel-wise attention for knowledge distillation. arXiv preprint arXiv:2006.01683 (2020)
140. Hu, J., et al.: Squeeze-and-excitation networks. In: CVPR, pp. 7132–7141 (2018)
141. Shu, C., et al.: Channel-wise knowledge distillation for dense prediction. In: ICCV, pp. 5311–5320 (2021)
142. Jiao, X., et al.: Tinybert: Distilling bert for natural language understanding. arXiv preprint arXiv:1909.10351 (2019)
143. Devlin, J., et al.: Bert: pre-training of deep bidirectional transformers for language understanding. arXiv preprint arXiv:1810.04805 (2018)
144. Goodfellow, I., et al.: Generative adversarial nets. NeurIPS **27** (2014)
145. Lopes, R.G., et al.: Data-free knowledge distillation for deep neural networks. arXiv preprint arXiv:1710.07535 (2017)
146. Yin, H., et al.: Dreaming to distill: data-free knowledge transfer via deepinversion. In: CVPR, pp. 8715–8724 (2020)
147. Nayak, G.K., et al.: Zero-shot knowledge distillation in deep networks. In: ICML, pp. 4743–4751. PMLR (2019)
148. Micaelli, P., Storkey, A.J.: Zero-shot knowledge transfer via adversarial belief matching. NeurIPS **32** (2019)
149. Fang, G., et al.: Contrastive model inversion for data-free knowledge distillation. arXiv preprint arXiv:2105.08584 (2021)
150. Fang, G., et al.: Up to 100x faster data-free knowledge distillation. In: AAAI, vol. 36, pp. 6597–6604 (2022)
151. Kimura, A., et al.: Few-shot learning of neural networks from scratch by pseudo example optimization. arXiv preprint arXiv:1802.03039 (2018)
152. Shen, C., et al.: Progressive network grafting for few-shot knowledge distillation. In: AAAI, vol. 35, pp. 2541–2549 (2021)
153. Liu, P., et al.: Ddflow: learning optical flow with unlabeled data distillation. In: AAAI, vol. 33, pp. 8770–8777 (2019)
154. Radosavovic, I., et al.: Data distillation: towards omni-supervised learning. In: CVPR, pp. 4119–4128 (2018)
155. Zhang, W., et al.: Reliable data distillation on graph convolutional network. In: SIGMOD, pp. 1399–1414 (2020)
156. Cazenavette, G., et al.: Dataset distillation by matching training trajectories. In: CVPR, pp. 4750–4759 (2022)
157. Fang, G., Song, J., Shen, C., Wang, X., Chen, D., Song, M.: Data-free adversarial distillation. arXiv preprint arXiv:1912.11006 (2019)
158. Liu, R., et al.: Teacher-student compression with generative adversarial networks. arXiv preprint arXiv:1812.02271 (2018)
159. Yoo, J., et al.: Knowledge extraction with no observable data. NeurIPS **32** (2019)
160. Zhai, M., et al.: Lifelong gan: Continual learning for conditional image generation. In: ICCV, pp. 2759–2768 (2019)
161. Belagiannis, V., et al.: Adversarial network compression. In: ECCV Workshops, pp. 0–0 (2018)
162. Liu, P., et al.: Ktan: knowledge transfer adversarial network. In: IJCNN, pp. 1–7. IEEE (2020)
163. Wang, W., et al.: Gan-knowledge distillation for one-stage object detection. IEEE Access **8**, 60719–60727 (2020)

164. Aguinaldo, A., et al.: Compressing gans using knowledge distillation. arXiv preprint arXiv:1902.00159 (2019)
165. Chen, H., et al.: Distilling portable generative adversarial networks for image translation. In: AAAI, vol. 34, pp. 3585–3592 (2020)
166. Li, M., et al.: Gan compression: efficient architectures for interactive conditional GANs. In: CVPR, pp. 5284–5294 (2020)
167. Zhang, L., et al.: Wavelet knowledge distillation: towards efficient image-to-image translation. In: CVPR, pp. 12464–12474 (2022)
168. Gou, J., Yu, B., Maybank, S.J., Tao, D.: Knowledge distillation: a survey. Int. J. Comput. Vis. **129**(6), 1789–1819 (2021)
169. Wang, L., Yoon, K.J.: Knowledge distillation and student-teacher learning for visual intelligence: a review and new outlooks. IEEE Trans. Pattern Anal. Mach. Intell. (2021)

A Geometric Perspective on Feature-Based Distillation

Ilias Theodorakopoulos and Dimitrios Tsourounis

Abstract Feature-based Knowledge Distillation (FKD) is a method for guiding the activations at the intermediate layers of a Convolutional Neural Network (CNN) during training. It has recently gained significant popularity as either a standalone or complementary method of Knowledge Distillation (KD). Most techniques however, handle the teacher-to-student knowledge transfer in a statistical or probabilistic manner. In this chapter, we propose a family of efficient loss functions that aim to transfer the geometry of activations from intermediate layers of a teacher CNN to the activations of a student CNN model at matching spatial resolutions through a novel Feature-based Knowledge Distillation (FKD) method. We discuss the challenges of geometric methods, provide connections with manifold-to-manifold comparison and regularization techniques, and draw some interesting connections to the field of random graphs. Experiments on benchmark tasks show evidence that by focusing on replicating the feature relationships only across local neighborhoods, results in better performance. Furthermore, the definition of neighborhoods important for sufficient performance, with neighborhoods defined over parsimonious graphs such as the Minimal Spanning Tree achieving better results that standard kNN rule. Finally, we present a case study on data-free Knowledge Distillation in the field of offline handwritten signature verification. The case study demonstrates a way to harness knowledge from an expert CNN model to enhance the training of a new model with different architecture, using only external (task-irrelevant) data. Results indicate that both geometric FKD and its combination with standard KD techniques can effectively create new models with better performance than the expert teacher model.

Keywords Knowledge distillation · Geometric regularization · Manifold dissimilarity · Data-free knowledge transfer · Offline signature verification

I. Theodorakopoulos (✉)
Electrical and Computer Engineering Department, Democritus University of Thrace, Komotini, Greece
e-mail: iltheodo@ee.duth.gr

D. Tsourounis
Physics Department, University of Patras, Patras, Greece
e-mail: dtsourounis@upatras.gr

W. Pedrycz and S.-M. Chen (eds.), *Advancements in Knowledge Distillation: Towards New Horizons of Intelligent Systems*, Studies in Computational Intelligence 1100,
https://doi.org/10.1007/978-3-031-32095-8_2

1 Introduction

From the advent of modern Deep Learning and the realization that deeper and larger model architectures are more capable of solving difficult vision tasks [1, 2], there is a constant quest for techniques to create efficient models for accurate inference with smaller computational footprint. This trend is particularly important in the applications domain, where system designers are often dealing with embedded hardware with resources limited in comparison to workstations and datacenters. Different routes were explored for solving these problems, including pruning and factorization techniques [3], architectural search [4], and Knowledge Distillation (KD) [5]. Most of the former techniques are aiming at eliminating redundancies within large trained models, hence creating networks with smaller footprint. The architectural search methods try to generate architectures just as big as necessary for the target task through iterative optimization processes. Knowledge Distillation methods on the other hand, are designed to harness the ability of large models to learn better representations of the training data, and transfer those directly into a smaller model.

Since the introduction of KD in the realm of deep learning by Hinton et al. [6], the focus of researchers has been shifted across different qualities of Neural Networks and data that may reflect the knowledge stored in a trained model. The original KD scheme proposed in [6] utilizes the responses (logits) of a large "teacher" neural network derived from the final layer of its architecture, and formulates the knowledge transfer as an additional objective, introduced in the training of a small "student" model. The distillation objective is a loss term, incentivizing the student to generate responses similar to those of the teacher model for the same input data, in conjunction to the optimization of the main task. This form of distillation is called response-based [5] and the idea behind it is that the teacher's knowledge is hidden in the representations generated at the final layer(s), thus by learning to mimic such responses the student model will gain much of the teacher's generalization qualities.

Shortly after response-based KD was introduced, feature-based distillation was proposed in [7], utilizing the activations in the intermediate layers of teacher model as additional guidance (hints) during the training of the student model. Following that, many researchers approached feature-based distillation from different perspectives [5], formulating loss functions based on probabilistic, statistical, contextual and other interpretations of the intermediate activations, producing a variety of methods that aim at instilling different qualities of the intermediate representations of the teacher, into the student model. In parallel, beginning with the work of Yim et al. [8] that proposed the inner product of activations from consecutive layers as the main knowledge carrier for KD, a variety of techniques for relational-based KD were presented in the literature, focusing on the student model to mimic certain qualities of the evolution of activations in consecutive layers, rather than the activations alone.

In this chapter we will focus on feature-based KD from a geometry perspective, discussing the challenges and possible approaches on formulating feature-based KD as a problem of learning similar manifolds of local activations in corresponding layers of teacher and student models. The rest of this chapter is organized as follows: In

Sect. 2 we formulate the problem of feature-based knowledge distillation (FKD) and present some indicative existing approaches. In Sect. 3 we present the connections with manifold-to-manifold comparison and regularization techniques and discuss challenges for the different approaches, drawing some interesting connections to the field of random graphs. In Sect. 4 we present a method to design geometric loss functions for feature-based KD. In Sect. 5 we present experiments using FKD on standard KD configurations and in Sect. 6 a case study on a practical problem from the field of offline signature verification (OSV). Finally, Sect. 7 discusses the scalability of the proposed FKD method while 8 presents the conclusions.

2 Prior Art on Feature-Based Knowledge Distillation

2.1 Definitions

Loosely defined, feature-based KD (FKD) is the process of using the activations in the intermediate layers of a trained "teacher" model to guide the training of the "student" model. Its main different with response-based KD is that in the latter, the activations involved in distillation are usually drawn from layers that generate global representations of the input multidimensional signal, thus lacking the spatially localized information that characterizes the activations in the output of intermediate convolutional layers. In FKD, for a typical input signal with 2D spatial domain such as images, the activations in each intermediate convolutional layer are treated as a 3D volume of multiple feature maps containing local features across the channel dimension, arranged in a regular 2D grid. More specifically, the output of a layer with spatial dimensions of $H \times W$ and C channels is considered as a collection of $H \cdot W$ feature vectors in $\mathbb{R}^C$, each representing the qualities of the input signal over a $H \times W$ regular grid.

The process of KD using such responses is accomplished via the addition of appropriate differentiable loss terms into the training objective function. Essentially FKD, as any other similar form of KD, is a regularization mechanism, that utilizes the teacher model's activations to regularize student's activations at corresponding positions of their architectures, using some criteria reflected in the chosen distillation loss terms. A schematic overview of this process is shown in Fig. 1. In the general case, the layer pairings between teacher and student models do not always have the same number of channels or the same spatial resolution, although the latter restriction is almost always being followed by FKD methods to avoid comparing signal qualities from incompatible scales.

In general, the distillation loss functions used in FKD methods fall under the following formulation:

$$L_{FKD}(X_S, X_T) = \mathcal{L}_F(f_S(X_S), f_T(X_T)) \tag{1}$$

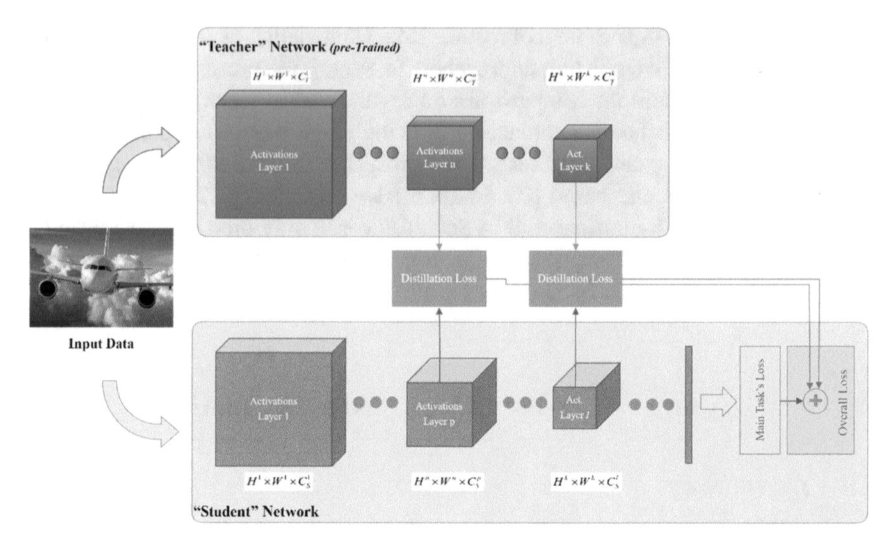

Fig. 1 Schematic overview of Feature-based Knowledge Distillation (FKD), for transferring qualities of teacher model's intermediate representations to the student model's feature maps having corresponding spatial dimensions (H × W) but arbitrary number of channels C

where X_S, X_T are the feature maps in the outputs of a student–teacher layer pair respectively and $\mathcal{L}_F$ is a similarity or dissimilarity function, measuring how well the two feature maps match. The functions f_S and f_T are used to transform the feature maps appropriately, to facilitate the measurement of their (dis)similarity according to the criteria of each method. The transformation functions can be anything from simple linear transformations aiming to project feature maps of different dimensionality on a common vector space for measuring divergence, up to kernel and domain transformations.

2.2 *Related Work*

Several methods have been proposed so far, approaching FKD from slightly difference angles. Since a detailed survey of the literature is out of scope for this chapter, we present some methods which are indicative of the different tools that researchers use to formulate distillation loss functions that follow the general scheme defined in Eq. (1). As stated earlier in the preceding section, FKD was first introduced in [7] were the FKD loss term was the sum of square Euclidean distances between corresponding feature vectors from teacher (hints) and student's activations, with the addition of a learnable linear projection layer introduced whenever needed to transform the teacher's activations to the same dimensionality as the student's. Later, in [9, 10] authors followed an approach of indirect comparison, utilizing some form of attention that weighted more some local features in expense of the less informative ones, also utilizing kernel transformations of the teacher's and student's activations in order to overcome the difference in features' dimensionality between the two models.

Following a probabilistic modeling approach, authors in [11] used kernel density estimation to model the probability distribution of activations in the features space of teacher and student, and subsequently utilized KL divergence as the distillations loss term. In [12], authors used a purely statistical approach, trying transfer the channel-wise statistics of the teacher into the student model, adding though a layer that creates affined-transformed instances of teacher features, which are evaluated by the teacher model to ensure that the transformed features do not alter the result. This mechanism is used as an online evaluator of whether the desired statistics are reliably transferred to the student. Recently, in a technique inspired by vision transformers and aiming to introduce a form of semantic context to the pairing between teacher and student layers, authors in [13] proposed a method where the activations from different layers of teacher and student are projected to a low dimensional space where using a query-key mechanism, the student model can focus on what it seems to be the most informative activations across all involved layers.

3 Geometric Considerations on FKD

3.1 Local Manifolds and FKD

In all methods described in the previous section, there is no consideration regarding the geometric properties of the feature representations taken into account during the formulation of the distillation loss functions. On one end, the direct comparison methods only rely on feature similarities in corresponding spatial locations of teacher and student's activations, while on the other end, methods that utilize the kernel trick take into account all pairwise similarities within the two feature maps. While the attention mechanisms induce some weighting on the contribution of each local feature, the criteria are usually statistical and not based the intrinsic geometry of the activations. The method of [13] although may be able to capture some of the geometric properties of the activations through the sparse nature of similarities induced by the learned query-key dynamics, the mechanism is not explicit and it is hard to interpret the information of these connections.

In this chapter we argue that the problem of FKD is in its essence a problem of matching the manifolds of local activations at the different layers of teacher and student models. Under this perspective, let's consider that the observed feature maps X_S and X_T drawn from an intermediate layer of the student and teacher model respectively for the same input, are the result of a sampling of the corresponding manifolds of local activations $\mathcal{M}_S$ and $\mathcal{M}_T$ respectively. Different inputs generate different samplings of the underlying manifolds, but the manifold structures themselves remain constant if the models are not updated. In this context, the target of FKD could be defined as the minimization of the manifold-manifold distance between the student's and teacher's manifolds of local activation, therefore

$$L_{FKD} = d(\mathcal{M}_S, \mathcal{M}_T) \approx \tilde{d}(X_S, X_T) \tag{2}$$

where $d()$ is an appropriate manifold-manifold distance function and $\tilde{d}()$ is a function that can approximate $d()$ from a sample of the underlying manifolds.

The manifolds of local sub-structures in signals and images, rely on the nature of the signals at the scale they are studied. The simpler the structure inside the scale of reference, the fewer the parameters needed to describe the structures, thus the smaller the intrinsic dimensionality of the manifold. For example, the manifold of local patches from smooth images with periodic boundaries, is a 2D surface in the form of a torus in 3 dimensions with self-intersecting segments [14]. In the case of binary images with smooth contours (cartoon images), the manifold of local patches for example is equivalent to a cylinder embedded in 3 dimensions. Such properties of signals in local scale are well studied [14, 15] in the past with applications in manifold regularization of signal restoration inverse problems [14].

Despite the simplicity of the idea to formulate FKD as a manifold-manifold distance minimization problem, there are some challenges intrinsic to the domain of FKD. First, natural images usually contain mixtures of smooth, structured and stochastic content, thus the underlying manifolds are more complex than those described above. Second, there are two opposing forces acting on the complexity manifolds: as the depth of representation increases, so does the non-linearity, potentially enabling models to learn representations with simpler structure but on the other hand, the receptive field of each kernel increases effectively representing larger and thus more complex areas of the initial signal. Third, as the depth of representation increases inside a deep model, the spatial resolution decreases, thus offering fewer samples of the underlying structure with each input in order to estimate the manifold-manifold distance. Finally, since in most realistic problems the input signals do not always include a uniform sample of all possible local content, a feature map generated by a single input cannot be—in general—considered as a uniform sample of the underlying manifold structure.

3.2 Manifold-Manifold Distance Functions

The use of conventional methods for estimating the distance between manifolds can be impractical due to various issues. The most common approach to measuring the similarity or dissimilarity between manifolds is based on the assumption that they consist of locally linear subspaces. The overall dissimilarity between two manifold structures is then calculated by comparing individual subspaces using concepts such as principal angles [16–18]. Other methods for constructing manifold-to-manifold dissimilarity functions include Tangent Distance (TD) [19], Grassmannian distances

[20], or reconstruction errors from Local Linear Coding performed in manifold subspaces [21]. However, these methods can be greatly affected by the limited and non-uniform sampling of the underlying manifolds in a FKD setting. Furthermore, in most cases, the distance functions are not differentiable and cannot be easily approximated in order to formulate an appropriate Distillation loss function with the desired behavior.

A more suitable approach to manifold-manifold distance estimation, designed for local manifolds and the challenges they pose, is the work presented in [22]. In that work, two compared manifolds $\mathcal{M}_1$ and $\mathcal{M}_2$ are represented through the respective neighborhood graphs $G_1(E_1, V_1)$ and $G_2(E_2, V_2)$, with E_n and V_n denoting the sets of nodes and edges respectively. The two graphs are constructed from two corresponding sets of features $\mathbf{X}_n \in \mathbb{R}^{N \times C}$ with N vectors in C dimensions each, using any relevant criterion. The weights between nodes i and j of the n-th graph are denoted as w_{ij}^n and are derived via a standard heat kernel. For each of the graphs, a permutation $\mathbf{p}$ of its nodes is computed using spectral reordering, from the indices of the sorted Fiedler vector of each graph's Laplacian. Each of these permutations has the property of minimizing the distance of strongly connected nodes on the permutation, or alternatively concentrating the strongest weights closest to the diagonal of the reordered graph's adjacency matrix. Using this property, the dissimilarity between $\mathcal{M}_1$ and $\mathcal{M}_2$ is defined via the notion of reordering efficiency, by applying to each graph the permutation $\mathbf{p}$ from the other, and quantifying how well the large weights are still concentrated along the main diagonal of the adjacency matric. The rationale behind this is that since the nodes of the two graphs are considered to have one-to-one correspondence, the more similar the underlying manifolds, the more compatible will be either of the permutations for reordering both graphs. Therefore, the reordering efficiency of a graph G with weights W, by a permutation $\mathbf{p}$ is defined as

$$E_r(G, \mathbf{p}) = \frac{\sum_{i,j} \tilde{w}_{ij} \cdot b_r(p_i, p_j)}{\sum_{i,j} \tilde{w}_{ij}} \tag{3}$$

where

$$b_r(i, j) = \begin{cases} 1, & |i - j| \leq r \\ 0, & |i - j| > r \end{cases}, \quad \tilde{w}_{ij} = w_{p_i p_j}, \quad r \in \mathbb{N}, \quad i, j \in \{1, 2, ..., N\} \tag{4}$$

and finally, the distance between $\mathcal{M}_1$ and $\mathcal{M}_2$ is defined as

$$d(\mathcal{M}_1, \mathcal{M}_2) = 1 - \frac{\left(\frac{E_r(G_1, \mathbf{p}_2)}{E_r(G_1, \mathbf{p}_1)} + \frac{E_r(G_2, \mathbf{p}_1)}{E_r(G_2, \mathbf{p}_2)}\right)}{2} \tag{5}$$

Subsequently, authors generalized the 1-dimensional ordering to a multi-dimensional ordering using Minimal Spanning Tree (MST) as the structure that defines neighborhoods between nodes through their geodesic distance over the MST.

The generalization offered best results when the MST was computed in the original feature space, eliminating the need to compute the Fiedler vector or any other mapping of the nodes' features.

The neighborhood mask M is defined according to Eq. (4), simply computed using the geodesic distance instead of distance over the 1D permutation $\mathbf{p}$, the dissimilarity between the manifold structures $\mathcal{M}_1$ and $\mathcal{M}_2$ expressed by the corresponding neighborhood graphs G_1 and G_2, with respective neighborhoods M_1 and M_2, can be defined as:

$$d(\mathcal{M}_1, \mathcal{M}_2) = 1 - \frac{\left(\frac{E_r(G_1,M_2)}{E_r(G_1,M_1)} + \frac{E_r(G_2,M_1)}{E_r(G_2,M_2)}\right)}{2} \tag{6}$$

Essentially, the distance measure in Eq. (6) is constructed by taking the average of the normalized bi-directional reordering efficiency of every neighborhood graph, which is reordered based on the opposite MST. The rationale behind this formulation is that the more similar the data manifolds $\mathcal{M}_1$ and $\mathcal{M}_2$ are, the more compatible would be the respective neighboring masks M_1 and M_2 with the opposite neighborhood graph G_2 and G_1. Thus, the reordering efficiencies $E_r(G_1, M_2)$ and $E_r(G_2, M_1)$ would become larger, approaching the values obtained by matching mask and graph pairs, with the overall distance approaching zero. Despite the efficiency of the distance measure defined in Eq. (6) in estimating dissimilarity between local manifolds, it has two major limitations when used as a regularization function in deep learning setting. First, even an unidirectional variation of Eq. (6) necessitates computing the MST and k-NN graphs for each compared layer of the student CNN at each iteration, since student model is constantly updated. Secondly, the derivation of a term $E_r(G_S, M_T)/E_r(G_S, M_S)$ involves factors that affect the connectivity of the M_S that are very difficult to approximate in a differentiable manner. In the next section we present the necessary modification in order to formulate a differentiable and computationally efficient geometric Distillation loss based on this mechanism.

3.3 Interpretation of Graph Reordering as a Tool Measuring Similarity

We can derive some interesting insights regarding the underlying mechanism of the dissimilarity measure of Eq. (5) from the field of range-dependent random graphs, and especially the work of Grindrod in [23]. In that work the author shows that the permutation of the nodes resulting from spectral reordering of a graph, is the solution to the inverse problem of defining the parameters of a model of random graphs that most likely generated the observed graph. Specifically, consider the family of range dependent graphs in which the probability of an edge connecting two nodes i and j is controlled by the distance k of these nodes on an enumeration of the nodes $\mathbf{p}$ and is provided by a probability function in power-law form

$$f(k) = a\lambda^{k-1}\ ,\ \ k = \left|p_i - p_j\right|,\ \ a, \lambda \in (0, 1] \tag{7}$$

Then, given a weighted graph with weights $w_{ij} \in [0, 1]$, these weights can be interpreted as edge probabilities, and the permutation resulting from the spectral reordering of this graph provides the maximum likelihood estimation for the generating permutation of this graph under the model of Eq. (7).

According to this and in the context of pair-wise signals comparison, using the criterion of Eq. (5) for assessing (dis)similarity of the underlying manifolds, equals to the assumption that the underlying graphs of local features can be modeled as power-law random graphs. Thus, Eq. (3) is just a simple function to evaluate the validity of the permutation $\mathbf{p}$ as a generator of the graph with weights w_{ij}. In fact, this function could be any function that quantifies properties of the graph that can distinguish a properly identified neighborhood from a random subset of the graph's nodes. Regarding the validity of the power-law graph assumption, despite there is no definitive answer of what type of random graph model is best to model graphs of local descriptors, there are significant evidence [24, 25] that the scale invariance occurring in (at least) natural images is connected to the power-law statistics of objects and region sizes. Given that the power-law range dependent graphs show both significant small-world characteristics of localized clustering and a natural hierarchy of edges across different scales, the hypothesis that such model may sufficiently describe the structure of graphs generated from local features in similar signals is not invalid.

Finally, the generalization followed in the criterion of Eq. (6) using MST of the respective features departs from the strict definition of Eq. (7) but does not change the nature of the criterion. To understand this, consider that the MST is essentially a connected acyclic graph connecting all nodes in a skeletonized form of the underlying structure, therefore is just a permutation of the nodes with the ability to have multiple branches. Additionally, due to its minimalistic nature is less prone to topological short circuits [26], thus less likely to connect nodes from (geodetically) distant parts of the underlying manifold with short paths. Therefore, neighbors defined via the MST are more likely to be real neighbors on the data structure than through other proximity graphs such as k-NN or ε-ball, even if not all neighbors are selected each time. This means that the use of MST essentially provides some additional degrees of freedom to the generator permutation, which is fact has proven to be beneficial in several tasks [22]. In fact, recent evidence [27] indicate that neighborhoods defined over non-parametric proximity graphs such as MST and Gabriel graphs, provide significant advantages for NN-based regression.

4 Formulating Geometric FKD Loss Functions

Drawing on the ideas presented in earlier sections, here we describe a way to formulate efficient loss functions for FKD, that utilize geometric criteria of dissimilarity based on the mechanisms outlined in Sect. 3.2. The objective here is to design loss functions that follow the general form of Eq. (1), facilitating vastly different CNN architectures between student and teacher, with the only requirement being the presence of matching spatial dimensions at the outputs of different layers at the two networks. In order to regularize the spatial activations $\mathbf{X}_S \in \mathbb{R}^{H \times W \times C_S}$ at a layer of the student CNN, we need to formulate a differentiable loss function that quantifies the affinity between $\mathbf{X}_s$ and a set of corresponding activations, $\mathbf{X}_T \in \mathbb{R}^{H \times W \times C_T}$ from the teacher CNN. The assumption here is that that the dimensionality of the local features' C_S and C_T is arbitrarily different in the student and teacher models, but the matching spatial dimensions offer a trivial one-to-one correspondence between the local descriptors from the two models. One way to overcome the limitations of conventional methods for estimating manifold-manifold distance is to use a loss function that enforces analogous affinity patterns between the corresponding vectors in both the teacher and student feature sets. This can be achieved by defining the affinity pattern of a feature vector as a function based on its distance to other vectors in the same set. Such an approach allows for regularization on the affinities within the student's vector space, enabling the teacher's activations to exist in a different dimensionality feature space. This method is similar to those used in other FKD approaches (e.g. [9–11]). In the present context, a general expression for the distillation loss is:

$$L_{FKD} = f(A_S(\mathbf{X}_S), A_T(\mathbf{X}_T)), \;\; A_i : \mathbb{R}^{HxWxC_i} \rightarrow \mathbb{R}^{N \times N}, N = H \cdot W \qquad (8)$$

where A_S and A_T are functions that measure pairwise similarities or distances inside the C_i-dimensional vector spaces at the student's and teacher's sides respectively. In the rest of this section, we will explore two methods for defining the similarity function. The first method is to directly compare the patterns of neighboring activation features between the student and teacher models. The second method is more lenient, allowing for more freedom in the student model's activations, by comparing only the ratio of the sum of distances to each feature's neighbors to the sum of distances to all features of the activation map.

4.1 Neighboring Pattern Loss

The relationship between sets of vectors can be quantified either as similarity or dissimilarity (distance) quantities. Methods for geometrical regularization often utilize similarity (e.g. [10, 22, 28]), since there is a theoretical link [29] between the Laplacian of data samples to the Laplace–Beltrami operator on their underlying manifold for Gaussian kernels. However, the disadvantage of using Gaussian kernel for similarity in the context of FKD is two-fold. Firstly, it involves computing exponential terms at each location and training iteration. Secondly, it incorporates adjustable parameters related to the kernel functions, associated with the dimensionality of the features' space in non-trivial manner, that can significantly affect the quality of the distillation.

To overcome these issues, we can use a simple definition for the neighboring patterns, by computing the pairwise squared Euclidean distances between each feature vector and other vectors and then, normalizing it by the sum of distances to all other vectors in the set. This method requires less computation, the value is bounded, and does not rely on any additional parameter that requires tuning. Similar approaches have also been used in previous works for relational knowledge distillation using global features [30] and local features [28]. Therefore, by momentarily disregarding the spatial distribution of the activation features, we define the matrix $\mathbf{D} \in \mathbb{R}^{N \times N}$ to hold the neighboring patterns of all vectors in a set of N activation features with C dimensions stored in $\mathbf{X} \in \mathbb{R}^{N \times C}$, as follows:

$$d_{ij} = \frac{\left\|\mathbf{x}_i - \mathbf{x}_j\right\|_2^2}{\sum_j \left\|\mathbf{x}_i - \mathbf{x}_j\right\|_2^2} \tag{9}$$

An important step here, is to promote the transfer of the geometric characteristics of the local features' distribution instead of their global statistics. To accomplish that, the loss function should specifically target the impact of regularization on local neighborhoods within the desired manifold structure, therefore we formulate the Neighboring Pattern Loss to take into account only the neighbors of each vector. An easy way to overcome the drawbacks explained in Sect. 3.2 and construct a distillation loss function that follows the spirit of Eq. (6), is to rely on the spatial correspondence between features and define the neighbors of each feature vector only in the teacher's side. This eliminates the need to estimate the neighborhood for every forward pass based on the constantly updating student model, as it can be computed once for each input datum based on the fixed (static) teacher model. Ultimately, the Neighboring Pattern loss (NP) can be expressed as:

$$L_{NP} = \left\|\tilde{\mathbf{D}}^S - \tilde{\mathbf{D}}^T\right\|_F \tag{10}$$

with

$$\tilde{\mathbf{D}}^n = \mathbf{D}^n \odot \mathbf{M}^T, \; \mathbf{M}^T \in \{0, 1\}^{N \times N} \tag{11}$$

and $\|\cdot\|_F$ is the Frobenius norm.

In Eq. (11), $\mathbf{M}^T$ is a binary mask indicating the neighborhood of each vector. Specifically, in the position (i, j) of the mask, the entry is one (1) if the jth vector of the teacher's feature set is among the neighbors of the ith vector, and zero (0) otherwise.

4.2 *Affinity Contrast Loss*

Another direction to formulate the loss function for geometric regularization can be derived from the analogy with Eq. (3) and the effectiveness of this simple criterion as a function to measure the dissimilarity of the local structures from different types of signals [22]. In the context of geometric FKD and the previous considerations, it is possible to estimate the geometric dissimilarity between sets of activation features by using a function that employs the normalized square Euclidean distances defined in Eq. (9) as the pairwise comparison measure instead of similarity. Again, the neighborhood is computed only on the teacher's side, according to Eq. (11). The loss function is constructed using the measure of Local Affinity Contrast, which is defined for a set of N feature vectors using a neighborhood mask $\boldsymbol{M}$ and the normalized pairwise distances $\boldsymbol{D}$, and finally, it is formulated as:

$$\mathbf{J}_D^M = \frac{\sum_{j=1}^{N} d_{ij} \cdot m_{ij}}{\sum_{j=1}^{N} d_{ij}}, \quad \mathbf{J} \in \mathbb{R}^N \tag{12}$$

where d_{ij} is provided by Eq. (9). This is essentially a measure of how close the neighbors of each feature are compared to the overall distribution of distances to it. After all, we can describe the Affinity Contrast Distillation loss (AC) for the activations of student model simply as:

$$L_{AC} = \left\| \mathbf{J}_{D^S}^{M^T} - \mathbf{J}_{D^T}^{M^T} \right\|_2 \tag{13}$$

In principle, the neighborhood for each feature can be defined by any relevant criterion such as proximity rules (e.g. k-NN, ε-ball etc.) or rules such Eq. (4) that bare some interesting qualities as discussed in Sect. 3.3. In the following investigations the primary criterion to define the neighborhoods is via the MST of each features set $\mathbf{X} \in \mathbb{R}^{N \times C}$, where the neighborhoods are specified via a fixed radius r of geodesic distance between nodes on the MST. Therefore, similarly to Eq. (4), the neighborhood of each feature can be determined by calculating the MST on the activation features of the teacher's model. Thus, the neighborhood mask could be defined as follows:

$$\mathbf{M}_r^T \in \{0, 1\}^{N \times N}, m_{ij} = \begin{cases} 1, & g_{ij}^{G_{MST}^T} \leq r \\ 0, & g_{ij}^{G_{MST}^T} > r \end{cases} \tag{14}$$

where $g_{ij}^{G_{MST}^T}$ is the geodesic distance between the ith and jth node on the MST computed on the teacher's activation features G_{MST}^T. The investigation in the following Experimental Section also includes an analysis on the effectiveness of the neighborhood definition method. So, we will compare the MST-based criterion to the k-NN rule that is constructed a neighborhood mask $\mathbf{M}_k^T$ when applied on the activation features of the teacher's model indicating the k-nearest neighbors of each feature in the feature space. Note that the teacher model remains fixed (not updated) during training and thus, the neighborhoods resulting from either the MST or k-NN can be computed offline for each training sample only once. Hence, the overall computational overhead of the presented FKD approach is reduced as the necessary extra computations in each training iteration are focused only on calculating pair-wise distances between the local features D^S.

5 Experimental Verification

In this section, we present experiments designed to validate that the described FKD losses can serve as an efficient regularization mechanism for improving the generalization of the student models in standard KD settings and basic commonly used datasets. We compare the above regularization criteria in terms of accuracy, demonstrating that FKD can be easily combined with response-based KD, improving the results in most cases. We also highlight the importance of utilized neighborhood rules to the effectiveness of the regularization.

5.1 Materials and Methods

This section aims to demonstrate FKD between models with different architectures and thus, many CNNs with various characteristics were used in the experiments. Table 1 provides an overview of the models' architectures used. The first model is a simple vanilla CNN that consists of 3 (three) convolutional layers and 2 (two) fully connected (dense) layers. It has a relatively small number of parameters, totaling around 146,000, and requires approximately 12.35 million Multiply-Accumulate (MAC) operations for inference when the input is a color image with resolution of 32×32 px. This model serves as an example of a lightweight architecture, which can be used as a student in KD schemes with the goal of improving accuracy for deployment on resource-constrained embedded devices.

The second model is the Network-in-Network (NiN) [31] architecture, with three convolutional layers using kernel sizes of 5×5 and 3×3, and two layers with 1×1 kernels. This model can be considered medium-to-large-sized, with approximately 967,000 learnable parameters while requiring 222.5 million MAC operations for inference on a 32×32 px image. Both the teacher and student models were built on this architecture since NiN models were used in both roles in the experimental investigation.

The third category of models is based on the ResNet [2] architecture, and the two representatives of the ResNet family being used are ResNet-20 and ResNet-32. ResNet-20 comprises six (6) residual blocks, divided into three (3) groups, each containing two convolutional layers with 3×3 kernels and batch normalization, operating in different spatial scales. On the other hand, ResNet-32 follows similar design with the ResNet-20 but having five (5) blocks in each group instead of three (3). ResNet-20 has 273,000 parameters and requires 41 million MAC operations for inference on a 32×32 image, while ResNet-32 has 468,000 parameters and

Table 1 CNN architectures and activations' size for the used models

Simple CNN		NiN		ResNet-20		ResNet-32	
Layer name	Activation size	Layer name	Activation size	Layer/block name	Activation size	Layer/block name	Activation size
Input	32 × 32 × 3	Input	32 × 32 × 3	Input	32 × 32 × 3	Input	32 × 32 × 3
Conv1	32 × 32 × 32	Conv1	32 × 32 × 192	Conv1	32 × 32 × 16	Conv1	32 × 32 × 16
Pool1	16 × 16 × 32	Cccp1	32 × 32 × 160	Group0_block0	32 × 32 × 16	Group0_block0	32 × 32 × 16
Conv2	16 × 16 × 32	Cccp2	32 × 32 × 96	Group0_block1	32 × 32 × 16	Group0_block1	32 × 32 × 16
Pool2	8 × 8 × 32	Pool1	16 × 16 × 96	Group0_block2	32 × 32 × 16	Group0_block2	32 × 32 × 16
Conv3	8 × 8 × 64	Conv2	16 × 16 × 192	Group1_block0	16 × 16 × 32	Group0_block3	32 × 32 × 16
Pool3	4 × 4 × 64	Cccp3	16 × 16 × 192	Group1_block1	16 × 16 × 32	Group0_block4	32 × 32 × 16
FC1	64	Cccp4	16 × 16 × 192	Group1_block2	16 × 16 × 32	Group1_block0	16 × 16 × 32
FC2	#Classes	Pool2	8 × 8 × 192	Group2_block0	8 × 8 × 64	Group1_block1	16 × 16 × 32
		Conv3	8 × 8 × 192	Group2_block1	8 × 8 × 64	Group1_block2	16 × 16 × 32
		Cccp5	8 × 8 × 192	Group2_block2	8 × 8 × 64	Group1_block3	16 × 16 × 32
		Cccp6	8 × 8 × 100	Pool	64	Group1_block4	16 × 16 × 32
		Pool	100	FC	#Classes	Group2_block0	8 × 8 × 64
		FC	#Classes			Group2_block1	8 × 8 × 64
						Group2_block2	8 × 8 × 64
						Group2_block3	8 × 8 × 64
						Group2_block4	8 × 8 × 64
						Pool	64
						FC	#Classes

requires 70 million MAC operations. Despite having fewer parameters than NiN models, ResNets can typically achieve higher accuracy taking advantage from their architectural characteristics, like the bypass (or skip) connections that equipped the residual blocks. As a result, these architectural features produce activations with different qualities at the output of each residual block. Hence, their scientific worth to this study is for assessing the ability of the investigated FKD mechanisms to transfer knowledge between teacher and student with very different architectures.

The image classification task is evaluated in all the experiments as it allows a more direct way to evaluate the effectiveness of the trained models. Three benchmark datasets, CIFAR10, CIFAR100 [32], and SVHN [33], were used, covering a wide range of difficulty levels. Each dataset consists of 32×32 px color images. CIFAR10 includes 10 visual categories, with a training set of 50 k images and a validation set of 10 k images. CIFAR100 contains 50 k training and 10 k validation images from 100 classes. The SVHN dataset contains 73,257 single-digit training and 26,032 validation images, extracted from real-world images of numerical digits. Table 1 shows the size of activations in the output of each individual layer or layer block of the utilized architectures for inputs of 32×32 pixels.

The experiments were run on a personal PC using two NVIDIA GTX 1080 Ti GPUs. The training process involved running 120 epochs and utilizing the Stochastic Gradient Descent (SGD) optimizer. The learning rate was set to 0.01 initially and was reduced by a factor of 0.1 every 40 epochs for the vanilla CNN models, while it was reduced once at the 100th epoch for NiN models. In the case of ResNet models, the learning rate was initialized to 0.1 and was also reduced by the same factor at the 100th epoch. To ensure fair comparisons, the same random seed was used in all training sessions to initialize the models with the same random parameters. Also, this option enables that the training data was randomly permuted in the same way across all training iterations. Furthermore, no image augmentation process was applied in order to remove external variables from the results.

5.2 Knowledge Distillation from Large Teacher to Small Student Models

The first experimental setting is the typical KD case where the goal is to improve the performance of a small student model using a single or an ensemble of bigger teacher models. To enable knowledge transfer from a more complex model to a simpler one, in these experiments, models with NiN architecture are considered as the complex models and were trained on all the classification tasks in order to be used as the teachers. In the role of the student was utilized a model with a Simple CNN architecture. The teacher models were able to achieve 86.19% classification accuracy on the CIFAR10 task, 63.24% accuracy on the CIFAR100 task, and 95.57% accuracy on the SVHN classification task.

Choosing the positions for FKD in the student's architecture between the different layers is straightforward. The first layer was excluded from the regularization process, as its purpose is to create a set of primitive filters, therefore no significant differences in behavior of different models at this layer are expected. Consequently, the output activation tensors from the Conv2 and Conv3 layers were subjected to geometric regularization with spatial grids of 16×16 and 8×8 respectively. Now the layers on the teacher's architecture are defined to estimate the neighborhoods and calculate the target affinity relations for the student's activations. The layers of the teacher were selected following the rule of using the deepest possible feature representations for each spatial grid, considering that the supervision of the student will be most efficient with the most informative representations that the teacher has to offer at each spatial scale. Specifically, the student's Conv2 layer was matched with the teacher's Cccp4 layer, and the student's Conv3 layer was paired with the teacher's Conv3 layer. Thus, the student's feature representations are regularized with the activations from the teacher in two positions. It is important to note here that the dimensionality of activation features differs between the two models, since the feature vectors from the teacher have dimensionality equal to 192 while from the student have dimensionality of 32/64.

The multi-loss function used in the training is a weighted sum of the regular multinomial logistic classification loss and a loss term for each of the regularized layers based on the applied function. Hence, the overall loss function can be written as:

$$L = \alpha_1 L_{\text{classification}} + \alpha_2 L_{NP/AC}^{Conv2} + \alpha_3 L_{NP/AC}^{Conv3} \tag{15}$$

where α_i are the contribution coefficients corresponding to each term of the overall loss. The coefficients α_1 and α_3 are set to 1, and α_2 is set to 0.1. To ensure training stability, we found that following the general rule of thumb of assigning a smaller contribution to earlier layers (such as Conv2) compared to deeper ones in the overall loss function is an effective solution. To achieve this, we tuned the weighting factor α_2, so that the contribution of the regularization on Conv2 layer was approximately half that for Conv3 layer in the beginning of the training.

In this set of experiments, a baseline student model was trained for all tested tasks without KD. Subsequently, the student model was trained with NP and AC KD losses as described above. The CNN models were initialized with random weights (i.e. random parameters) in all the training settings. The neighborhoods were defined using the MST (Minimum Spanning Tree) criterion with respective masks given by Eq. (14), and varying the radius r via testing three different values. After each training, the best accuracy of the respective trained model on the corresponding test set is presented in Table 2.

In the conducted experiment, it was observed that the geometric regularization has a beneficial effect on the performance of the student model, especially for the most challenging classification task like CIFAR100, where the accuracy improved by 2.22%, compared to SVHN, where there was a modest 0.14% improvement. The AC function yielded better results than the NP function for all tested configurations and

Table 2 Comparison between the student model (Simple CNN) from regular training and from FKD with AC or NP functions using a NiN model as the teacher. The neighborhoods are defined based on a radius r on the teacher activations' MST

Classification accuracy (%)	Reference training	FKD loss	Neighborhood radius			
Dataset			r = 2	r = 5	r = 10	r = ∞
CIFAR10	76.24	AC	77.30	77.98	**78.34**	–
		NP	77.03	77.21	78.00	76.14
CIFAR100	44.49	AC	45.14	**46.71**	46.1	–
		NP	44.90	45.42	46.03	45.74
SVHN	92.76	AC	92.8	92.84	**92.9**	–
		NP	92.79	92.8	92.82	92.8

also, using the AC criterion for regularization consistently resulted in significantly improved accuracy when compared to the reference model.

It is also noteworthy that the direct solution of regularizing all the neighboring relationships using the NP loss (r $= \infty$) performs worse when compared to its purely geometric and neighborhood-oriented counterparts. This finding highlights the significance of locality in geometric regularization, in comparison to the straightforward approach of targeting all pairwise relations of the local activations. This is particularly evident for the NP loss, which delivers its maximum benefits when r = 10 across all datasets but decreases performance in full-graph regularization (r $= \infty$), suggesting that attempting to replicate the most distant relationships could have detrimental effects on the generalization of the student model. Additionally, the results indicate that both loss functions provide better performance for large- to medium-sized neighborhoods, as opposed to solely focusing on the close vicinity of each vector in the feature space.

5.3 Comparison with Vanilla Knowledge Distillation

In order to gain some perspective on the results of Table 2, we compare the performance of the geometric FKD losses to the standard KD approach of [6] which softens the network's outputs at its last layer using temperature scaling. To ensure direct comparability, we used the same teacher model as well as the same training procedure for the student models, with the only difference being the addition of the new distillation loss term into the objective function. After exploring various configurations, we found that the optimal distillation result was achieved using a temperature scaling factor of $T = 6$ and we define the loss weight to T^2, as recommended in [6], and this was kept consistent throughout all experiments involving the standard KD.

Table 3 includes the experimental results, where the performance of the standard KD approach is compared with that of the geometric FKD methods. The best

Table 3 Comparison of geometric FKD methods and standard KD approach as well as combinations between both

Accuracy (%) / Dataset	Regular training	AC loss (best)	NP loss (best)	Standard KD [6]	AC (r = 5) and standard KD	NP (r = 10) and standard KD
CIFAR10	76.24	78.34	78.00	78.25	**80.03**	78.61
CIFAR100	44.49	**46.71**	46.03	44.71	46.10	46.22
SVHN	92.76	92.9	92.82	92.98	**93.05**	92.91

result for the geometric FKD methods is also reported on the same table for easy comparison. Since standard KD and geometric FKD aim to regulate different parts of the CNN graph, it is possible to combine the two schemes to achieve more effective regularization of the student model, since the combination could regularize the student more intensely on additional levels. To investigate this, the geometric FKD and standard KD losses were combined with the typical classification loss, using the same weights α_i, as in the cases that they were trained individually with each method. The teacher models and training hyperparameters remained the same for all experiments. To ensure a consistent setting across different tasks, the geometric losses were applied with the radius that produced the largest improvement over the baseline. These experimental results are also reported on Table 3, with the highest accuracy for each task indicated in bold.

The results indicate that the geometric FKD, specifically using the AC loss function, achieved higher accuracy values in the majority of the experimental evaluations. It is also noteworthy that the benefit of the AC loss over standard KD was even more pronounced for more difficult classification tasks. Furthermore, the accuracy of the combined FKD and standard KD training was better than any individual technique in most settings, with the exception of the CIFAR100 where the lower performance of standard KD resulted in inferior accuracy value compared to the AC loss alone.

5.4 Knowledge Distillation Between Large Models

Beyond model compression, KD has practical importance in use cases involving different architectures that exhibit similar performance, but with diversity in their error profiles. In such scenarios, all the models can be regarded as experts, and KD can be utilized to merge their distinct “perspectives” to the task obtained by the different architectures, into one model with enhanced performance. To assess the effectiveness of geometric FKD in a similar context, we employed the NiN architecture as the student, and teachers with ResNet architecture.

Initially, ResNet-20 and ResNet-32 models were traines on the two CIFAR datasets, following the regular training described in the previous section. The accuracies achieved for both tasks are provided in Table 4. Notably, the student and

Table 4 Comparison with different teachers in the FKD scheme via AC loss since the teacher models are ResNet CNNs and the student model always follows a NiN architecture

Accuracy (%)	Teacher architecture	Teacher accuracy	Regular training	FKD with AC loss	
Dataset				r = 5	r = 10
CIFAR10	ResNet20	91.40	86.19	88.06	88.75
	ResNet32	92.48		88.35	88.82
CIFAR100	ResNet32	64.95	63.24	65.80	66.40

teacher models have similar capacities based on their respective baseline performance. Particularly for the CIFAR100 task, the accuracy gap between the student's reference training and the architecture of the teachers is about 1.7%. Hence, this setup falls under a transfer-between-experts scenario rather than KD from an expert into a smaller student, as in the previous experiments.

Following a similar rationale as described in the former section, the deepest layers for each spatial resolution from each architecture were matched. Thus, the activations at the output of the final residual block of group1 in the ResNet models were employed for the regularization of the Cccp4 layer's activations in the NiN student model, while the activations from the last block of group2 were used to regularize the activations of Cccp6 layer. The experimental results of the NiN student models for geometric FKD via AC loss are also presented in the following Table 4.

The above results demonstrate that the tested geometric FKD approach can effectively enhance the performance of already capable models. Additionally, it suggests the benefit of utilizing different models' architectures as an extra source of knowledge even when working with the same data. Results on the CIFAR10 dataset reveal minimal variations between the ResNet20 and ResNet32 architectures when serving as teachers. Our hypothesis is that this behavior suggests that in similar scenarios, the improvements primarily stem from the contrasting architectural characteristics between the student and teacher, rather than the teacher's depth and accuracy. A key takeaway from the study is that the student models' accuracy surpasses both the teacher's and student's reference training for both tested radii in the case of CIFAR100, highlighting the usefulness of KD as a technique for knowledge transfer in situations beyond model compression.

5.5 *Effects of Neighborhood*

Determining the neighbors of each feature vector is a crucial aspect of geometric methods, and it remains to be seen to what extent the criterion of Eq. (14) influences the efficiency of regularization, or whether a more plain k-NN neighborhood rule would produce similar outcomes. To explore this, experiments were conducted on the more challenging datasets, i.e. the CIFAR10 and CIFAR100 tasks, to compare

Table 5 Comparison of various neighborhood criteria for the geometric FKD with AC loss. The NiN model is used as the teacher and a 3-layer CNN (Simple CNN) as the student

Accuracy (%)	MST-based neighborhood			k-NN neighborhood		
Dataset	r = 2	r = 5	r = 10	Equiv. to r = 2	Equiv. to r = 5	Equiv. to r = 10
CIFAR10	77.30	77.98	**78.34**	76.92	77.07	78.23
CIFAR100	45.14	**46.71**	46.1	43.87	44.87	45.05

the accuracy of a student model trained with geometric regularization via the AC function, using either the k-NN rule or the MST r-radius criterion. The teacher and student models, regularization weighting factors, and learning hyperparameters were the same to those of the previous Sect. 5.2. Because the radius determines the neighborhood on MST criterion, the neighbors' number is not fixed. To ensure fairness, the number of neighbors, k, in the k-NN rule was selected independently for each layer and it was set to the average number of neighbors at each corresponding radius. Table 5 includes the results.

The efficiency of the geometric FKD scheme depends on the use of the MST-based criterion according to the results. In all settings that were tested, the accuracy of the student models improved when using the MST-based neighborhoods. Furthermore, for the challenging CIFAR100 task, the k-NN rule reduced the efficiency of KD for all radii. A look at the Table 2 indicates that, in most experiments, the NP loss with MST-based neighborhoods was better than the AC loss with k-NN neighborhoods. These results suggest that a more complex neighboring criterion is preferable, as it provides more information about the underlying geometry of activation features. Additionally, the findings are consistent with those of [22], which showed that in more difficult and fine-grained tasks requiring smaller radii, the advantage of the MST-based neighborhoods is even greater than that of other methods.

6 Case Study: Geometric FKD in Data-Free Knowledge Transfer Between Architectures. An Application in Offline Signature Verification

Offline Signature Verification (OSV) is the binary behavioral biometric problem of distinguishing between genuine and forgery signatures by analyzing only the shape information of the signature after the writing process, typically using digitized version of the signing document. In this manner, a query -static- signature along with the claimed identity of the user are the inputs to an OSV system, and the output is a decision on acceptance if the query signature is classified as genuine or rejection if the query signature is regarded as forgery. A common OSV system consists of three stages, the preprocessing stage to remove noise due to digitization process, the feature extraction stage that encodes the content of the image into an appropriate representation, and the decision stage for classifying the image producing the binary result. The main challenges for an OSV system are (a) the high intra-class variability between signatures of the same user, (b) the partial knowledge during the design of the system and registration of a user since there is access only to genuine signatures of the user while its skilled forgeries are not practically available, and accordingly (c) the limited number of available samples for each user.

These limitations pose particularly significant challenges to the incorporation of CNNs in the OSV problem, following the current efficient trend of supervised deep learning. However, in previous years the availability of a large offline signature dataset, namely GPDS-960 corpus, containing about five hundred writers with 24 genuine along with 30 forgeries signatures per writer [34], allowed training deep models into the similar task of writer identification [35]. Due to the large size of GPDS-960 dataset, these CNN models are considered as good universal functions for producing image-level feature descriptors for signature images, thus their adoption as feature extractors in the OSV pipeline. Unfortunately, this dataset, is no longer available due to the implementation of the General Data Protection Regulation (EU) 2016/679 ("GDPR"), thus hindering the efforts of the research community to develop new models and complex methods that require more training data. The approaches trying to address the lack of adequate signature training data are summarized to the generation of synthetic signatures using geometrical transformation or generative learning models [36–39], the augmentation on feature space (after the feature extraction stage) to artificially populate samples for improving the classifier' performance [40], and the utilization of images from a relative domain, such as the handwritten text documents [41, 42]. Since the generation of synthetic signatures has not yet been validated for its efficacy on creating signatures with realistic intra-subject variability, none of these efforts cannot contribute to the creation of new and more efficient CNN models for OSV. In this case study we demonstrate how to utilize KD and especially geometric FKD, to train new architectures that inherit the knowledge of the older benchmark models whose training data are unavailable, using external data of similar nature.

6.1 Problem Formulation

The condition where an effective teacher model exists but there is no access to its training data (or to a meaningful amount of data for the target task) is the domain of data-free KD [5], where the distillation process uses only external or artificial data to perform the knowledge transfer. In the OSV case, an appropriate teacher is one of the benchmark CNN models in the field, such as SigNet [35] which is trained with the genuine signature images of 531 writers from GPDS-960 corpus and the trained model is publicly available from the official repository.[1] In this case, the teacher model can provide valuable feature representations for any input image, but not a

[1] https://drive.google.com/file/d/1l8NFdxSvQSLb2QTv71E6bKcTgvShKPpx/view.

meaningful classification response, since the training classes are person IDs which are not relevant without access to their data samples. The student model could be any CNN architecture that can be utilized under the described geometrical FKD scheme. The data which act as information carriers for distillation, can be either synthetic or external. In the following experiments we chose to utilize images of handwritten text since they possess a similar data structure as signatures (thin pen strokes on a paper) and most importantly, there is an abundance of data available from public sources. The alternative of using synthetically generated signatures was rejected since, given the poor image quality of these methods in general (e.g., [38]), there is no real benefit for utilizing such techniques.

The training of a feature extraction model for OSV is a learning task different than the main verification task, since the identity and data of the users involved in the operational phase are not available during the model's training phase (often called development phase). The decision stage, which consumes the feature representation of a signature image and decides upon its validity, is a classifier trained with only the samples of the users registered during operation, with either a separate model trained for each user (writer dependent—WD) or a universal model used for all users (writer independent—WI). Since the purpose of this section is to illustrate the effectiveness of geometric FKD in a data-free KD setting, we will follow the most straightforward WD approach using WD Support Vector Machine classifiers as the final decision stage at the operational phase.

The training of the feature extraction model will include KD, assessing both geometric FKD as also a regularization loss based on the global feature generated at the penultimate layer of teacher model, similarly to standard KD. Since the global feature incorporates the encoded knowledge of the CNN just before the final classification layer, we can treat it as the response of the network independently of the classification task. Consequently, we can incorporate the minimization of cross-entropy between the two temperature-scaled features (T-CE) extracted from the teacher and student, as a response-based KD loss utilized as additional term in the overall loss function in the following experiments. A graphical overview of the training scheme is presented on Fig. 2.

6.2 Experimental Setup

The teacher model used in this section is SigNet, a CNN based on a vanilla AlexNet architecture without residual skip connections. Our goal in this case study is to transfer the knowledge from the teacher's model into a deeper, more complex and efficient student's architecture, thus we chose ResNet-18 as the basis architecture for the student model. The student model is slightly modified to provide activations with the same spatial resolutions as SigNet (by varying the stride parameter to some layer) and also, a fully connected layer with 2048 neurons was added as a penultimate layer to facilitate the feature extraction before the final classification layer. A summary of the architectures of teacher and student CNNs is provided in For the

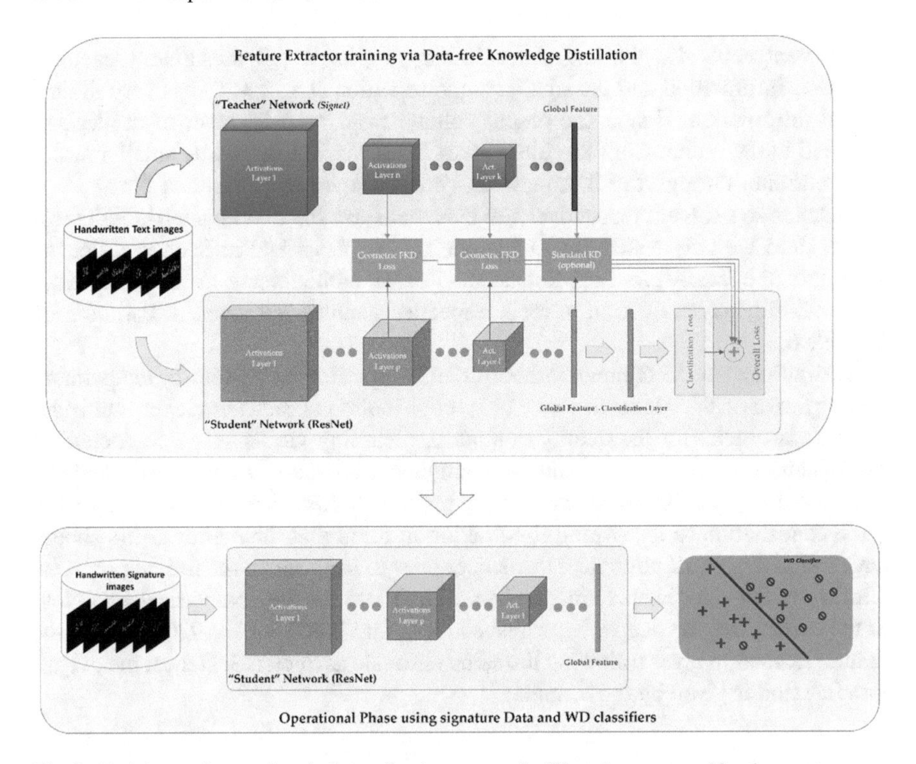

Fig. 2 Training and operational phases for the proposed offline signature verification system

connections between teacher and student architectures we followed the same rationale as in previous experiments, assuming that the deepest representation for each spatial resolution is the most informative in order to incorporate the optimal encoding. Considering the four common spatial sizes through the teacher and student networks, the largest resolution (i.e., 35×53) presumably captures low-level information and the smallest representation (i.e., 3×5) has very limited spatial information. Hence, the output volume from "conv2" layer of the teacher is linked to the output from "residual block 2" of the student, while smaller scales the contribute through the link between "conv5" layer and "residual block 3" of the teacher and student respectively, to eventual form the two geometric FKD loss terms. Here we only tested the AC geometric loss described in Sect. 4.2. For the T-CE loss, the extracted features from both architectures, SigNet and ResNet, originated from the 2048-dimensional feature at the penultimate layer of the two architectures. An interesting note here is that the learnable parameters are roughly equal in the two CNNs that made up the S-T pair.

For the connections between teacher and student architectures we followed the same rationale as in previous experiments, assuming that the deepest representation for each spatial resolution is the most informative in order to incorporate the optimal encoding. Considering the four common spatial sizes through the teacher and student networks, the largest resolution (i.e., 35×53) presumably captures low-level information and the smallest representation (i.e., 3×5) has very limited spatial information. Hence, the output volume from "conv2" layer of the teacher is linked to the output from "residual block 2" of the student, while smaller scales the contribute through the link between "conv5" layer and "residual block 3" of the teacher and student respectively, to eventual form the two geometric FKD loss terms. Here we only tested the AC geometric loss described in Sect. 4.2. For the T-CE loss, the features for both SigNet and ResNet architectures, are extracted from the 2048-dimensional output at their respective penultimate layer, as can be seen in Table 6.

The radius of the MST neighborhood criterion is set to $r = 5$ while the temperature scaling factor of T-CE is set to $T = 10$. The contribution of each term in the multi-loss function is crucial for the quality of student's learning via KD. The coefficient of classification loss is set to 1, and the coefficients for the KD terms were selected such that, in the beginning of the training process all loss terms having roughly the same contribution to the overall loss, taking in mind that the earlier layers should have a smaller contribution than the deeper ones to keep stable the training process. Ultimately, the coefficients are tuned to 10 and 100 for the geometrical FKD loss at the earlier and the deeper layer respectively, and 0.001 for the T-CE loss at the feature extraction layer following the same rationale as Sect. 5.3. Hence, the overall loss function is given by the formula:

Table 6 Architectures and activations' size for teacher (SigNet) and student (ResNet18) models

SigNet		ResNet-18	
Layer/block name	Activation size	Layer/block name	Activation size
Input	$150 \times 220 \times 1$	Input	$150 \times 220 \times 1$
Conv1	$35 \times 53 \times 96$	Conv	$72 \times 107 \times 64$
Pool1	$17 \times 26 \times 256$	Pool	$35 \times 53 \times 64$
Conv2	$17 \times 26 \times 256$	Block1	$35 \times 53 \times 64$
Pool2	$8 \times 12 \times 256$	Block2	$17 \times 26 \times 128$
conv3	$8 \times 12 \times 384$	Block3	$8 \times 12 \times 256$
Conv4	$8 \times 12 \times 384$	Block4	$3 \times 5 \times 256$
Conv5	$8 \times 12 \times 256$	fc1	2048
Pool5	$3 \times 5 \times 256$	fc2	#Classes
fc6	2048		
fc7	2048		
fc8	#Classes		

$$L = L_{classification} + 10 \cdot L_{NS}^{conv2/block2} + 100 \cdot L_{NS}^{conv5/block3} + 0.001 \cdot L_{T-CE}^{fc7/fc1} \tag{16}$$

Training was executed by minimizing the overall loss with Stochastic Gradient Descent (SGD) using initial learning rate of 0.01, which is divided by 10 every 20 epochs for a total of 60 epochs, and Nesterov Momentum with momentum factor given by 0.9. The mini-batch size was 64 images (as the maximum capacity of the utilized GeForce RTX 2070) and each S-T training executed on approximately 30 h.

The external data used for training were images of handwritten text from CVL dataset [43], cropped and preprocessed via the procedure proposed by Tsourounis et al. in [42] to simulate the distribution of signature images. In this work, the utilized set of text images were obtained using random aspect-ratio cropping and sampling one text image for each canvas. Finally, the dataset for the feature extraction model contained about sixty thousand training and twenty-five thousand validation text images, with the ID of the writer used for the classification loss.

The evaluation of the trained ResNet as the feature extractor in OSV setting is implemented using CEDAR dataset [44], which is a popular and publicly available offline signature dataset. CEDAR database includes 55 writers with 24 genuine and 24 forgeries signatures per writer, while the forgeries are a mixture of random, simple, and skilled simulated signatures. The preprocessing steps as well as the implementation of Writer Dependent Support Vector Machine classifiers (WD SVMs) are based on the work of Hafemann et al. [35]. In fact, to establish a fair comparison with results of SigNet, we used the implementation downloaded from the official repository,[2] including the same image preprocess, partition into training and test sets, classifiers' training, selection of hyper-parameters, and calculation of evaluation metrics. Therefore, for each writer's SVM, the positive training class consists of 12 reference genuine signatures of the writer while the negative training class composed from the reference signatures of all other writers of the dataset. The test set of each writer includes 10 genuine signatures of the writer (different from the references) and 10 skilled forgeries of the writer. The equal size of two test populations lead to better estimation of evaluation metrics using the ROC curve (Receiver Operating Characteristic curve) and especially, the reported value of EER (Equal Error Rate) with user-specific threshold. For each experiment, ten repetitions of WD classifiers are performed using randomly selected splits of data (with different signatures) and the results are presented in terms of the average and standard deviation values of EER.

6.3 Results

The aim here is to transfer the knowledge from a frozen teacher model to a student CNN with different architecture utilizing geometric FKD and/or global distillation,

[2] https://github.com/luizgh/sigver.

Table 7 EER metric for various FKD schemes evaluated on the offline signature verification (OSV) task using the CEDAR dataset

S-T	EER
Teacher (SigNet)	3.21 ± 0.33
Student (random weights)	9.12 ± 0.91
KD (GEOM FKD & T-CE)	2.50 ± 0.38
CL (w/o KD)	2.28 ± 0.60
CL + GEOM FKD	2.05 ± 0.28
CL + T-CE	2.19 ± 0.44
CL + (GEOM FKD and T-CE)	$\mathbf{1.99 \pm 0.39}$

and assessing how the student performs as a feature extractor in an OSV system. The results for the various experiments including various KD schemes, are summarized in Table 7. In each experiment, the overall loss function consists of only the relevant terms i.e., geometric FKD (GEOM FKD) and/or the global feature KD (T-CE) term and/or the classification loss (CL). In order to provide some perspective on the accuracy differences, we also included the performance of the student model as a feature extractor initialized with random weights.

The obtained performance of the student model trained only using text classification loss (CL w/o KD) is better compared to utilizing only KD (GEOM FKD & T-CE) terms in the overall loss function, something that is consistent with the findings in [6]. The KD losses alone though, still managed to produce a model that is better than the teacher in terms of effectiveness for the generated features in OSV. The importance of KD as a regularization mechanism though, is highlighted from the fact that the synergy between the classification loss (CL) and any of the tested KD losses produces a more effective model than the straightforward training with CL loss. An important aspect in the context of this chapter is that the combination of CL with geometrical FKD (CL + GEOM FKD) achieves better performance compared to the combination of CL with response-based KD (CL + T-CE), demonstrating that the approaches presented in the previous sections are valid and very efficient mechanisms for KD, despite their simplicity. The regularization of local activations controls the training along the data path of the student network in a more explicit way, avoiding the divergence of the learning process such could be induced from using KD only in deeper stages of the network, relying only on global feature information. Finally, the best results overall, and significantly better than those achieved by the teacher model, were obtained when using both local and global based KD along with CL loss. In conclusion, the proposed approach to geometric FKD was able to exploit the teacher's knowledge without the need to utilize data related to the primary task, and transfer the knowledge to a different CNN architecture, generating an even more effective model.

Table 8 includes the performance of state-of-the-art methods in CEDAR dataset, thus providing a general overview of the WD OSV field since the many differences in the systems' stages make infeasible to compare all of them fairly. We can argue that

Table 8 Summary of the state-of-the-art methods for offline signature verification (OSV) problem using CEDAR dataset

WD OSV systems in CEDAR dataset			
References	Feature extraction method	NREF	EER
[35]	SigNet	12	4.76
[35]	SigNet-F	10	4.63
[45]	SigNet-SPP	10	3.60
[46]	PDSN	10	4.37
[47]	SR-KSVD/OMP	10	0.79
[48]	Hybrid Texture	10	6.66
[49]	CNN-Triplet and Graph edit distance	10	5.91
[50]	HOCCNN	12	4.94
[51]	Visibility motif profiles	10	0.51
[40]	SigNet-F classifier Gauss augments	3	0.82
[52]	MSDN	–	1.75
[53]	Micro deformations	–	2.76
[42]	CNN-CoLL	10	1.66
Proposed	S-T KD (CL + GEOM FKD + T-CE)	12	1.99

the best of the above models achieves a low verification error that is at least comparable to the other methods. Furthermore, the presented feature extraction method exhibits SoTA result even it is trained without any signatures while the WD classifiers do not utilize skilled forgeries during their training, which is a significant benefit in real-life conditions.

7 Discussion

In this chapter, we demonstrated that by transferring geometric characteristics of the data manifold from a teacher CNN to a student model, the performance of the latter can be improved. Our study focused on CNNs, but the concept can be easily extended to other types of deep neural networks, such as Vision Transformers [54]. Recent research [55] has demonstrated that using distillation with neural architectures featuring Transformers, can enhance performance in visual tasks such as semantic segmentation. In the various stages of such architectures, the visual information is represented either in the form of patch embeddings, or as typical 3D volumes of feature maps reconstructed from the patch embeddings. In both cases, since a (spatial) correspondence between the intermediate embeddings at the student and a teacher models' sides can be established, the dissimilarity measures of Eq. (6) and the criteria developed in Sect. 4 are directly applicable. Thus, the geometry of the intermediate embeddings of a teacher vision transformer can be distilled to a student transformer that follows the same tokenization scheme of the processed multi-dimensional signal, using the same methods.

Regarding the computational overhead imposed to the training process from the presented methods, it is dominated by the computation of features' pairwise distances in each layer that contributes to the distillation. Since the teacher is not updated and the computation of all neighborhoods on the Minimal Spanning Tree are performed only on teacher, this intensive process can be performed offline without contributing to the main training loop. Therefore, the extra training workload is linear to the number of layers used for knowledge transfer, and proportional to the area of the input image. Furthermore, since only a small fraction of the pairwise distances actually contributes to the distillation losses (only neighbors of each feature are considered in Eq. (10) and (13)), an implementation that computes only the necessary pairwise distances can greatly reduce the amount of computations needed for this step. Finally, the potential computational gain by employing the distillation, is rather difficult to estimate since the performance advantage in the student relies heavily on the task, and the estimation of an equivalent network's size for the same performance is non trivial.

8 Conclusions

In this chapter, we focused on developing a family of computationally efficient loss functions that aim to transfer the geometry of activations from intermediate layers of a teacher CNN to the activations of a student CNN model at matching spatial resolutions. In contrast to most feature-based distillation techniques, which handle the teacher-to-student knowledge transfer in a statistical or probabilistic manner, we argue that the problem of FKD is in essence a problem of matching the manifolds of local activations at the different layers of teacher and student models. Therefore, by formulating a geometric FKD loss, the student model can efficiently harness the teacher's knowledge. We introduced a family of loss functions that have their roots in manifold-manifold distance functions, designed to overcome the inherent problems of sampling local manifolds. Experimental results on benchmark tasks show evidence that the formulated geometric FKD losses can indeed transfer knowledge from teacher to student efficiently, and that neighborhoods defined over parsimonious graphs such as Minimal Spanning Tree achieve better performance compared to the standard kNN rule or fully connected neighborhood graphs.

Finally, we presented a case study from the challenging field of offline handwritten signature verification, in which a publicly available CNN model was used as a teacher to enhance the training of a new model of different architecture using only external (task-irrelevant) data. Results showed that the geometric FKD alone or combined with standard KD along with the external classification task, managed to produce a model with significantly better performance as a feature extractor for OSV compared to the teacher model. Such training regime can be useful beyond OSV, in any situation where trained models are available for a particular task, but training data is scarce.

References

1. Ba, J., Caruana, R.: Do deep nets really need to be deep? In: Advances in neural information processing systems, vol. 27. Curran Associates, Inc. (2014)
2. He, K., Zhang, X., Ren, S., Sun, J.: Deep residual learning for image recognition. In: Proceedings of the IEEE conference on computer vision and pattern recognition, pp. 770–778 (2016)
3. Goel, A., Tung, C., Lu, Y.-H., Thiruvathukal, G.K.: A survey of methods for low-power deep learning and computer vision. In: 2020 IEEE 6th World Forum on Internet of Things (WF-IoT), pp. 1–6 (2020). https://doi.org/10.1109/WF-IoT48130.2020.9221198
4. Elsken, T., Metzen, J.H., Hutter, F.: Neural architecture search: a survey. J. Mach. Learn. Res. **20**, 1997–2017 (2019)
5. Gou, J., Yu, B., Maybank, S.J., Tao, D.: Knowledge distillation: a survey. Int. J. Comput. Vis. **129**, 1789–1819 (2021)
6. Hinton, G., Vinyals, O., Dean, J.: Distilling the knowledge in a neural network (2015). https://doi.org/10.48550/arXiv.1503.02531
7. Romero, A., et al.: Fitnets: hints for thin deep nets. In: International Conference on Learning Representations (ICLR) (2015)
8. Yim, J., Joo, D., Bae, J., Kim, J.: A gift from knowledge distillation: fast optimization, network minimization and transfer learning. In: 2017 IEEE Conference on Computer Vision and Pattern Recognition (CVPR), pp. 7130–7138 (2017). https://doi.org/10.1109/CVPR.2017.754
9. Huang, Z., Wang, N.: Like what you like: knowledge distill via neuron selectivity transfer (2017). https://doi.org/10.48550/arXiv.1707.01219
10. Zagoruyko, S., Komodakis, N.: Paying More Attention to Attention: Improving the Performance of Convolutional Neural Networks via Attention Transfer (2022)
11. Passalis, N., Tefas, A.: Learning Deep Representations with Probabilistic Knowledge Transfer, pp. 268–284 (2018)
12. Yang, J., Martinez, B., Bulat, A., Tzimiropoulos, G.: Knowledge distillation via adaptive instance normalization (2020). https://doi.org/10.48550/arXiv.2003.04289
13. Chen, D., et al.: Cross-layer distillation with semantic calibration. In: Proceedings of the AAAI Conference on Artificial Intelligence, vol. 35, pp. 7028–7036 (2021)
14. Peyré, G.: Manifold models for signals and images. Comput. Vis. Image Underst. **113**, 249–260 (2009)
15. Carlsson, G., Ishkhanov, T., de Silva, V., Zomorodian, A.: On the local behavior of spaces of natural images. Int. J. Comput. Vis. **76**, 1–12 (2008)
16. Bjorck, A., Golub, G.: Numerical methods for computing angles between linear subspaces. Math. Comput. **27**, 123 (1973)
17. Kim, T.-K., Arandjelović, O., Cipolla, R.: Boosted manifold principal angles for image set-based recognition. Pattern Recogn. **40**, 2475–2484 (2007)
18. Wang, R., Shan, S., Chen, X., Gao, W.: Manifold-manifold distance with application to face recognition based on image set. In: 2008 IEEE Conference on Computer Vision and Pattern Recognition, pp. 1–8 (2008). https://doi.org/10.1109/CVPR.2008.4587719
19. Vasconcelos, N., Lippman, A.: A multiresolution manifold distance for invariant image similarity. IEEE Trans. Multimed. **7**, 127–142 (2005)
20. Hamm, J., Lee, D.D.: Grassmann discriminant analysis: a unifying view on subspace-based learning. In: Proceedings of the 25th International Conference on Machine Learning, pp. 376–383. Association for Computing Machinery (2008). https://doi.org/10.1145/1390156.1390204

21. Lu, J., Tan, Y.-P., Wang, G.: Discriminative multimanifold analysis for face recognition from a single training sample per person. IEEE Trans. Pattern Anal. Mach. Intell. **35**, 39–51 (2013)
22. Theodorakopoulos, I., Economou, G., Fotopoulos, S., Theoharatos, C.: Local manifold distance based on neighborhood graph reordering. Pattern Recogn. **53**, 195–211 (2016)
23. Grindrod, P.: Range-dependent random graphs and their application to modeling large small-world Proteome datasets. Phys. Rev. E **66**, 066702 (2002)
24. Turiel, A., Mato, G., Parga, N., Nadal, J.-P.: Self-similarity properties of natural images. In: Advances in Neural Information Processing Systems, vol. 10. MIT Press (1997)
25. Ruderman, D.L.: Origins of scaling in natural images. Vis. Res. **37**, 3385–3398 (1997)
26. Carreira-Perpiñán, M.Á., Zemel, R.S.: Proximity graphs for clustering and manifold learning. In: Proceedings of the 17th International Conference on Neural Information Processing Systems, pp. 225–232. MIT Press (2004)
27. İnkaya, T.: Parameter-free surrounding neighborhood based regression methods. Exp. Syst. Appl. **199**, 116881 (2022)
28. Lassance, C., et al.: Deep geometric knowledge distillation with graphs. In: ICASSP 2020—2020 IEEE International Conference on Acoustics, Speech and Signal Processing (ICASSP), pp. 8484–8488 (2020). https://doi.org/10.1109/ICASSP40776.2020.9053986
29. Belkin, M., Niyogi, P.: Towards a theoretical foundation for Laplacian-based manifold methods. J. Comput. Syst. Sci. **74**, 1289–1308 (2008)
30. Park, W., Kim, D., Lu, Y., Cho, M.: Relational knowledge distillation, pp. 3967–3976 (2019)
31. Lin, M., Chen, Q., Yan, S.: Network in network (2014). arXiv:1312.4400 [cs]
32. Krizhevsky, A.: Learning Multiple Layers of Features from Tiny Images. University of Toronto (2012)
33. Netzer, Y., et al.: Reading digits in natural images with unsupervised feature learning. In: NIPS Workshop on Deep Learning and Unsupervised Feature Learning 2011 (2011)
34. Vargas, F., Ferrer, M., Travieso, C., Alonso, J.: Off-line handwritten signature GPDS-960 corpus. In: Proceedings of the Ninth International Conference on Document Analysis and Recognition, vol. 02, pp. 764–768. IEEE Computer Society (2007)
35. Hafemann, L.G., Sabourin, R., Oliveira, L.S.: Learning features for offline handwritten signature verification using deep convolutional neural networks. Pattern Recogn. **70**, 163–176 (2017)
36. Diaz, M., Ferrer, M.A., Eskander, G.S., Sabourin, R.: Generation of duplicated off-line signature images for verification systems. IEEE Trans. Pattern Anal. Mach. Intell. **39**, 951–964 (2017)
37. Parmar, M., Puranik, N., Joshi, D., Malpani, S.: Image processing based signature duplication and its verification. In: SSRN Scholarly Paper. https://doi.org/10.2139/ssrn.3645426
38. Yapıcı, M.M., Tekerek, A., Topaloğlu, N.: Deep learning-based data augmentation method and signature verification system for offline handwritten signature. Pattern Anal. Appl. **24**, 165–179 (2021)
39. Yonekura, D.C., Guedes, E.B.: Offline handwritten signature authentication with conditional deep convolutional generative adversarial networks. In: Anais do Encontro Nacional de Inteligência Artificial e Computacional (ENIAC), pp. 482–491 (SBC, 2021). https://doi.org/10.5753/eniac.2021.18277
40. Maruyama, T.M., Oliveira, L.S., Britto, A.S., Sabourin, R.: Intrapersonal parameter optimization for offline handwritten signature augmentation. IEEE Trans. Inf. Forensics Secur. **16**, 1335–1350 (2021)
41. Mersa, O., Etaati, F., Masoudnia, S., Araabi, B.N.: Learning representations from Persian handwriting for offline signature verification, a deep transfer learning approach. In: 2019 4th International Conference on Pattern Recognition and Image Analysis (IPRIA), pp. 268–273 (2019). https://doi.org/10.1109/PRIA.2019.8785979
42. Tsourounis, D., Theodorakopoulos, I., Zois, E.N., Economou, G.: From text to signatures: knowledge transfer for efficient deep feature learning in offline signature verification. Exp. Syst. Appl. **189**, 116136 (2022)
43. Kleber, F., Fiel, S., Diem, M., Sablatnig, R.: CVL-DataBase: an off-line database for writer retrieval, writer identification and word spotting. In: 2013 12th International Conference on

Document Analysis and Recognition, pp. 560–564 (2013). https://doi.org/10.1109/ICDAR.2013.117

44. Kalera, M.K., Srihari, S., Xu, A.: Offline signature verification and identification using distance statistics. Int. J. Pattern Recogn. Artif. Intell. **18**, 1339–1360 (2004)
45. Hafemann, L.G., Oliveira, L.S., Sabourin, R.: Fixed-sized representation learning from offline handwritten signatures of different sizes. IJDAR **21**, 219–232 (2018)
46. Lai, S., Jin, L.: Learning discriminative feature hierarchies for off-line signature verification. In: 2018 16th International Conference on Frontiers in Handwriting Recognition (ICFHR), pp. 175–180 (2018). https://doi.org/10.1109/ICFHR-2018.2018.00039
47. Zois, E.N., Tsourounis, D., Theodorakopoulos, I., Kesidis, A.L., Economou, G.: A comprehensive study of sparse representation techniques for offline signature verification. IEEE Trans. Biom. Behav. Identity Sci. **1**, 68–81 (2019)
48. Bhunia, A.K., Alaei, A., Roy, P.P.: Signature verification approach using fusion of hybrid texture features. Neural Comput. Appl. **31**, 8737–8748 (2019)
49. Maergner, P., et al.: Combining graph edit distance and triplet networks for offline signature verification. Pattern Recogn. Lett. **125**, 527–533 (2019)
50. Shariatmadari, S., Emadi, S., Akbari, Y.: Patch-based offline signature verification using one-class hierarchical deep learning. IJDAR **22**, 375–385 (2019)
51. Zois, E.N., Zervas, E., Tsourounis, D., Economou, G.: Sequential motif profiles and topological plots for offline signature verification. 13248–13258 (2020)
52. Liu, L., Huang, L., Yin, F., Chen, Y.: Offline signature verification using a region based deep metric learning network. Pattern Recogn. **118**, 108009 (2021)
53. Zheng, Y., et al.: Learning the micro deformations by max-pooling for offline signature verification. Pattern Recogn. **118**, 108008 (2021)
54. Dosovitskiy, A., et al.: An image is worth 16x16 words: transformers for image recognition at scale (2021). arXiv: https://doi.org/10.48550/arXiv.2010.11929
55. Liu, R., et al.: TransKD: transformer knowledge distillation for efficient semantic segmentation (2022). arXiv: https://doi.org/10.48550/arXiv.2202.13393

Knowledge Distillation Across Vision and Language

Zhiyuan Fang and Yezhou Yang

Abstract Recent years have witnessed the fast development of Vision and Language (VL) learning with deep neural architectures, benefiting from large amounts of unlabeled or weakly labeled, and heterogeneous forms of data. A notable challenge arises when deploying these cross-modal models on an edge device that usually has the limited computational power to be undertaken. It is impractical for real-world applications to exploit the power of prevailing models under a constrained training/inference budget, especially when dealing with abundant multi-modal data that requires significant resources. As an important technique for deep model compressing, Knowledge Distillation (KD) has been widely applied to various tasks that aim to build a small yet powerful student model from a large teacher model. In the furtherance of gaining a more real-world practical compact Vision-language model, it is essential to exploit knowledge distillation techniques to improve VL learning on compact models. KD + VL, is of great practical value but is less explored in the previous literature. This inspires the studies of KD on different modalities (Vision, Language); learning tasks (Self-supervised learning, Contrastive Learning); and also will serve as a vital role for many sub-disciplines of cross-modal tasks, such as image/video captioning, visual-question answering, image/video retrieval, etc. This chapter summarizes and discusses in-depth this emerging area of research. We first retrospect the existing KD algorithms comprehensively, and mainly focuses on their applications for Vision-Language tasks as well as the recent coming-up works leverage the self-distillation technique for training joint VL representation. The chapter also features recent works at the forefront of research on this topic leveraging KD for Vision-Language representation learning, as well as its application to downstream VL tasks, captioning, visual question answering, and so forth. The chapter further discusses and studies variations of KD applied to Vision and Language-related topics.

Z. Fang (✉)
Applied Scientist of Alexa AI-Natural Understanding Team, Amazon, USA
e-mail: zyfang@amazon.com

Y. Yang
Faculty of School of Computing and AI, Active Perception Group, Arizona State University, Tempe, USA
e-mail: yz.yang@asu.edu

W. Pedrycz and S.-M. Chen (eds.), *Advancements in Knowledge Distillation: Towards New Horizons of Intelligent Systems*, Studies in Computational Intelligence 1100,
https://doi.org/10.1007/978-3-031-32095-8_3

Keywords Knowledge distillation · Vision and language learning · Representation learning

1 Introduction

Knowledge Distillation (KD), one of the key techniques for deep neural model compression, has been intensively studied and applied to a variety of applications in recent years. The fact that the deployment of the most powerful AI models on edge devices is problematic due to resource limitation necessitates the development of compression techniques that preserve model performance. In the foreseeable future, AI will be confronted with complex and diverse data in various forms, such as Vision, Language, Audio, and even Tactile, posing a persistent obstacle. In pursuit of a more competitive, practical, and compact AI model, we investigate the use of knowledge distillation techniques to enhance representation learning on compact architectures, particularly at the intersection of Vision and Language (VL) modalities.

Representation learning aims to train a deep architecture and acquire "omni" representations that can be transferred to a variety of tasks through pre-training and fine-tuning. This learning can take various forms (e.g., supervised, self-directed, and unsupervised) and utilize diverse data sources. In essence, representation learning enables the deep model to fully exploit the power of scaling data while removing its reliance on prohibitively expensive data annotations, as pre-training is now conducted flexibly on the basis of multi-modalities. Most existing self- or unsupervised learning can be categorized as generative or contrastive learning, where the former emphasizes the model's ability to "reconstruct and predict" (e.g., the prevalent auto-regressive modeling and masked language modeling are formulated as pre-predicting the masked or imperceptible future words in language model), whereas the latter adheres to the "instance discrimination" principle, encouraging the instances to be distinguished from one another. This chapter introduces the application of a contrastive learning based distillation technique for Vision-Language representation learning, which shows great performances on small VL architectures.

The contribution of this chapter includes providing an comprehensive overview of the KD's application to contrastive representation learning and its derivatives for self-supervised visual representation learning/VL representation learning, which is rarely discussed in prior works. We discuss in detail how KD facilitates learning from a theoretical perspective, highlighting its performance and inference advantages across downstream tasks. For instance, with our VL distillation technique, we are able to compress the prevailing VL model to its 70% in regards of learnable parameters but with comparable performances, largely alleviating the computational burdens for VL models. This chapter's comprehensive coverage of Vision and Language also aims to provide readers with necessary and foundational knowledge, as well as demonstrate the significance of KD for VL learning.

The remaining sections of this chapter are structured as follows: To provide readers with the necessary context for Vision and Language Learning, we will first examine

Sect. 2's common VL Tasks. On the basis of this, we examine the frequently employed contrastive learning for representation learning and the fundamental formulations of knowledge distillation. Then, we present SEED [9], a recently developed method that combines contrastive learning with unsupervised visual representation learning, followed by its application when extended to VL representation learning [10, 11]. Section 4 discusses our experimental results for this method following a large-scale pre-training, including Visual Object Recognition, Visual Question Answering, Image Retrieval by Text, Image Captioning, etc. To this end, we discuss the potential and broader effects of multi-modal learning on the future of KD in uncharted regimes.

2 Vision Language Learning and Contrastive Distillation

The advancements in deep learning led to great development in computer vision and natural language processing. Unprecedented improvements have been achieved in computer vision benchmarks, e.g., image classification [37], object detection [27], and so in classical NLP tasks, e.g., named entity recognition [53], question-answering [35, 38], etc. Yet, to fully understand and enable human–machine interaction, capabilities of cross-modal association, comprehension, and reasoning at a higher level are to be involved as an important stepping stone. Numerous applications of AI systems, including visual search/localization by language, and assistive robots, would be largely enhanced by giving the AI a better understanding of human instructions and their connections with the observable environment. The studies of Vision and Language, thus, are no longer carried on individually but are to be explored more complementarily. This chapter firstly reviews the prevailing Vision and Language pre-training technique, then introduces the formulations of multiple VL tasks. To this end, some preliminary issues about contrastive learning and knowledge distillation are to be discussed.

2.1 Vision and Language Representation Learning

The emergence of pre-training techniques pushes computer vision and natural language processing into a new era, and the key lies in the substantial data it leverages. A great amount of work has shown that the pre-trained weights from large-scale training obviously benefit downstream tasks when they serve as initialization for fine-tuning. The goal of Vision and Language representation learning follows a similar schema and aims to train the VL architecture from a large VL corpus (**C**) consisting of mage-text pairs. This process can be defined as:

$$\boldsymbol{R} = \text{VL ENC}(\boldsymbol{I}, \boldsymbol{T}), <\boldsymbol{I}, \boldsymbol{T}> \epsilon \, \mathbf{C}, \tag{1}$$

where VL-ENC is a VL architecture that takes the input of an image-text pair, and the image and text are encoded by an offline object detector as vector representations (***I***, ***T***). For optimization, image-text matching (ITM) modeling and masked language modeling (MLM) are frequently used as pre-training losses. In particular, ITM encourages the VL model to distinguish the matched (or un-matched) pairs via a binary classification task:

$$\boldsymbol{L}_{ITM} = y \log \mathrm{P}_{\theta}(\boldsymbol{I}, \boldsymbol{T}) + (1 - \mathbf{y}) \log \mathrm{P}_{\theta}(\boldsymbol{I}, \boldsymbol{T}), \tag{2}$$

where $\mathbf{y}$ represents the binary ground-truth label for the (un)-matched image-text pair and P_{θ} $(\boldsymbol{I}, \boldsymbol{T})$ denotes the probability score outputs by a task-specific head on the top of VL architecture for ITM training. Intuitively, ITM modeling introduces the discriminativeness to the VL learning that it learns to associate the semantics across modalities. MLM originates from large language model (LLM) training and is constructed in a "*mask and predict*" fashion where certain ratios of the words are randomly masked. The model is optimized to predict them, giving the contextualized observations:

$$\boldsymbol{L}_{\mathrm{MLM}} = -\log \mathrm{P}_{\theta}(\boldsymbol{w}_k | \boldsymbol{I}, \boldsymbol{w}_0 \ldots \boldsymbol{w}_{k-1}, \boldsymbol{w}_{k+1} \ldots \boldsymbol{w}_N), \tag{3}$$

with $\boldsymbol{w}_i$ representing the $\boldsymbol{i}$-th word token and $\boldsymbol{w}_k$ being the masked one to be predicted. In LLM, this modeling can be auto-regressive (single directional, the model predicts the words of the unseen future given historical words) or bidirectional (the model predicts masked words given both previous and future words), depending on the applications it is dealing with (e.g., auto-regressive modeling is mostly used on generative models like GPT for sentence generation, whilst bidirectional one is adopted on BERT for discriminative tasks like sentiment classification).

2.2 *Contrastive Learning and Knowledge Distillation*

Self-supervised Learning (SSL) shows superior results on tasks for both the vision and language domains. Many techniques for learning latent distributions involve contrastive learning, which is based on noise-contrastive estimation [14]. These techniques typically involve generating random or artificial noises to estimate the latent distribution. Info-NCE [30] was one of the first methods to propose using an autoregressive model to predict the future in order to learn image representations in an unsupervised manner. There have been numerous follow-up works that aim to improve the efficiency of this approach [17] or use multi-view as positive samples [44]. However, these methods are limited by their access to only a limited number of negative instances. To overcome this problem, Wu [52] suggested employing a memory bank to store previously observed random representations (to be used as negative samples for contrastive learning), considering each of them as distinct discrimination categories. Nevertheless, this strategy might be troublesome since the stored vectors may be inconsistent with the freshly calculated representations during the early phases of pre-training. Chen [4] sought to alleviate this problem by sampling negative samples from a big batch, while He [16] proposed employing a momentum updated encoder to decrease representation inconsistency. Misra [29] combines a pretext-invariant objective loss with contrastive learning, and Khosla [20] decomposes the contrastive loss into alignment and uniformity objectives:

$$L_{contrastive} = -\log\left(\exp\left(z \cdot z^{+}\right) \Big/ \sum_{i}^{N} \exp\left(z \cdot z^{-}\right)\right), \tag{4}$$

with z represents the embedding of the target instance, and $z^{+/-}$ is the embedding of positive and negative instance w.r.t. the target instance.

Hinton initially proposed the concept of knowledge distillation back in 2015 for neural network training: Knowledge distillation is a method for transferring information (called as dark knowledge according to Hinton) from a strong teacher network to a student network. The process of training may be described as follows:

$$\boldsymbol{L} = \sum_{i}^{N} L_{supervised}(x_i, \theta_S, y_i) + L_{distillation}(x_i, \theta_S, \theta_T), \tag{5}$$

where x_i is the input signal (in the form of a visual image or textual sentence), θ_S is the learnable parameter set for the student network, while θ_T represents the parameters for the teacher network. In the process of distilling knowledge, the student network attempts to learn the knowledge from a network of pre-trained teachers. To achieve so, Hinton proposes to leverage the Kullback–Leibler divergence to estimate the distance of output logit (or class probability) between the Teacher-Student networks. Ebrahimi et al. [8] attempts to directly minimize the feature maps' distances via L2 distance instead and also found performances improvement. Knowledge distillation has been shown to be successful in both supervised and unsupervised scenarios, and is not limited to specific tasks. For example, in image classification tasks, this loss is typically the cross-entropy loss. In object detection task, it may also include bounding box regression instead as complementary loss term. In supervised settings, the loss $L_{\text{supervised}}$ represents the discrepancy between the network's prediction and the annotated data, while in self/un-supervised setting, it can be a pseudo-label constructed manually (to be discussed at later section).

On these basis, several follow-ups continue to propose the contrastive distillation, which reformulates the distillation process as a contrastive learning alike schema, with the target instance/negative instances encoded by a pre-trained teacher architecture:

$$\boldsymbol{L} = \sum_{i}^{N} L_{supervised}(x_i, \theta_S, y_i) - \log\left(\exp\left(z^{S} \cdot z^{T}\right) \Big/ \exp\left(z^{S} \cdot z^{-T}\right)\right). \tag{6}$$

Such design largely alleviates the challenge during distillation where only positive samples are leveraged, whilst it remains un-charted about how this can be inherited for self-supervised learning and in multi-modalities.

2.3 Contrastive Distillation for Self-supervised Learning

Contrastive learning for self-supervised learning, in which a small model learns representations that can be used for downstream tasks by imitating a large model without the need for labeled data, has not been widely explored until the work of [9] on SEED. Inspired by this approach, we propose a simple method for distillation based on the distribution of instance similarity over a contrastive instance queue. The overall pipeline for this method is shown in Fig. 1. This approach differs from supervised distillation, in which labeled data is used.

In our method, we follow the approach of MoCo [16] and generate a view x_i, that is produced from some randomly augmentations (say random rotation and etc.) of an input image. The Teacher and Student encoders transform and normalize this view into feature vectors. The instance queue, D, is represented as $D = [d_1 ... d_k]$, where K represents the the length of this constructed queue and d is a vector of representations obtained from the Teacher encoder. Like in the traditional schema of contrastive learning, D is updated progressively using a "first-in, first-out" updating way as the process of distillation continues. This signifies that the visual features of the current batch, as deduced by the Teacher, are added to the queue and the earliest observed samples are eliminated after each iteration. It is essential to note that the majority of the samples in the queue are random and unrelated to the target instance. To overcome this issue, we add the Teacher's embedding (z_i^T) to the queue to guarantee there's always a positive target, resulting in $D^+ = [d_1 ... d_k, d_{k+1}]$ with $d_{k+1} = z_i^T$. In essense, by minimizing the cross-entropy between the similarity score distributions produced by the Student and Teacher, the network compares x_i with

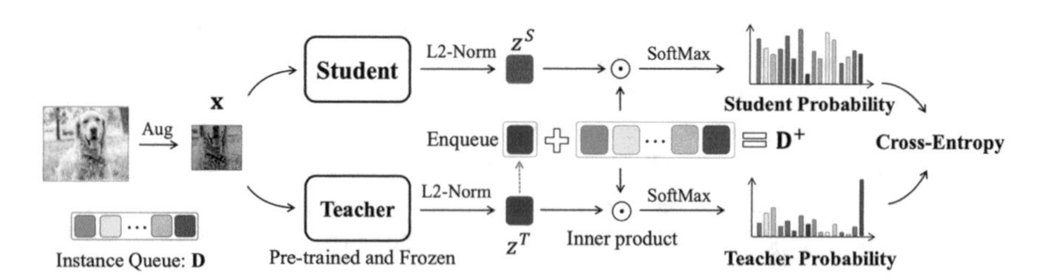

Fig. 1 SEED is a distillation pipeline for self-supervised learning, as illustrated in above figure. The SEED teacher encoder is pre-trained using SSL and is frozen throughout the first step. The student encoder is then taught by reducing the difference in error between the teacher and student encoders for an augmented picture view. To provide negative samples for contrastive distillation, we propose maintaining a dynamically updated queue. In this way, the student can learn from the teacher without requiring labeled data

randomly chosen samples for contrasting and meanwhile aligning it directly with the Teacher encoder's embedding.

Let $p^T(x_i, \theta_T, \text{D})$ denotes the score of similarity between the feature z_i^T produced by the Teacher, and d_j's (j = 1 ... K + 1) computed by the teacher model with SoftMax normalization:

$$p^T(x_i, \theta_T, \text{D}^+) = [p_1^T ... p_{K+1}^T],\ p_j^T = \left(\exp\left(z_i^T \cdot d_j/\tau^T\right)\right) \Big/ \sum_{d \sim D^+} \exp\left(z_i^T \cdot \text{d}/\tau^T\right)), \tag{7}$$

The parameters of the so called "temp" (temperature) for the Teacher is represented by τ. $()^{\mathrm{T}}$ denotes the feature output from the teacher network, and the inner product between two features is represented by (·). The student model computes a similarity score for each instance, denoted as $p^S(x_i, \theta_S, \text{D}^+)$, using the same process as the Teacher. The objective of self-supervised distillation is then to minimize the cross entropy between the Teacher's similarity scores and the Student for all instances, which can be expressed as:

$$L_{SEED} = \sum_{i}^{N} -p^T(x_i, \theta_T, \text{D}^+) \cdot \log p^S\left(x_i, \theta_S, \text{D}^+\right). \tag{8}$$

The Teacher network's queued features are de facto fixed during training because it is pre-trained and frozen. The student network uses these features to compute similarity scores denoted as p^S. With larger values of p_j^T, it results in more weight being placed on p_j^S. L2 normalization keeps the similarity score between c and d_{k+1} constant at **1** before it is normalized by the softmax function, which is the highest value among p_j^T. Hence, the weight for p_{K+1}^S is the greatest and may be modified by varying τ^T. By minimizing the loss enables the z_i^S feature to align with z_i^T and meanwhile contrast with unrelated image features in D.

Note that, during the experiment, when using a relatively small temperature parameter, the p^T becomes a spiky distribution similar to a one-hot vector. In such an extreme case, the SEED objective is equivalent to the general contrastive learning formulation as in Eq. 4, but the positive target is obtained from Teacher network. Intuitively, such a learning objective consists of two sub-goals: aligning the encoding supplied by the student model with the encoding produced by the teacher model; and gently contrasting random samples preserved in the D.

As was shown in [10, 11], SEED obtains great performance improvement over small architectures on a series of visual downstream tasks, including visual recognition, object detection, and instance segmentation. Yet, it is still unknown how this contrastive distillation can be applied to the multi-modal inputs, especially the Vision and Language representation learning. Next, we describe DistilLVM, a technique where we investigate a similar contrastive distillation schema but for Vision Language Learning.

3 Contrastive Distillation for Vision Language Representation Learning

Significant progress has been made in the pre-training of visual linguistic (VL) models, also known as omni VL representation models, which can be applied to a variety of downstream tasks including image captioning, visual question answering, and image retrieval [25, 26, 28, 42, 55]. Its achievement can be largely ascribed to the self-attention transformer architecture, such as BERT [7], which is effective at learning from vast quantities of image-text pairings. Nevertheless, many of these models are big and have significant latency and memory needs, limiting their applicability on resource-constrained edge devices in practical applications. To solve this problem, knowledge distillation (KD) has been suggested as a method for compressing big models by transferring knowledge from a powerful Teacher model to a smaller Student model without affecting generalizability. Such knowledge transfer may be accomplished using techniques like as mimicking the output logits, minimizing the divergence of intermediate layerwise (or averaged) feature maps. KD has been shown to be useful for compressing language models, with DistillBERT [23] using a cosine embedding loss and a soft-target probability loss to decrease the size of the BERT-base model by 40%. TinyBERT [19], MobileBERT [41], and MiniLM [48] have highlighted the significance of conducting distillation by aligning the attention distributions across networks. Particularly, Clark [5] illustrates clearly that attention maps in BERT capture significant distillation-relevant language information and syntactic relationships.

These developments have not been applied to the compression of VL models as of yet. We have discovered numerous obstacles that make direct use of these approaches to VL distillation challenging. Several current VLP efforts use pre-trained or obtained from other domain object detectors, such as Faster-RCNN [36]. The object detectors are utilized extraction of regional representations, which are concatenated as visual tokens for input to the multi-modal encoder network for Vision-Language pre-training. A smaller/efficient VL model often has computational resources limitation so they employ a lightweight visual encoder like EfficientNet [43] as object detector for prompt inference, such as the MiniVLM, which may be distinct from the Teacher's detector. It is challenging to establish semantic connection between the two sets of object produced by the two distinct detectors, since they are often rather dissimilar. This makes it hard to conduct distillation on the top of the Student and Teacher's attention maps or the hidden embeddings from intermediate layers.

To overcome these challenges, we suggest a series of training pipelines to enable the VL model distillation using contrastive distillation. First, we adopt the identical list of the objects, generated by the Student's efficient object detector. This is for the visual token formulation of both the Teacher network and Student network (T-S) (as is shown in Fig. 2). This assures that the visual tokens of the both side coincide semantically. Second, we use a loss term to make the Student learn from the Teacher's distribution of self-attention from the final layer of the transformer encoder. Third,

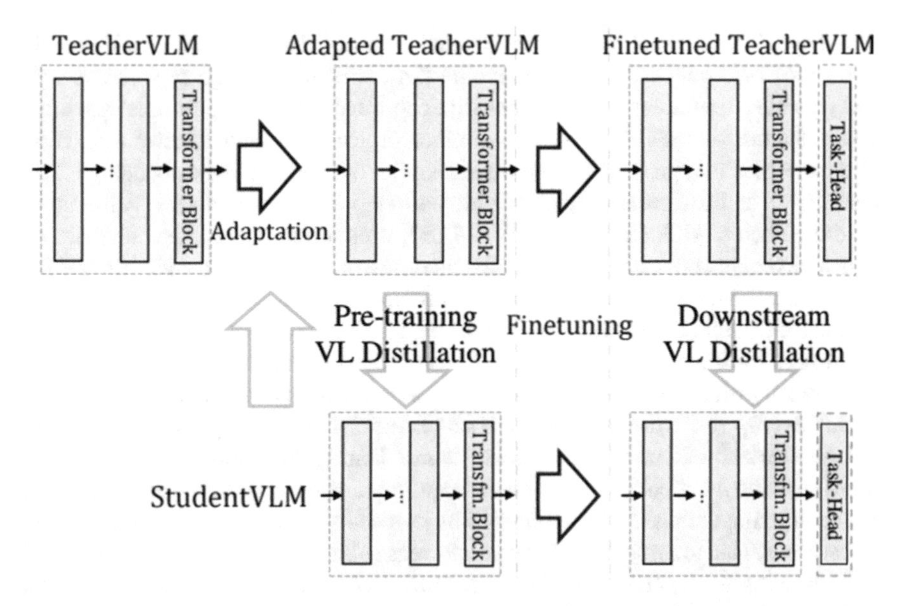

Fig. 2 In our tests, we use a VL model with a region feature extraction module and a multi-modal transformer module. The Teacher VL model is changed to include region proposals from the Student's region feature extractor in order to guarantee accurate alignment of the input. During both the pre-training and fine-tuning phases, the VL distillation process happens

we additionally extract information from the transformer layers (i.e., the embeddings of hidden-states). We discover that distillation from the Teacher's layer-wise embeddings is insufficient supervision for the distillation, so we apply a contrastive loss to keep the token embeddings consistent by making them contrasting with some chosen negative embeddings maintained in a sample queue. Figure 2 offers an illustration of our suggested VL distillation scheme, which may be used during both the pre-training and fine-tuning phases. To evaluate the effectiveness of our proposed distillation schema, we use the identical transformer network, in its compact form, as in MiniVLM and the same lightweight object detector [48], but leverage knowledge distillation techniques to train a small VL model, dubbed DistillVLM [10, 11]. We show that our DistillVLM performs comparably to a larger architecture and outperforms its the previous compact model, MiniVLM, as non-distilled counterpart.

3.1 DistillVLM

Visual token alignment is a key aspect of VL pre-training. In the process of visual token alignment for VL pre-training, a tuple pair <image-caption> is inputted as part of the triple <*w*, *q*, *v*> consisting of the caption embeddings (*w*), the word embeddings of the object tags output from detector in the text (*q*), and tokens (*v*) extracted from the image. These visual tokens are obtained by using detector such as a Faster R-CNN [36] pre-trained on Visual Genome [24], to extract image regional vectors and concatenating them with their position coordinates to form positional-sensitive region feature vectors. These vectors are then projected through a linear projection to ensure that they have the same dimension as the caption/tag embeddings. The resulting visual tokens are typically ordered based on the confidence score output by the detector. Alternatively, MiniVLM [50] uses a lightweight detector with an EfficientNet backbone and a BiFPN [43] module to generate multi-scale features for visual token extraction. However, using different tokens between the T-S networks during the training of distillation can make the direct use of the distillation objectives ineffective for transferring knowledge. Overall, the VL pre-training process aims to semantically align the image regions with the textual input words. To align T-S's visual tokens, we use the exact identical set of detected object proposals and maintain the same order based on their confidence scores. During distillation, both the T and S VLMs use the identical regional objects' tags, predicted by the lightweight detector. While utilizing the visual tokens derived by proposals of the Student's light detector directly may lead to a minor performance decrease, this concern may be circumvented by re-training the T-VLM with the unseen novel visual tokens (we call this process as the Teacher network adaptation) (Fig. 3).

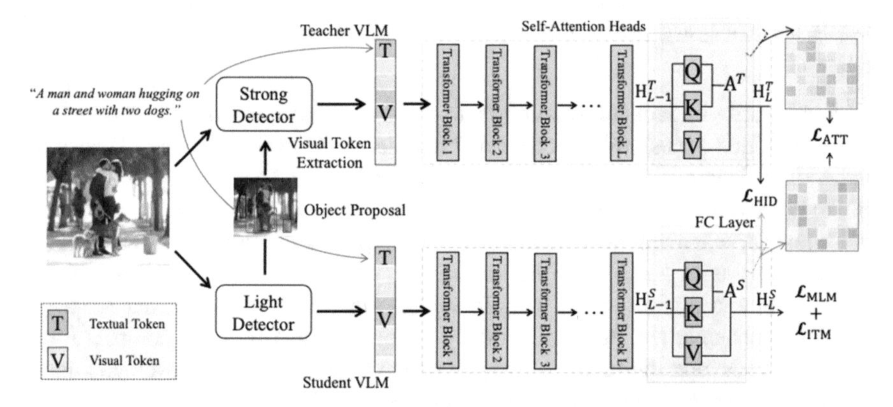

Fig. 3 The illustration depicts the our proposed architecture (dubbed DistillVLM). The lightweight detector firstly obtain regional representations as visual tokens, and the region proposals are sent to the Teacher's large detector to ensure alignment of the region features between both the Teacher network and the Student network. Before distillation, the Teacher transformer network is updated to take the new input tokens for further adaptation. Techniques such as hidden-states alignment and matching-up of attention distribution are used to stepwisely condense the Student VLM

3.2 Attention Distribution Distillation

We first introduce the so-called multi-head attention block, allows contextualized knowledge to be captured from an input sequence, is a crucial component of the transformer block [46]. This module produces a set of attended values as output.

$$Attention(\boldsymbol{Q}, \boldsymbol{V}, \boldsymbol{K}) = \text{softmax}\left(\boldsymbol{K}, \boldsymbol{Q} \Big/ \sqrt{d_k}\right) \cdot \boldsymbol{V}, \tag{9}$$

In this process, $\mathrm{H_i}$ denotes the hidden-states embedding from the median transformer layer is transformed into query (dubbed as Q), key (denoted as the K), and V the value representations through three independent linear transformations. Key (d_k) is the dimension of these vectors and it serves as a scaling factor. The attention matrix is obtained by taking the dot product operation between the and key-query, which is then normalized using a softmax function (as shown in Fig. 4):

$$\boldsymbol{A} = \text{SoftMax}\left(\boldsymbol{K}, \boldsymbol{Q} \Big/ \sqrt{d_k}\right). \tag{10}$$

A transformer block includes a sequence of linear transformations, such as a multi-head attention module, a two-layer feed-forward network, a normalization layer. It is believe that residual connection bridges the layers and allows gradient pass to contour gradient vanishing issue as layers scale so it is adopted in transformer block as well. In the past, it has been shown that transferring self-attention matrices is important for language model distillation [19], as they are believed to include the hidden linguistic knowledge such as syntactic relations (or co-reference relations) of input tokens. It has been found that using just the attention distribution map from the last transformer block is sufficient, even if the T-S networks have different numbers of layers. For VL pre-training, certain attention matrices of pre-trained VL

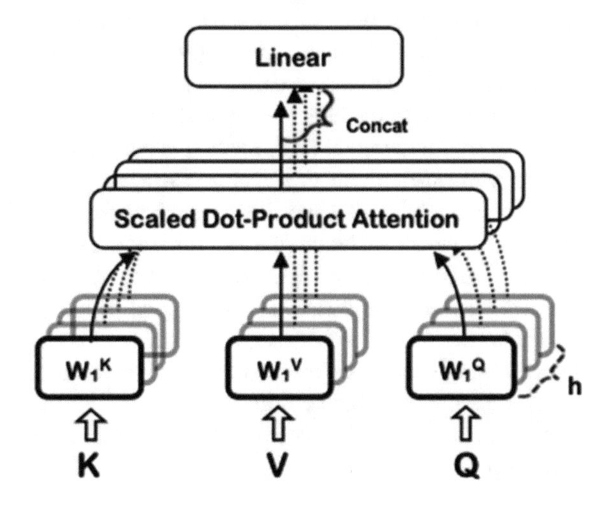

Fig. 4 Example of multi-head attention block. Each block has L layers of attention blocks

models have been shown to contain valuable visual-linguistic knowledge, including extensive intra-modal and cross-modal co-reference relations, which can be used for VL distillation [2]. Following this intuition, we propose to similarly formulate the distillation loss in the form of:

$$L_{ATT} = \frac{1}{T \cdot H} \sum_{i=1}^{T} \sum_{j=1}^{H} MSE(\boldsymbol{A}_{i,j}^{S}, \boldsymbol{A}_{i,j}^{T}), \tag{11}$$

where T represents the total tokens during distillation, H denotes the number of attention heads in a transformer. $\boldsymbol{A}_{i,j}^{S/T}$ denotes Student and Teacher's normalized attention metrics on the i-th token at the j-th head. In essence, we perform knowledge transfer by reducing the self-attention matrices' divergence of the Teacher and Student's final layer. We investigate further the impact of distillation on the distribution of attention in later ablative studies with detailed comparisons.

3.3 Hidden Representation Distillation

During distillation, we also leverage the hidden token embedding for aligning the T and S networks. This was achieved in earlier efforts by focusing on minimizing the divergence of the hidden embedding ($\boldsymbol{H}$) of each transformer block, with the following objective:

$$L_{HID-MSE} = \frac{1}{T \cdot L} \sum_{i=1}^{T} \sum_{j=1}^{L} MSE(\boldsymbol{H}_{i,j}^{S} \boldsymbol{W}_{h}, \boldsymbol{H}_{i,j}^{T}), \tag{12}$$

To align the hidden representations of the Teacher and Student, such a layer-to-layer alignment method is used widely in language model compression, where the divergence of the hidden embedding ($\boldsymbol{H}$) of each transformer layer is minimized. $\boldsymbol{W}$ is a learnable linear transformation that maps the Student's hidden embedding to the same dimension as the Teacher's embedding.

However, limitation of such method is also self-evident: for instance, TinyBERT uses a uniform-function to select a subset aggregation of transformer layers for distillation; MobileBERT has the prerequisite that the T-S need to have same number of layers for layerwise learning. Additionally, the noisy nature of visual tokens during VL distillation can make alignment more challenging.

To circumvent these issues, CoDIR [41] proposes using contrastive loss (NCE or also known as noise-contrastive estimation) loss to align the T-S's hidden states' embedding by contrastive objective. The target instance is encouraged to discriminate with negative samples of random instances and meanwhile aligned with the positive sample from the Teacher network. A pre-defined instance queue, $([h_0^T ... h_K^T])$, is used to store K randomly sampled hidden embeddings and just one embedding as the positive instance from the Teacher network. So the final objective of it can be expressed as follows:

$$L_{HID} = -\log exp(h_i^S \cdot h_i^T/\tau) \Big/ \sum_{j=0}^{K} exp(h_i^S \cdot h_j/\tau), \qquad (13)$$

In this approach, the temperature hyperparameter is represented by τ and the cosine function is denoted as $\langle\cdot\rangle$. There're various methods for obtaining hidden representations, such as using mean-pooled (or known as averaged) token representations as a so-called "summarized" embeddings at each layer (as proposed in [41]). However, we have found that using the contrastive loss with token-wise embeddings leads to improved results, as we will discuss in later experiments. To transform the dimensions, we likewisely introduce a linear mapping as previous work.

In order to update the instance queue with more negative samples, we update the queue in iteratively: we first add the Teacher-produced representation of the target sample (h^T) and remove the oldest stored samples' representation after each iteration. Such queue updation design allows for batch-size independent distillation as historical representations are leveraged for contrastive distillation and so allows for comparison with a larger number of contrasting samples using limited computational resources. In our ablation studies, we investigate the impact of increasing the queue size and using different distillation methods. We keep representations from the pre-trained and frozen Teacher network in the sample queue, which stay constant throughout training. This obviates the need for a momentum encoder as in MoCo [16].

3.4 Classification Distillation

During pre-training, we use task-agnostic distillation to reduce losses. In the fine-tuning stage, we also use knowledge distillation to improve the performance of certain visual language (VL) downstream tasks. These tasks typically involve classification using labels, such as image captioning or visual question answering. By continuing the distillation process at the downstream stage, we can reduce the gap between the pre-training VL corpus and the downstream task. We minimize the softmax prediction of the student and teacher networks using cross-entropy loss as a measure:

$$L_{CLS} = \text{Cross} - \text{Entropy}\ (\mathbf{z}^S/\tau_d, \mathbf{z}^T/\tau_d), \qquad (14)$$

and the temperature parameter, represented as τ_d, is kept as a constant of and the soft label outputs from the Student/Teacher network are represented by $\mathbf{z}^{S/T}$. For the training process, we use the original pre-training objective from VL which consists of two parts: the masked language modeling loss and the image-text (contrastive) matching (*ITM*) loss. In the masked language modeling loss, 15% of the textual tokens are replaced with a special token called $[MASK]$ and the VL model must classify these tokens. The ITM loss requires the model to predict whether the image and text pair match.

$$L = L_{VLP} + \alpha \cdot L_{ATT} + \beta \cdot L_{HID}, \tag{15}$$

where α denotes the weights for attention loss and β is the weights of hidden distillation loss terms. We observe in experiments that L_{CLS} will not affect the results to the pre-training, and we just use this loss term for certain fine-tuning stage at downstream (e.g., for the task of image captioning, or the task of VQA):

$$L = L_{CE} + L_{CLS} + \alpha \cdot L_{ATT} + \beta \cdot L_{HID}, \tag{16}$$

where L_{CE} denotes the specific downstream, classification related objective. We conduct ablative study the impacts of various loss items and report results in later chapter. We summarize the specific overall pipeline of our algorithm in the following part.

4 Experiments

In this study, we examine the effectiveness of a proposed method and evaluate the performance of proposed method on the challenging image captioning and VQA tasks, providing the results and ablation studies.

4.1 Datasets

We create a VL dataset for pre-training using Conceptual Captions (CC) [3], SBU Captions [31], Flicker30k [27, 34], and VQA-2.0 [13]. VL-7M has 4 million unique images and 7 million image-text pairs. Our teacher model and DistillVLM are VL-7M-trained and applied to downstream VL tasks. We assess downstream COCO Captions and VQA-2.0 image captioning and visual question answering. Like Karpathy, we split image captioning into 11,000 training images and 5,000/5,000 validation/testing images. To fine-tune and test the visual question answering task, the VQA-2.0 dataset uses 83,000 images and 444,000 questions for training and 41,000 images and 214,000 questions for validation. The online evaluation server compares test-dev split ablation results.

4.2 Implementation Details Visual Representation

We implement MiniVLM using an EfficientNet-based lightweight object detector (TEE) to extract regional visual features. The input token sequence is the regional feature concatenation. TEE reduces total inference time by 90% and has 91% fewer parameters than the R101-F-based Faster R-CNN model [36]. (86.9M for R101-F vs. 7.5M for TEE). Before extracting visual representations, we pre-train the TEE detector on Object365 [40] and Visual Genome [24]. R101-based Faster-RCNN and TEE detected proposals extract regional visual representations for the teacher model. This semantically aligns teacher and student model input tokens. VL pre-training with object tags has been shown to improve performance. TEE detects object tags and visual tokens of 15 and 50 lengths during distillation for both Teacher and Student networks.

4.3 VL Pre-training and Distillation

We compare VL distillation and regular VLP on compact transformer architecture. Our compact transformer has 12 layers, 12 attention heads, and 384 hidden dimensions. The larger Teacher model has 12 transformer blocks with 12 attention heads and 768 hidden dimensions. The AdamW optimizer pre-trains Student and Teacher architectures on the VL-7M corpus for 1 million steps. Initial LR = 5e-5 and batch size 768 for learning rate. 34.5 million learnable parameters make our compact transformer 70% smaller than Oscarb. After retraining the Teacher with the new visual tokens, we freeze the model and continue VL distillation. Unlike previous work [26], DistilVLM weights perform better when randomly initialized rather than inherited from BERT. Distillation uses 2e-4 learning rate and 768 batches. In ablative experiments, we set $d = 1$ and $\alpha = 10$, $\beta = 10$ and discuss hyperparameter effects.

4.4 Transferring to Downstream Tasks

We also apply our pre-trained model to downstream classification tasks like image captioning, as well as the task requires more comprehending (VQA). These tasks allow direct task-specific distillation and comparison. The VL distillation can easily be applied to other VL tasks as well in future. We use the fine-tuned Teacher model output logit as soft labels with the classification distillation loss for downstream distillation.

To better summarize our algorithm, we give the pseudo-implementation as follows:

```
# Q: Maintaining queue of random history representations: (K X DT)

# teacher_model , student_model: Large/small Teacher/Student VL
model.

# FC: Linear layer for dimension transformation of Student hidden
representation.

# temp: Temperatures of the NCE loss. lamda , beta: Weights for
loss terms.

# activate evaluation mode for Teacher to freeze BN and updation.
with torch.no_grad():

    # t_atts: 1 X N X N, t_hids: 1 X N X DT
    t_atts, t_hids = teacher_model(**teacher_inputs)

    # s_atts: 1 X N X N, s_hids: 1 X N X DS
    mlm_loss, itm_loss, s_atts, s_hids = student_model(**stu-
dent_inputs)

     # dimension transformation on student hidden, s_hids: 1 X N X
DT
    s_hids = FC(s_hids)

# attention distillation loss
att_loss = mse_loss(s_atts , t_atts)

# NCE hidden states distillation loss

# l2-norm the hidden representation
t_hids , s_hids = L2_norm(s_hids), L2_norm(t_hids)

# positive logit: Nx1
l_pos = torch.einsum('nd,nd->n', [s_hids.squeeze(0), t_hids.
squeeze(0)])

# negative logit: NxK
l_neg = torch.einsum('nd,dk->nk', [s_hids.squeeze(0), Q.de-
tach() .T()])

# logit: Nx(1+K)
logits = torch.cat([l_pos, l_neg], dim=1)

# apply the temperature for NCE loss
logits /= temp

# NCE label: N
labels = torch.zeros(logits.shape[0], dtype=torch.long).cuda()
```

Image Captioning. Our VL distillation learning is firstly evaluated by applying it to the challenging captioning task. As general VL pre-training, we train our model by masking 15% words at random and imposing a classification head to estimate the masked, optimized with the CE loss. As [7], we cut and extend textual phrases to 20 characters for the sake of a consistent length input. During inference, we input in [MASK] tokens iteratively and predict successive captions with a beam search, in the size of 1. Metrics BLEU@4 [32], METEOR [6], CIDEr [47], and SPICE [21] are employed to determine the effectiveness of our distilled model.

VQA. The Visual Question Answering challenge requires the model to choose the best answer from multiple options when presented with an image and textual question. We fine-tune the model on the [13] dataset and report DistilLVLM accuracy on the two evaluation splits. A multi-way prediction task with cross-entropy loss trains the VQA model [1]. The VQA problem is mildly combinatorial parameter searched under these constraints.

4.5 Experimental Results

Table 1 shows DistillVLM's Oscar Teacher model. Top lines have VLP baselines with larger transformer structures and better detector. DistillVLM w/o distillation yields 34.0 BLEU@4 and 115.7 CIDEr on TEE (masked language prediction and IMT losses). Suboptimal hyper-parameters may explain MiniVLM pre-trained on VL-7M scoring 116.7 CIDEr and our reproduction 115.7. VLP models. Downstream distillation improves COCO captioning results on CIDEr, and BLEU@4 as can be observed from table. VQA downstream distillation scored 69.2 against 69.0. This is likely because YES/NO or counting-type questions do not improve advice and VQA job answers are mostly irrelevant/mutually exclusive. However, pre-training with distillation benefits the DistillVLM on captioning and VQA tasks across all metrics: 1.2% at B@4, 4.4% at CIDEr, and 0.7% higher VQA. We run ablations for several designing alternatives to understand DistillVLM and examine the benefits of our method at various training iterations and data utilization (e.g., leveraging only partial VL corpus for training).

Table 1 DistillVLM, derived from a more robust model (Teacher), performs well on COCO captioning regardless of the small visual encoder. Even with 50% of the training corpus (# I-T pairs) pre-trained, our model outperforms MiniVLM. The proposed distillation enhances pre-training and fine-tuning (F.D.). For fair comparison, all captioning methods use cross-entropy optimization without CIDEr training. V. Feat. denotes the visual features used by methods

Method	# Param	# *I-T pairs*	V-Feat.	P.D.	F.D.	COCO captioning				VQA	
						B@4	M	C	S	Std.	Dev.
UVLP	111.7M	3M	ResNeXt101	–	–	36.5	28.4	116.9	21.2	70.7	–
OSCARB	111.7M	7M	R101-F	–	–	36.5	30.3	123.7	23.1	73.4	73.2
MiniVLM	34.5M	7M	TEE	–	–	34.3	28.1	116.7	21.3	–	–
MiniVLM	34.5M	14M	TEE	–	–	35.6	28.6	119.8	21.6	69.4	69.1
DistillVLM	34.5M	7M	TEE	–	–	34.0	28.0	115.7	21.1	69.0	68.8
				–	Y	34.5	28.2	117.1	21.5	69.2	69.0
				Y	–	35.2	28.6	120.1	21.9	69.7	69.6
				Y	Y	35.6	28.7	120.8	22.1	69.8	69.6

Table 2 Comparing pre-training distillation effects based various losses. We show results of the model after 20 epochs pretraining and downstream fine-tuning (using only the cross-entropy optimization here)

L_{VLP}	L_{ATT}	L_{HID}	COCO captioning				VQA
			B@4	M	C	S	Dev.
Y	–	–	36.5	28.4	116.9	21.2	68.5
Y	Y	–	36.5	30.3	123.7	23.1	68.9
Y	–	Y	34.3	28.1	116.7	21.3	69.2
Y	Y	Y	35.6	28.6	119.8	21.6	69.4

4.6 Distillation over Different Losses

Table 2 shows the distillation loss contribution based on VL pre-training. VL Pre-training/Distillation uses the same hyperparameters for same epochs. The table yields these conclusions: The non-distilled method scores 110.6 for picture captioning and 67.2 for VQA accuracy (see first row in Table). Simulating attention distribution boosts CIDEr and VQA scores by 1.2 and 0.4 points, respectively. VLP and hidden embedding distillation produce a similar pattern. Hidden-states embedding distillation outperforms the VLP baseline in all metrics, demonstrating the schema's efficacy. Combining all loss terms yields the best performance, confirming that attention and hidden embedding distillation losses are complementary. Distillation alone performs well, suggesting that knowledge transfer via distillation is similar to VL pre-training loss.

4.7 Different Distillation Strategies

Table 3 shows distillation results using different learning techniques (layerwise vs. final-layer) and different losses (Mean Square Error vs. Contrastive). First, we investigate the impact of the visual token alignment we suggested by applying different losses purely on the textual token component, using the "T2T" attention sub-matrices and the textual token embedding that corresponds to each sub-matrix. Table 3's second line shows the textual distillation's marginal quality improvement over the VLP baseline. Following on from previous language distillation studies, we conduct layer-to-layer attention and hidden-states distillation between the Teacher-Student and find that performance is worse than using just the final layer for distillation technique.

The layer-by-layer method is also limited by architectural structures (e.g., when models have different architectures). Our contrastive objective function that uses negative samples for alignment learning yields "NCE + Last-layer" results from DistillVLM. Contrastive learning outperforms MSE loss slightly. For contrastive

Table 3 Ablation of DistillVLM using various distillation procedures, i.e., layerwise distillation or final-layer distillation, using MSE or NCE loss. "Text Distillation" denotes the process of applying the distillation just to the textual tokens rather than the visual ones. The results on captioning task is presented after training and distillation on VL-7M for 20 iterations using CE as objective

Methods	COCO captioning				VQA
	B@4	M	C	S	Dev.
VL pre-training	33.0	27.3	110.6	20.4	68.5
Textual distill	34.1	27.7	114.3	20.9	69.0
MSE + layerwise	34.2	27.8	114.8	21.1	69.2
MSE + last-layer*	33.3	27.6	112.4	20.7	68.5
MSE + layer-layer	34.3	27.8	115.3	21.2	69.4
NCE + last-layer*	34.3	27.9	115.4	21.2	69.3
NCE + last-layer	34.6	27.9	115.6	21.3	69.4

* denotes the result with averaged embedding distillation

learning, we compare token-wise and averaged embedding across layers for distillation. We observe that token-wise embedding improves results, which is contrary to previous findings [41]. The mean-pooled/averaged embedding with NCE loss solves this problem and produces results comparable to the token-wise NCE technique (as suggested in last two lines of Table 3).

In Table 3, we investigate what occurs to training with NCE loss when additional negative samples are used. We discovered that gradually increasing the size of the sample queue may make a significant contribution to the VL models' overall performance. The model obtains a score of 112.5 CIDEr, even when we just use one negative sample, which is consistent with the MSE findings (a score of 112.4 CIDEr) shown in Table 3. When the number of negative sample is brought up to 4,096, the model exhibits the greatest performance across all metrics. In spite of the fact that continuing to make use of a greater number of negative samples may result in improved results, the size of the sample queue that was used in our studies was just 4,096 for limited computing resources. It is important to note that the pre-constructed D holds the random sample representations from Teacher VLM, and these representations do not change at any point throughout the distillation process. This also suggests that it is possible to use samples from same batch for the training, while using the D for sample storation frees the model from the limitations regarding the size of the batches to be used and enables the usage of more negative examples. The length of the queue has a significant impact on the performance of the distillation process, an issue that has been highlighted in a number of earlier research. Even if using a queue that is as big as feasible assures optimal performance for contrastive learning, doing so inevitably brings up substantial computing burdens. In Table 4, we conduct an analysis of the results of the VL distillation process while using various queue sizes. We have found that using a queue size of 4,096 produces the best results possible when used to the captioning and VQA jobs.

Table 4 The effect that the amount of negative samples has on the distillation sing noise contrastive estimate loss. A bigger queue size makes a small but incremental contribution to the performance of the distillation. When the size of the queue is getting closer to one, the NCE loss is essentially the same as the MSE loss, with the Teacher serving as the lone positive anchor. After all of the trials have been trained for twenty iterations on VL-7M with sample queues of varying sizes, they are sent downstream to be evaluated

# Neg	COCO captioning				VQA
	B@4	M	C	S	Dev.
1	33.3	27.6	112.5	20.7	68.5
128	33.6	27.7	112.7	20.9	68.9
512	33.7	27.8	113.3	21.0	68.8
1,024	34.1	27.9	114.7	21.2	69.1
4,096	34.3	27.9	115.4	21.2	69.3

As also brought up frequently by previous works, the size of the queue affects the distillation performances to a large extent. Though using as large queue as possible guarantees the contrastive learning performance, it brings heavy computational burdens naturally. In Table 4, we study the results of VL distillation leveraging the different sizes of the queue. We observe that using a 4,096 queue leads to optimal results on both tasks.

4.8 *Is VL Distillation Data Efficient?*

One important features of VL distillation in the real applications is its capacity for efficiently train efficient VL models at a lower cost, that is, with a smaller VL corpus (from the perspective of data scarcity) and fewer converging epochs (from the perspective of training efficiency). This is a key benefit of VL distillation.

We train models with various epochs and compare results with those of VLP in order to get a greater understanding of whether or not VL distillation is capable of overcoming these obstacles (data scarcity and training efficiency). We propose conducting VL training utilizing equally sampled partial data. This is in accordance with Lu et al., which suggested that some partial data could be essential enough for training so studying the performances of DistillVLM leveraging random partial data is necessary. In addition, they assist to test whether or not DistillVLM is improved by having more converging epochs and more VL data. From the aforementioned findings, one might conclude a few different take-aways. First, VL distillation results in a steady increase in CIDEr over the course of more training epochs. The non-distillation approach only receives a score of 99.8 on the CIDEr after one period of training, but the DistillVLM method receives a score of 103.1 (see Fig. 5). It should be noted that the CIDEr of our method improves progressively when epochs scale, which reveals that our distillation is more successful. A similar pattern emerges for us when we consider the use of varying proportions of the VL data. In the edge case,

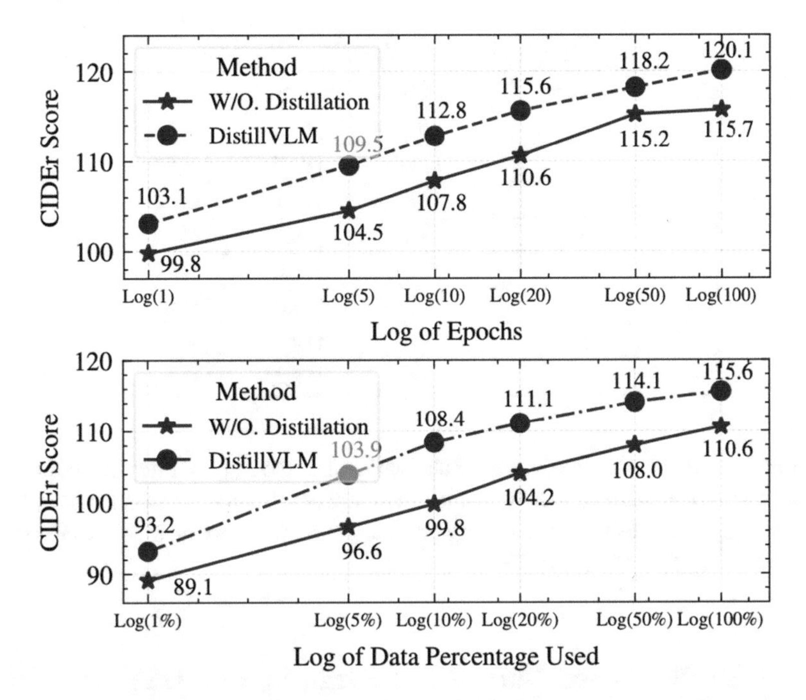

Fig. 5 The performances of DistilLVLM and its non-distilled counterpart under different training epochs and partial data. Top: Captioning CIDEr score of DistillVLM and non-distilled model with different optimization epochs (1, 10, 20, 50, 100) on VL-7M; Bottom shows the results part of data with fixed number of training time

when just 1% of the training data is accessible, the CIDEr score produced by regular pre-training is 89.1, which is 4.1 points lower than the value produced by our method. When there are more training data to work with, our distillation method naturally produces results that are even more impressive. This demonstrates our strategy offers supervision that is far more effective than the regular technique, that in turn leads to improved results. Additional information on the outcomes of the distillVLM process that were transmitted to the image retrieval job is provided by us. In conclusion, we give a qualitative study of other distillation losses, as well as an in-depth discussion of our method and the wider influence it will have on future research.

4.9 Results for Captioning

During the stage that is dedicated to the task-specific training, we additionally carry out ablations. Table 4 provides a summary of the results of the fine-tuning phase of the picture captioning job, which made use of many various distill. losses. On the VL-7M dataset, each experiment goes through 20 iterations of pre-training and distillation. In contrast to the findings obtained during the training stage of the VL

Table 5 Detailed effects based on various loss weights

L_{VLP}	L_{ATT}	L_{HID}	COCO captioning			
			B@4	M	C	S
0	*0*	*0*	34.6	27.9	115.6	21.3
1	*0*	*0*	34.6	28.1	116.3	21.3
1	*1*	*0*	34.6	28.1	116.3	21.3
1	*10*	*0*	34.5	28.0	116.1	21.3
1	*0*	*1*	34.5	28.0	116.0	21.3
1	*0*	*10*	34.3	28.0	115.2	21.2
10	*10*	*10*	34.6	28.1	116.4	21.3

distillation process, our findings indicate that neither the L_{HID} nor the Latt contribute in any way to the performances of the downstream distill. process. Solely using L_{CLS}, as opposed to both sets of data, clearly produces superior outcomes; the CIDEr score rises from 115.6 to 116.3 (Table 5).

5 VL Distillation on Unified One-Stage Architecture

As was mentioned earlier, the intricate two-stage architectures of VL are primarily to blame for the complicated first phases of VL distillation, which include Teacher adaptation and VL token alignment across Teacher and Student. This is because VL distillation requires a Teacher adaptation in order to proceed. The uncoupled representations of Vision and Language each respond in their own unique way to different pre-trained feature extractors that come from a variety of different domains. ViLT [22], Grid VQA [18], ViTCAP [12], and UFO [49] are some examples of recent works that attempt to build up the unified VL architectures, with the goal of using joint VL architecture to encode VL inputs. These are just some of the examples of recent works that attempt to build up the unified VL architectures. These works are an attempt to circumvent this problem that has been identified by employing an unified multimodal encoder w.o. object detector as intermediate representor or simply use general visual encoders with light computations as replacement. In particular, ViTCAP and EfficientVLM [49] make an effort to make use of a more powerful VL model for distillation on a unified architecture, which greatly simplifies the process of distillation. This was done in an effort to make the distillation process more efficient. The author believes that VL distillation would most effectively facilitate the training of a "one-stage" model, and that this learning schema will extend further to the on-device and mobile VL models. This is because an increasing number of interesting applications are becoming possible with the use of VL models. This chapter takes a cursory look at the recently developed one-stage VL models and discusses the application of VL distillation within the context of this newly conceived architectural framework. In the following sections, we will first discuss the benefits that a one-stage VL architecture brings for computational headway, and then we will introduce their fundamental formulations under for representation learning.

5.1 One-Stage VL Architecture

Figure 6 demonstrates that, despite these significant advancements, the majority of mainstream VL models still rely primarily on a large detector to give regional visual tokens for multi-modality fusion/interaction. This is the indeed one of the major challenge that blocks the VL models from been deployed in real-world, despite the fact that there have been many other advancements in this field. To achieve an optimum trade-off between performances and fast inference speed, it becomes more valuable to construct an efficient VL architecture with competing performance. There are still difficulties that come about as a result of the following: (1) feature extraction that needs to be done that requires regional operations (see pictures). These intermediary actions will invariably result in training inefficiency as well as an excessive amount of inference delay during the prediction stage; (2) it requires additional box annotations, which severely restricts both training and the ability to adapt to new circumstances. A growing body of research [18, 51, 54] proposes removing the detector from the VL pre-training procedure from the perspective of architecture in order to address these concerns. Figure 6 illustrates a detector-free system in which visual encoder acts as a replacement of the object detector. The grid features are extracted for subsequent cross-modality interaction from these encoders. These grid features can be obtained from vision transformers or as simply as intermediate ResNet layers. Few of these efforts have shed light on the general VL representation learning tasks, while the majority of these efforts have concentrated primarily on the specific VL task, which is frequently framed as a classification issue. It is possible that adopting a unified cross-modal encoder will prevent the model from effectively capturing the fine-grained cross-model correspondence, which will in turn result in less representative cross-modal features. This is an additional difficulty associated with the process of building a one-stage VL architecture. We ascribe this to the inconsistent information density across the modalities: visual inputs from nature are typically encoded as raw pixels and are predominantly repetitive and continuous (most pixels for the same neighboring objects are in similar RGB values and change moderately), whereas textual inputs from the language side encode highly condensed semantic information that is discretized and pre-defined by human beings. During the process of learning the cross-modal representations, such conflicts present the untrained multi-modal encoder with a significant challenge. Because of this, a few pioneering works such as ViLT and Grid VQA, despite significantly improving the inference speed by moving away from the object detector, suffer significantly when transferred to challenging VL tasks such as captioning and VQA that require highly comprehending. The image in [15] is encoded with a pre-trained ResNet for the

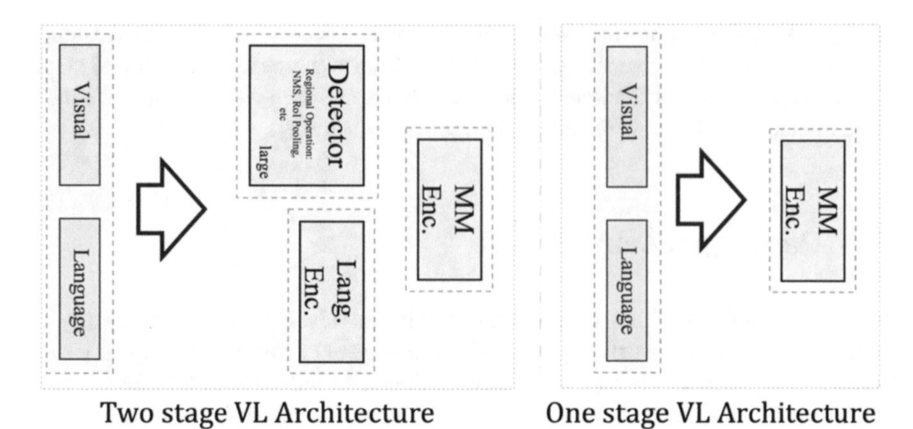

Fig. 6 Two-stage VL architecture versus one-stage VL architecture: typical two-stage VL architecture (e.g., VinVL, Oscar) reply on a pre-trained object detector for visual representation extraction. Such module largely increases the training/inference computational headways as they involve time-consuming regional operations like NMS, RoI and etc. To avoid such design, one-stage VL architecture bridge the Visual and Language domains via a direct multi-modal encoder (see right side of figure). Such design of course largely reduce the computational burdens required for VL models, but also unavoidably sacrifice the performances of VL architectures. For this reason, it becomes important to introduce VL distillation to assist the cross-model learning on one-stage VL architectures

captioning task; however, its performance (117.3 CIDEr on COCO) is still a long way behind that of the leading detector-based method (129.3 CIDEr with VinVL-base). Concerning how to create a detector-free VL model that is more robust, the terrain is still unknown and has only been poorly explored. The following section further delve deep into the details of one-stage VL architecture like ViLT.

ViLT's [22] design is minimalistic, taking the form of a VL model with a rudimentary visual encoding process and a singlestream architecture for multi-modal interaction. Figure 7 gives a direct overview of ViLT. In specific, it simply initializes the multi-modal interaction transformer weights with pre-trained ViT pre-trained on ImageNet rather than BERT from language corpus, as empirically they discover that a good initial visual representations are essential for successful training. In the lack of a separate deep visual embedder (e.g., object detector), such an approach utilizes interaction layers to interpret visual information. ViT consists of stacked blocks consisting of an MSA (multi self-attention) layer and an MLP layer. The only difference between ViT and BERT is the position of layer normalization (LN): in BERT, LN comes after MSA and MLP, whereas in ViT, it comes before. ViLT also attempt to initializing the layers with BERT weights and then simply use the pre-trained patch projection from ViT for image encoding, but neither approach was effective according to reported experiments in [22]. This is expected as using language initialization may lead to noisy visual representations and largely hinders the VL interaction.

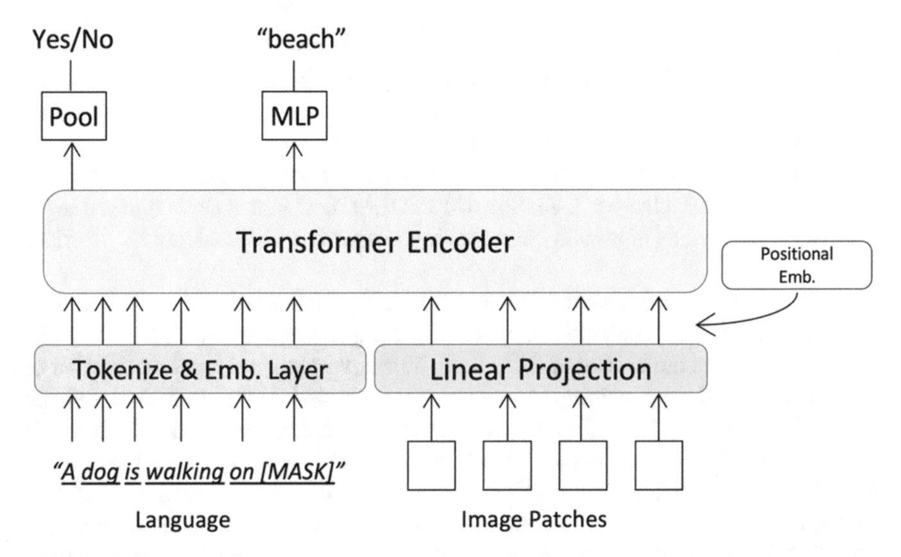

Fig. 7 Overview of ViLT, a typical one-stage VL architecture that exploits vision transformer as the backbone. ViLT encodes images and language concurrently by a simple linear projection function and tokenization and embedding layer. ViLT is pre-trained with traditional VL tasks: image-text matching loss and masked language prediction loss. Positional embedding are applied on VL embedding before fed into transformer encoders

The input caption is embedded using the word embedding layer after tokenization operation, and the position are applied on before concatenate with visual embedding. A raw image input is firstly sliced into non-overlapping square patches, and then flattened to vector. After linear projection and the position embedding application, visual embedding are then concatenated with textual embedding, with multi-type embedding added to them. This helps the modal to identify the different types of embedding during training This contextualized vector is updated iteratively through the D depth transformer layers until the final contextualized sequence is obtained for training objectives: image-text matching loss, and masked language prediction loss.

As for the detailed architectural setting: ViLT follows the ViT/32 setting 12 attention heads in each MSA with the hidden size = 768. There are 12 layers in ViLT and each small image patch size is 32 × 32. The MLP dimension is 3,072. ViT-B/32 for ViLT has been pretrained on ImageNet-21K and refined on ImageNet-1 K for image classification for better initialization. ViLT mentioned weights trained on larger datasets (such as JFT-300M or LAION5B) may further improve results but is out of the scope of ViLT.

More about Objectives of pre-training ViLT: ViLT is trained with 4M image-captions with traditional objectives for VLP models. That is the image-caption matching (ICM or ITM) and masked language modeling (MLM). For encouraging alignment, ViLT replaces the aligned image-text pairs with a different image or text with a probability of 0.5 randomly. As for ITM loss, it is simply computed by a negative log-likelihood loss. In addition, motivated by the textual-word and image-region alignment goal of Chen et al., ViLT also leverages a novel word patch alignment (WPA), which computes the alignment score between textual/visual subsets, utilizing the approximate proximal point method for optimal transports. While WPA, according to following experiments, is less effective. Masked Language Modeling: Following the heuristics of [7], ViLT randomly masks inout texts at random with a 15% chance. A shallow MLP MLM head that accepts masked input token's embedding as input and generates logits over the whole vocabulary for loss computation.

Whole textual-word masking is another training trick worthy of notice. WWM is a masking trick conceals all sub-word of a complete textual-word. It is revealed that future actions are enhanced. It is thought that whole-word masking will be of particular value for pre-training to maximize the utilization of data from the visual modal. With the pre-trained bert-base-uncased tokenizer, the word "elephant" is tokenized into three word piece tokens *["ele", "##ph", and "##ant"]*. Then, when not all of the words/tokens are selected to be masked, as in *["ele", "[MASK]", "##ant]*, the VL architecture can only depend on contextualized surrounding tokens *["ele", "##ant"]* to infer the intermediate masked "*##ph*" rather than picture data. Such a strategy is advantageous for VL learning because it drives the model to focus on a portion of the word that contains more semantic information.

5.2 VL Distillation on One-Stage Architecture

This section briefly presents how simple the VL Distillation can be formulated on one-stage VL architecture, taking image captioning task as the example. For captioning training, the multi-modal encoder now serves as a modal translator module: it accepts the image input embedding concatenating the masked textual-words after tokenization, and masking probability is set to 15%. The tokens that have been masked are substituted with a specially designed *[MASK]* token. Then, the prediction at location t depends on the previous visible words (caption generation process is auto-regressive process, which means only previous captions are visible to the masked token) visual representations. Using equation, this process is expressed as:

$$L_{cap} = -E\left[\log\left(\prod_{c_t \sim C_M} P_\theta(c_t \,|\, c_{<t}, v)\right)\right], \tag{17}$$

where C_M denotes the groundtruth aggregation of masked tokens $C_M = \{c_1 \dots c_M\}$, and θ is the learnable parameters of the VL model. $c_{<t}$ refers to the aggregation of all preceeding tokens' embedding and v is the visual embedding in inputs. By using the knowledge distillation technique, a pre-trained Teacher can be used to improve the VL model in comparison to its non-distilled counterpart. In ViTCAP [12], it employs a well-trained detector-based captioning model as the Teacher (parameterized by θ_T), i.e., VinVL to help the training ViTCAP. Surprisingly, per results of ViTCAP, by

simply adopting VinVL's make token logit as target for distillation already obviously improve the captioning results to an obvious extent (classification distillation). Note that, VinVL is indeed two stage VL model with a detection module for regional features extraction (ViTCAP and VinVL all use object tags as additional inputs for performance improvement, but we ignore this here for simplicity). This unavoidably result in disparate visual features with ViTCAP, and therefore the distillation objectives such as attention-map match-up loss and hidden embedding alignment loss are usable as introduced previously. As masked token prediction is classification task in essense, the classification distillation loss is formulated as the KL divergence between T-S:

$$L_{cls} = -E\big[\Sigma_{c_t \sim C_M} \mathrm{KL}(P_{\theta_T}(c_t), P_\theta(c_t))\big], \tag{18}$$

where $P_{\theta/\theta_T}(c_t)$ is the logit probability of token c_t from Student model and Teacher model respectively. Intuitively, by mimicking the soft probability of masked token, VL model learns from the Teacher the "relation" knowledge where certain tokens correlates with contextualized visual/textual inputs. Note that this classification distillation is also applicable on image-text matching modeling and masked language modeling during pre-training status. In addition, if a one-stage VL Teacher architecture is utilized (e.g., CLIP), it's also viable to apply averaged or layerwise hidden states alignment loss as in two-stage VL models but with much flexibility in contrastive format:

$$L_{hid} = -\log exp(h_i^S \cdot h_i^T/\tau) \Big/ \sum_{j=0}^{K} exp(h_i^S \cdot h_j/\tau), \tag{19}$$

with $h_i^{S/T}$ denotes the ith token's layer (or averaged across multiple layers) hidden embedding from Student/Teacher. Noticeably, such contrastive formulation now differentiates with Eq. (13) that it includes also the visual hidden embedding of image patches so knowledge encoded in Teacher model also transfers to the Student VL model during distillation compared with two-stage VL architectures. Similarly, attention map match-up loss can be formulated as:

$$L_{att} = -\mathrm{E}\left[MSE\big(\boldsymbol{A}_{i,j}^S, \boldsymbol{A}_{i,j}^T\big)\right], \tag{20}$$

that $\boldsymbol{A}_{i,j}^{S/T}$ refers to attention map for layer i head j. Then the overall distillation can be expressed as the concatenation of above-mentioned:

$$L_{overall} = L_{cap} + L_{cls} + L_{att} + L_{att}. \tag{21}$$

Interestingly, in ViTCAP [12], the authors also propose a novel sub-module: concept token network (CTN), where CTN predicts the image-level concepts (e.g., existed objects, adjectives of subject) as tokens. It is proved that concept tokens play as equivalent effects as object detector tags as in OSCAR [25] that helps to connect modalities. ViTCAP leverages the OD tags from OSCAR's detector as pseudo-labels for learning. This in essense can be viewed as a novel distillation schema, where Student mimics the projected masking token probability as well as learning the OD tags output from Teacher detector. ViTCAP also adopts the masked token distillation strategy during VL pre-training. The success of ViTCAP also verifies the importance of include VL distillation, which brings substantial improvements.

6 Conclusion and Future Works

This chapter summarizes the knowledge distillation and Vision-Language representation learning tasks, as well as the contrastive distillation technique. We discuss in depth the challenges of using knowledge distillation for multi-modal tasks and identify several points that impede VL distillation. In fact, knowledge distillation is used in multiple concurrent works for visual and textual representation learning, such as [33], DEiT [45], and DistillBERT [39]. This chapter studies the VL distillation schema, which involves distillation on a cross-modal model: VL distillation uses the knowledge distillation to successfully compress visual-linguistic models. Several aspects of empirical results verified the effectiveness of our proposed distillation technique: VL distillation not only produces better results, but it is also more data efficient, as better results can be obtained with a smaller amount of pre-training data. Our broad evaluations confirm that VL distillation has broad applicability to a wide range of VL tasks and domains. As results, we anticipate that VL distillation will become an essential technique for the task of large-scale representation learning, as well as for addressing other cross-modal learning challenges especially on mobile devices in foreseeable future.

References

1. Anderson, P.: Bottom-up and top-down attention for image captioning and visual question answering. In: Proceedings of the IEEE Conference on Computer Vision and Pattern Recognition (2018)
2. Cao, S.: Multilingual Alignment of Contextual Word Representations (2020)
3. Changpinyo, S.: Conceptual 12M: pushing web-scale image-text pre-training to recognize long-tail visual concepts. In: Proceedings of the IEEE/CVF Conference on Computer Vision and Pattern Recognition (2021)
4. Chen, T.: A simple framework for contrastive learning of visual representations. In: International Conference on Machine Learning (2020)
5. Clark, K.: What Does BERT Look at? An Analysis of BERT's Attention (2019)
6. Denkowski, M.: Meteor universal: language specific translation evaluation for any target language. In: Proceedings of the Ninth Workshop on Statistical Machine Translation (n.d.)
7. Devlin, J.: BERT: Pre-training of Deep Bidirectional Transformers for Language Understanding (2018)
8. Ebrahimi, Chassang, A., Romero, A., Ballas, N., Kahou, S.E., Chassang, A., Gatta, C., Bengio, Y.: Fitnets: Hints for Thin Deep Nets

9. Fang, Z.: SEED: Self-supervised distillation for visual representation. In: ICLR (2020)
10. Fang, Z.: Compressing visual-linguistic model via knowledge distillation. In: Proceedings of the IEEE/CVF International Conference on Computer Vision (2021)
11. Fang, Z.: Seed: self-supervised distillation for visual representation. In: ICLR (2021)
12. Fang, Z.J.: Injecting semantic concepts into end-to-end image captioning. In: Proceedings of the IEEE/CVF Conference on Computer Vision and Pattern Recognition (2022)
13. Goyal, Y.: Making the V in VQA matter: elevating the role of image understanding in visual question answering. In: Proceedings of the IEEE Conference on Computer Vision and Pattern Recognition (2017)
14. Gutmann, M.: Noise-contrastive estimation: a new estimation principle for unnormalized statistical models. In: Proceedings of the Thirteenth International Conference on Artificial Intelligence and Statistics (2010)
15. Haiyang Xu, M.Y.: E2E-VLP: End-to-End Vision-Language Pre-training Enhanced by Visual Learning (n.d.). arXiv:2106.01804
16. He, K.: Momentum contrast for unsupervised visual representation learning. In: Proceedings of the IEEE/CVF Conference on Computer Vision and Pattern Recognition (2020)
17. Henaff, O.: Data-efficient image recognition with contrastive predictive coding. In: International Conference on Machine Learning (2020)
18. Jiang, H.: In defense of grid features for visual question answering. In: Proceedings of the IEEE/CVF Conference on Computer Vision and Pattern Recognition (2020)
19. Jiao, X.: TinyBERT: Distilling BERT for Natural Language Understanding (2019)
20. Khosla, P.: Supervised contrastive learning. In: Advances in Neural Information Processing Systems (2020)
21. Kilickaya, M.: Re-evaluating Automatic Metrics for Image Captioning (2016)
22. Kim, W.B.: ViLT: vision-and-language transformer without convolution or region supervision. In: International Conference on Machine Learning. PMLR (2021)
23. Kim, Y.: Sequence-Level Knowledge Distillation (2016)
24. Krishna, R.: Visual genome: connecting language and vision using crowdsourced dense image annotations. Int. J. Comput. Vis. (2017)
25. Li, G.: Unicoder-VL: a universal encoder for vision and language by cross-modal pre-training. In: Proceedings of the AAAI Conference on Artificial Intelligence (2020)
26. Li, X.: Oscar: object-semantics aligned pre-training for vision-language tasks. In: European Conference on Computer Vision (2020)
27. Lin, T.-Y.: Microsoft COCO: common objects in context. In: European Conference on Computer Vision (2014)
28. Lu, J.: ViLBERT: Pretraining task-agnostic visiolinguistic representations for vision-and-language tasks. In: Advances in Neural Information Processing Systems (2019)
29. Misra, I.: Self-supervised learning of pretext-invariant representations. In: Proceedings of the IEEE/CVF Conference on Computer Vision and Pattern Recognition (2020)
30. Oord, A.: Representation Learning with Contrastive Predictive Coding (2018)
31. Ordonez, V.: Im2Text: Describing images using 1 million captioned photographs. In: Advances in Neural Information Processing Systems (2011)
32. Papineni, K.: BLEU: a method for automatic evaluation of machine translation. In: Proceedings of the 40th Annual Meeting of the Association for Computational Linguistics
33. Peng, Z.: BEiT v2: Masked Image Modeling with Vector-Quantized Visual Tokenizers (2022)
34. Plummer, B.A.: Flickr30k entities: collecting region-to-phrase correspondences for richer image-to-sentence models. In: Proceedings of the IEEE International Conference on Computer Vision
35. Reddy, S.: CoQA: a conversational question answering challenge. Trans. Assoc. Comput. Linguist.
36. Ren, S.: Faster R-CNN: towards real-time object detection with region proposal networks. In: Advances in Neural Information Processing Systems
37. Russakovsky, O.: ImageNet large scale visual recognition challenge. Int. J. Comput. Vis. (2015)

38. Saeidi, M.: Interpretation of Natural Language Rules in Conversational Machine Reading (2018)
39. Sanh, V.: DistilBERT, a Distilled Version of BERT: Smaller, Faster, Cheaper and Lighter (2019)
40. Shao, S.: Objects365: a large-scale, high-quality dataset for object detection. In: Proceedings of the IEEE/CVF International Conference on Computer Vision (2019)
41. Sun, Z.: MobileBERT: A Compact Task-Agnostic BERT for Resource-Limited Devices (2020)
42. Tan, H.: LXMERT: Learning Cross-Modality Encoder Representations from Transformers (2019)
43. Tan, M.: EfficientNet: rethinking model scaling for convolutional neural networks. In: International Conference on Machine Learning (2019)
44. Tian, Y.: Contrastive Representation Distillation (2019)
45. Touvron, H.: Training Data-Efficient Image Transformers & Distillation Through Attention (2021)
46. Vaswani, A.: Attention is all you need. In: Advances in Neural Information Processing Systems (2017)
47. Vedantam, R.: CIDEr: consensus-based image description evaluation. In: Proceedings of the IEEE Conference on Computer Vision and Pattern Recognition
48. Wang, J.: MiniVLM: A Smaller and Faster Vision-Language Model (2020)
49. Wang, J.X.: UFO: A Unified Transformer for Vision-Language Representation Learning (2021). arXiv:2111.10023
50. Wang, T.W., Wang, T., et al.: EfficientVLM: Fast and Accurate Vision-Language Models via Knowledge Distillation and Modal-Adaptive Pruning (2022). arXiv:2210.07795
51. Wonjae Kim, S.B.: ViLT: vision-and-language transformer without convolution or region supervision. In: International Conference on Machine Learning, 2021 (n.d.)
52. Wu, Z.: Unsupervised feature learning via non-parametric instance discrimination. In: Proceedings of the IEEE Conference on Computer Vision and Pattern Recognition (2018)
53. Yadav, V.: Deep Affix Features Improve Neural Named Entity Recognizers (n.d.)
54. Zhicheng Huang, Z.Z.: Pixel-BERT: Aligning Image Pixels with Text by Deep Multi-modal Transformers (n.d.). arXiv:2004.00849
55. Zhou, L.: Unified vision-language pre-training for image captioning and VQA. In: Proceedings of the AAAI Conference on Artificial Intelligence

Knowledge Distillation in Granular Fuzzy Models by Solving Fuzzy Relation Equations

Hanna Rakytyanska

Abstract A knowledge distillation method is proposed that uses inverse inference to deploy a granular-fuzzy classifier on devices with limited computing resources. The System of Fuzzy Relation Equations (SFRE) serves as the carrier of teacher knowledge. The self-organized set of rules is integrated into the hierarchical distillation structure based on granular solutions of the SFRE. At the first stage, knowledge is transferred from the granular teacher model to a compact relational model to identify semantic trends in data. At the second stage, knowledge is transferred from the relation-based teacher model to the granular student model. To control the granularity level, knowledge is distilled in the form of interval or constrained solutions of the SFRE. The completeness and accuracy of the rule set is ensured by the genetic-neural technology for solving the SFRE. The sequential data and knowledge distillation ensures a competitive compression performance. After compressing the training data, the relational model is trained using the distilled expert dataset. Incorporation of expert knowledge in the form of interval rules makes it possible to replace the granular teacher model with a compact relational model. The multitask distillation scheme ensures transferring distributed knowledge into the granular student model. Deployment of the student model can be parallelized when solving the SFRE for each output class. The self-organized set of rules based on solutions of the SFRE is autonomously identified using the incremental learning scheme. Compression of the student model is achieved through quantization at the level of granular or constrained solutions of the SFRE.

Keywords Knowledge distillation · Granular knowledge · Granular fuzzy classifier · Solving fuzzy relation equations · Interval solutions · Constrained solutions

H. Rakytyanska (✉)
Soft Ware Design Department, Vinnytsia National Technical University, Vinnytsia, Ukraine
e-mail: rakit@vntu.edu.ua

W. Pedrycz and S.-M. Chen (eds.), *Advancements in Knowledge Distillation: Towards New Horizons of Intelligent Systems*, Studies in Computational Intelligence 1100,
https://doi.org/10.1007/978-3-031-32095-8_4

1 Introduction

In machine learning, Knowledge Distillation (KD) is a well-established technique for transferring knowledge from a pre-trained teacher model to a student model [1, 2]. KD is a model compression technique aimed at training a more compact student model which is capable to imitate the predictions of the teacher model [3]. KD has become a paradigm for compressing and accelerating Deep Neural Networks (DNNs), since the high computational complexity and storage requirements make it problematic to deploy deep models in real-time applications [2]. The quality of distillation is determined by the structure of the teacher and student models and the knowledge transfer scheme between them [2]. Given the teacher model, the problem consists in searching for the optimal student model. Most of KD techniques use the predefined architecture based on a simplified or quantized structure of the student model [3].

This work addresses the problem of KD in granular fuzzy models, where understanding of KD mechanisms under uncertainty requires interpretable models of decision making and knowledge transferring [4, 5]. Granular fuzzy classifiers based on relations and rules are used to ensure the understanding of mutual interaction of the teacher and student during distillation [6]. Fuzzy relational calculus [7, 8] provides a powerful theoretical background for KD in granular fuzzy models. Combination of both paradigms stipulated for the development of a new hybrid KD technique based on solving the System of Fuzzy Relation Equations (SFRE) [9, 10].

The hierarchical teacher-student architecture is trained in two stages. At the first stage, knowledge is transferred from the granular teacher model to a relation-based model which describes semantic trends in data using primary fuzzy terms (*decrease, increase*). We consider the relation-based teacher model in the form of the SFRE with extended *max–min* composition [6]. The granular student model is built for significance measures of the primary terms. The modified granules in the form of intervals (*increase to* 56–74%) are described using linguistic modifiers (*moderate increase*). At the second stage, knowledge is transferred from the relation-based teacher model to the granular student model by solving the SFRE [9, 10]. Knowledge is distilled by simulating inverse inference, where a set of explanations for each class of the student model is generated with the help of the teacher model [9, 10].

For each output class, the set of granular solutions of the SFRE is determined using the genetic-neural algorithm [9, 10]. The use of the teacher model in the form of the SFRE, the solutions of which ensure the optimal partition at the level of rules, is a feasible solution to the problem of finding the optimal student structure [9, 10]. The proposed approach simplifies the process of rule generation, as it allows eliminating rule selection. However, in [9, 10] there are no mechanisms for linguistic interpretation of the interval solutions. In the general case, the distillation scheme requires solving the SFRE under granularity constraints. Therefore, it is important to develop the hybrid technique focused on transferring knowledge with controllable level of detail in rule-based solutions of the SFRE.

The remainder of this chapter is organized as follows. In Sect. 2 related works are discussed. In Sect. 3 the objective of the research is formulated. In Sect. 4 the hierarchical teacher-student architecture based on granular solutions of the SFRE is introduced. The problem of training the teacher-student architecture is defined in Sect. 5. In Sect. 6 we define the problem of structural optimization of the student model as the problem of solving the SFRE under granularity constraints. We also describe the genetic algorithm for parallel deployment of the student model. The method for training the neuro-fuzzy relational system is described in Sect. 7. Section 8 reviews the experimental results in time series forecasting for the benchmark problem of predicting the number of comments after post publication over Facebook pages. Model compression estimations are given in Sect. 9. Conclusions and suggestions for future work are offered in Sect. 10.

2 Related Works

Recent research shows a strong trend towards incorporating hybrid distillation schemes to tackle challenging problems in KD.

2.1 Knowledge Granularity in Transfer Learning

In [11, 12], the multi-granularity property of knowledge is introduced for transferring knowledge with different level of abstraction to students that have different abilities of knowledge understanding. In the human beings learning, experienced teachers not only summarize the knowledge, but also control the level of detail when facilitating students. When analyzing the essential characteristics, the multi-granularity distillation mechanism ensures supervised dimensionality reduction [11, 12]. In this case, knowledge quantization through information granulation can be considered as the mechanism underlying model compression [3]. The self-analyzing module focuses on feature compression by optimizing the encoders for the branches of abstracted and detailed knowledge. The low-dimensional distillation scheme allows transferring abstracted coarse-grained knowledge, focusing on the global characteristics, while the high-dimensional distillation scheme allows transferring the fine-grained knowledge, focusing on the detailed features [11, 12].

Incorporation of the multi-granularity classification mechanism into distillation frameworks provides extensive and detailed explanations in teacher-student learning. For understanding deep learning models, the methods which generate post-hoc explanations of the pre-trained model and methods with incorporated interpretability have been used [13]. Coarse-grained explanations are provided by highlighting the areas that are considered significant based on class activation maps [13, 14]. Following the principle of justifiable granularity [4], to explain the classification mechanism in detail, the highlighted areas are divided into commonality and specificity saliency maps [14]. In [15], it is proposed to construct a transparent and explainable student model by distilling structured knowledge from the pre-trained black-box model. When extracting features from the structured data, most Graph Neural Networks (GNN) use neighbor selection strategy, where representations of nodes are learned by randomly aggregating neighbors' features. For constructing an explainable GNN, a shallow network is trained using explicit contribution weights. In this case, the node selection strategy ensures the embedded interpretability.

In [16], the hyperbox fuzzy sets are introduced into the Convolutional Neural Network (CNN) to capture bounds between classes. In [17, 18], it was proven that the generalization bound, which is influenced by data geometry in the class separation, determines the quality of distillation, i.e., robust learning of the student network on the soft labels of the teacher network. The fuzzy *min–max* CNN is a principal solution when transferring geometric properties of the data distribution from the teacher to the student [16]. The CNN requires a complete re-training using the previous and new datasets for recognizing unseen classes from incoming data. When the new data related to unknown classes is obtained, the fuzzy *min–max* CNN is updated without the need of re-training the already acquired knowledge [16].

Injection of linguistic knowledge with the predefined granularity into deep learning is a principal way to reduce the amount of training data and distillation time. Instead of learning from scratch, the rule-based prior knowledge is integrated into the teacher-student distillation framework [19]. The logic statements for the detailed patterns which correspond to the predicates in the knowledge base are translated into a set of constrained blocks to be transferred into the student network [19]. In [20, 21], fuzzy decision trees are used to generate explanations of deep neural solutions at the level of rules recognized by domain experts. To explain the KD mechanism, in [22, 23] knowledge is distilled from the DNN into a neuro-fuzzy system. In [22], the finalized DNN architecture facilitates the training of the membership functions for each feature transferred to the rule-based classifier. In [23], the neuro-fuzzy model which replaces the final layers of the CNN is used for deriving explanations in the form of rule-based saliency maps obtained through transfer learning.

2.2 Evolutionary Neural Architecture Search

In transfer learning, the evolutionary Neural Architecture Search (NAS) focuses on finding the optimal subnetwork of the teacher model for the student task [24, 25]. To compress the model and simultaneously improve the testing accuracy, the NAS problem is reduced to a multiobjective optimization problem [24]. The multiobjective evolutionary algorithm concurrently minimizes the prediction error and model size when transferring knowledge from the source domain to the target domain. To optimize the network architecture, the coarse-grained global search for the suitable subnetworks is enhanced by the fine-grained local search [24]. In [25], the transferred and retrained modules are adaptively adjusted according to the dataset similarity of the source and target tasks. In [26], the multiobjective evolution strategy is proposed for filter pruning the CNN architecture, where convolution filters are eliminated to reach a trade-off between performance and computational complexity. In [27], the multiobjective evolutionary algorithm is proposed for automatically shallowing the DNN architecture by pruning less informative blocks. Performance is recovered when incorporating a prior knowledge into the distillation scheme to improve the exploration ability of the evolutionary search [27].

Evolutionary NAS through augmenting topologies results in advanced increasingly complex solutions that allow incrementally growing from the minimal structure, crossing-over of different topologies and inheriting structural innovations [28]. Since neuroevolution is computationally expensive, genetic algorithms are used to construct block-based evolutionary models with minimum computation [29–31]. When creating offspring architectures, most evolutionary NAS methods use only the mutation operator [29]. As a result, the generated models differ from their parent architectures and cannot inherit modular information to accelerate the learning process. In [29, 30], the evolutionary NAS methods use the modular cross-over operation which enables the offspring architectures to inherit and transfer encoded blocks from the parent architectures. The encoding strategy consists in mapping the fixed-length genotype to the structure with variable depth [30, 31].

In [32], a gradient-limited regularization is introduced into the pruning scheme, and the genetic algorithm is used for acquiring the optimal subnetwork from the sparse model. In [33, 34], the genetic algorithm is proposed for multi-objective pruning the DNN models with different trade-off between prediction error, computation, and sparsity. Fitness evaluation is based on the pre-defined criteria weights by laying emphasis on model compression [33]. In [34], it is shown that the compression ratio is influenced by the inconsistent distribution of the memory footprint and workload of the DNN model among different layers. When exploring the pruning structure space, the cross-layer constraints are imposed to find a judicious balance between the model size and workload [34]. In [35], the genetic algorithm is proposed for automatic selection of the effective layers for transfer learning. The genotype encodes layers whose weights are updated or fixed, and individuals with high validation accuracy are selected during evolutionary search [35]. When the first layers of a pre-trained architecture are tied to the fully connected layers to adapt them to a new task, the configuration of these layers affects the model performance [36]. For transfer learning based DNNs, an evolutionary pruning model replaces the last fully connected layers with sparse layers optimized by a genetic algorithm.

2.3 Deep Neuro-Fuzzy Networks

To deal with uncertainty in feature properties, in [37, 38] the deep rule-based classifier with transparent and interpretable hierarchical structure is proposed. The prototype-based fuzzy rules are integrated into the parallel multi-layer structure. The self-organized set of rules is autonomously identified by attracting similar data samples which form clusters with high data density. Feature descriptors imbedded into the multilayer architecture are distributed on three levels (low, medium, high) based on their descriptive abilities. The paper [38] focuses on aggregating less informative prototypes into highly generalized ones and self-arranging them into the hierarchical prototype-based structures. The multi-layer neuro-fuzzy model can start classification from a prior set of rules and can self-evolve without the need of re-training. Due to the prototype-based module structure, the training process can be parallelized, that makes it suitable for deployment in real-time applications.

In [39], a novel self-organizing Deep Fuzzy Neural Network (DFNN) is proposed. When detecting changes in the feature space, the structure of the DFNN can be deepened by stacking additional layers [39]. The DFNN evolves via feature augmentation guided by the stacked generalization principle [39]. The depth of the DFNN is controlled using the drift detection method, while the network structure can be simplified by merging the hidden layers [39]. To optimize the number of layers, in [40] an ensemble of rule-based units is integrated into the hierarchical structure. In deep integrated learning, multiple units are learned simultaneously, and compact and accurate units are selected to transfer their knowledge to the next layer according to the stacked generalization principle [40]. The work [41] focuses on embedding a deep learning scheme into the neuro-fuzzy system confronted with uncertainty in high-dimensional data. The deep implication operator serves as a feature extractor based on the concept of fuzzy association [41]. In [42], compression of the deep fuzzy network based on the stacked modules is carried out by eliminating redundant rules using singular value decomposition. In [43], the evolution algorithm is proposed for feature selection and subsequent reduction of the rule base.

3 Problem Statement

The type of knowledge, the teacher-student architecture, and the distillation method are the key issues in developing the KD framework based on solving the SFRE [2].

3.1 Granular Solutions of the SFRE

The SFRE serves as the carrier of teacher knowledge in the form of significance measures of the input and output primary terms. The output terms of the student model are selected as distillation spots to directly match the response-based knowledge when solving the SFRE. To control the level of granularity when passing messages from the teacher model, significance measures of the primary terms are described by interval values or linguistic modifiers. Following [44], the linguistic modifiers are associated with constraints imposed on significance measures of the primary terms. In this case, knowledge is distilled from the relation-based teacher model into the granular student model in the form of interval or constrained solutions of the SFRE.

Under granularity constraints, the structure of the solution set is modeled using binary relations interpreted as weights of linguistic modifiers, where the weights take the values 1(0) if the modifier is present (absent) in the linguistic description of the interval rule. In this case, the constrained solutions for significances measures of the primary terms are replaced by the crisp solutions for the weights of linguistic modifiers. To provide the complete linguistic description of the interval partition, the set of constrained solutions is determined by the set of maximum solutions for the vectors of binary weights, that is, the set of complete crisp solutions. Thus, the set of interval rules is replaced by the reduced set of linguistic rules that allow transferring expert knowledge in the form of the set of explanations.

3.2 Genetic Search for the Optimal Student Model

This paper proposes a teacher-student architecture based on granular solutions of the SFRE. For fuzzy rule-based classifiers, the method of generating rules by solving the SFRE provides the optimal partition at the level of rules [45, 46]. The number of rules in the class is determined by the number of solutions, and the granularity is determined by the interval solutions for significance measures of the primary terms. The method does not require rules pruning when generating candidate rules.

Following [9, 10], the problem of finding the optimal student model for the given relation-based teacher model is formulated as follows. For each class of the student model, significance measures of the input primary terms should be found which provide the maximum proximity to the teacher knowledge in the form of significance measures of the output primary terms. For the predefined granularity, solving the SFRE is reduced to identifying unknown relations in the constrained solutions. Thus, knowledge is transferred simultaneously with the search for the optimal structure of the student model under constrained granularity. The completeness and accuracy of the rule set is ensured by the genetic technology for solving the SFRE [9]. The distillation loss follows from approximate solvability of the SFRE.

The multitask learning scheme allows transferring the decomposed knowledge for each class of the student model. The deployment process can be parallelized when solving the SFRE for each output class. Besides, properties of the solution set of the extended *max–min* SFRE allow us to parallelize the search for the lower and upper solutions of the SFRE. It is proven that the solution set of the extended *max–min* SFRE can be decomposed into the lower and upper subsets corresponding to subsystems with *max–min* and dual *min–max* composition [47, 48]. The lower (upper) subset is defined by the unique greatest (least) aggregating solution and the set of minimal (maximum) solutions [7, 8]. Under granularity constraints, the properties of the set of interval solutions are generalized to the set of constrained ones. Thus, the set of constrained solutions is decomposed into the lower and upper subsets of complete crisp solutions bounded by the greatest (least) aggregating solutions. As a result, deployment of the student model is reduced to incremental learning by generating lower and upper solutions of the SFRE.

3.3 Self-Organizing Fuzzy Relational Neural Networks

The quality of distillation is determined by the relational teacher model focused on capturing semantic trends in data to transfer detailed knowledge to the student model. The hierarchical distillation structure can be represented as a system of granular and relational fuzzy models with an emphasis on evaluating significance measures of the primary terms using linguistic modifiers. This allows embedding the relation-based model into the special neuro-fuzzy network which is capable to acquire knowledge from the granular teacher model and transfer knowledge into the granular student model. The fuzzy relation matrix and rule-based solutions of the SFRE are embedded into the fuzzy *min–max* neural network so that the self-organized set of rules is autonomously identified using the incremental learning scheme. Feature space is controlled by the minimal solutions of the SFRE which define the subset of features in each rule. After imposing granularity constraints, the layer of interval rules is replaced by the fuzzy relational neural network that allows embedding the linguistic rules.

When used on devices with limited computing resources, the hierarchical distillation scheme ensures sequential model compression, as shown in Fig. 1. After training the granular teacher model, the training dataset is compressed to a set of interval rules. The relational model is trained using the compressed dataset by transferring expert knowledge from the granular teacher model. At this stage, the granular teacher model is replaced by a compact relational model. To reduce the structure complexity of the granular fuzzy classifier, stepwise deployment of the student model is carried out by generating interval or constrained solutions of the SFRE. At this stage, model quantization through knowledge granulation underlies the student model compression. The distillation loss is estimated after training the student model on experimental data. The quality of distillation shows that the method ensures a competitive compression performance while preserving completeness and accuracy of the rule set.

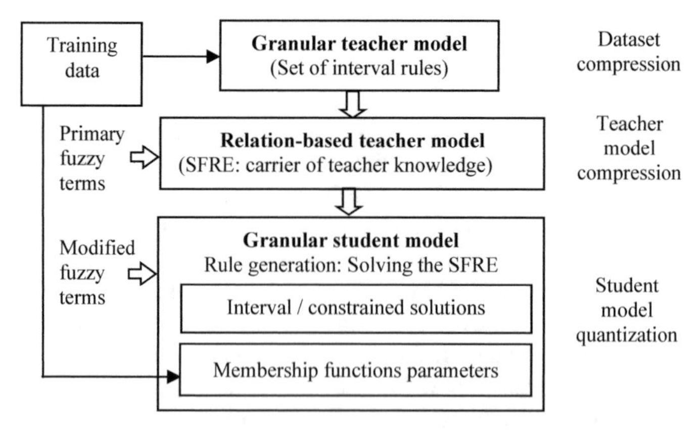

Fig. 1 Hierarchical distillation scheme based on solving the SFRE

4 Hierarchical Teacher-Student Architecture Based on Granular Solutions of the SFRE

4.1 Relation-Based Teacher Model

For an object of the form $y = f(\boldsymbol{X})$ with n inputs $\boldsymbol{X} = (x_1, \ldots, x_n)$ and the output y, the teacher model can be represented as a system of granular fuzzy rules [4, 9]:

$$\bigcup_{s=1}^{z_j^*}\left[\bigcap_{i=1}^{n}\left(x_i \in \left[\underline{\chi}_i^{js}, \overline{\chi}_i^{js}\right]\right)\right] \rightarrow y = d_j, j = 1, \ldots, m, \tag{1}$$

where $\underline{\chi}_i^{js}\left(\overline{\chi}_i^{js}\right)$ are the lower (upper) bounds of the values of the input variables x_i in the rule js of the teacher model; $d_j \in \left[\underline{y}_j, \overline{y}_j\right]$ is the output class for estimating the variable y; m is the number of output classes; $\underline{y}_j\left(\overline{y}_j\right)$ are the lower (upper) bounds of the output variable y in the class d_j; z_j^* is the number of interval rules in the class d_j of the teacher model.

The following system of fuzzy logic equations corresponds to the granular teacher model (1) [9]:

$$\mu_T^{d_j}(y) = \bigvee_{s=1}^{z_j^*}\left[\bigwedge_{i=1}^{n}\left[\mu^{T_i^{js}}(x_i)\right]\right], j = 1, \ldots, m, \tag{2}$$

where $\mu_T^{d_j}(y)$ is the membership function of the variable y to the class d_j of the granular teacher model; $\mu^{T_i^{js}}(x_i)$ is the membership function of the variable x_i to the term T_i^{js} associated with the interval $[\underline{\chi}_i^{js}, \overline{\chi}_i^{js}]$ in the rule js.

We shall redenote:

$\left\{c_{11}, \ldots, c_{1u_1}, \ldots, c_{n1}, \ldots, c_{nu_n}\right\} = \{C_1, \ldots, C_N\}$ is the set of primary fuzzy terms for estimating the variables x_i, $i = 1, \ldots, n$, where $N = u_1 + \ldots + u_n$;

$\{E_1, \ldots, E_M\}$ is the set of primary fuzzy terms for estimating the variable y;

$\left\{\alpha_{I1}, \ldots, \alpha_{Ig_I}\right\}$ is the set of linguistic modifiers for estimating the significance measure μ^{C_I}, $I = 1, \ldots, N$;

$\left\{a_{11}, \ldots, a_{1g_1}, \ldots, a_{N1}, \ldots, a_{Ng_N}\right\} = \{A_1, \ldots, A_L\}$ is the set of modified fuzzy terms $\left(C_I, \mu^{C_I} = \alpha_{IK}\right)$ for estimating the variables x_i, $i = 1, \ldots, n$, where $L = g_1 + \ldots + g_N$.

Examples of the use of primary and modified fuzzy terms in different knowledge domains are presented in Table 1.

Semantic trends described by the primary terms can be identified using the system of fuzzy relation matrices $\boldsymbol{R}_i \subseteq c_{il} \times E_J = [r_{il}^J]$, $i = 1, \ldots, n$, $l = 1, \ldots, u_i$, $J = 1, \ldots, M$, which is equivalent to the matrix $\boldsymbol{R} \subseteq C_I \times E_J = [r_{IJ}]$, $I = 1, \ldots, N$, $J = 1, \ldots, M$.

Table 1 Examples of the use of primary and modified fuzzy terms

Domain	Linguistic variable	Primary terms	Modified terms
Engineering	Consumed power	Increase	Strong, weak
Medicine	Blood pressure	Hypertension	I–III degree
Economics	Purchasing power	Inflation	Moderate
Sociology	Consumer demand	Growth	Rapid, slow
Biology	Demand of oxygen	Saturation	Insufficient
Ecology	Air condition	Contamination	Significant

Given matrices $\boldsymbol{R}_i$, $i = 1, \ldots, n$, the relation-based teacher model can be represented in the form of the extended *max–min* SFRE [6]:

$$\mu^{E_J}(y) = \bigwedge_{i=1}^{n}\left(\bigvee_{l=1}^{u_i}\left[\wedge\left(\mu^{c_{il}}(x_i), r_{il}^{J}\right)\right]\right), J = 1, \ldots, M, \tag{3}$$

where $\boldsymbol{\mu}^{C} = (\mu^{c_{11}}, \ldots, \mu^{c_{1u_1}}, \ldots, \mu^{c_{n1}}, \ldots, \mu^{c_{nu_n}}) = \left(\mu^{C_1}, \ldots, \mu^{C_N}\right)$ is the vector of significance measures of the primary terms C_I; $\boldsymbol{\mu}^{E}(y) = \left(\mu^{E_1}, \ldots, \mu^{E_M}\right)$ is the vector of significance measures of the primary terms E_J.

4.2 Student Model based on Granular Solutions of the SFRE

For the given output classes $y = d_j$, $j = 1, \ldots, m$, the granular student model can be represented as a set of interval solutions of the SFRE (3) [10, 45, 46]:

$$\bigcup_{k=1}^{z_j}\left[\bigcup_{I=1}^{N}\left(x_i \in \left[\underline{x}_I^{jk}, \overline{x}_I^{jk}\right]\right)\right] \rightarrow \bigcup_{k=1}^{z_j}\left[\bigcup_{I=1}^{N}\left(\mu^{C_I} \in \left[\underline{\mu}_I^{jk}, \overline{\mu}_I^{jk}\right]\right)\right] \rightarrow y = d_j, \tag{4}$$

where $\underline{x}_I^{jk}$ ($\overline{x}_I^{jk}$) are the lower (upper) bounds of the values of the input variables x_i in the rule jk of the granular student model;$\underline{\mu}_I^{jk}$ ($\overline{\mu}_I^{jk}$) are the lower (upper) bounds of the significance measure μ^{C_I} in the rule jk; z_j is the number of interval rules in the class d_j of the granular student model.

The linguistic student model can be represented as a set of constrained solutions of the SFRE (3):

$$\bigcup_{p=1}^{q_j}\left[\bigcup_{I=1}^{N}\left[\bigcup_{K=1}^{g_I}\left(x_i = a_{IK} \text{ with weight } v_{IK}^{jp}\right)\right]\right] \to y = d_j,\ j = 1, \ldots, m, \tag{5}$$

where v_{IK}^{jp} is the weight of the modified term a_{IK} for the rule *jp* in the class d_j of the linguistic student model; q_j is the number of linguistic rules in the class d_j.

The following system of fuzzy logic equations corresponds to the granular and linguistic student models (4), (5):

$$\mu_g^{d_j}(y) = \bigvee_{k=1}^{z_j}\left[\bigwedge_{i=1}^{n}\left(\bigvee_{l=1}^{u_i}\left[\mu^{b_{il}^{jk}}(x_i)\right]\right)\right],\ j = 1, \ldots, m, \tag{6}$$

$$\mu_c^{d_j}(y) = \bigvee_{p=1}^{q_j}\left[\bigwedge_{i=1}^{n}\left(\bigvee_{K=1}^{g_{i1}+\ldots+g_{iu_i}}\left[\wedge\left(\mu^{a_{i,K}}(x_i), v_{i,K}^{jp}\right)\right]\right)\right],\ j = 1, \ldots, m. \tag{7}$$

Here $\mu_g^{d_j}(y)$ ($\mu_c^{d_j}(y)$) is the membership function of the variable y to the class d_j of the granular (linguistic) student model; $\mu^{b_I^{jk}}(x_i)$ is the membership function of the variable x_i to the term b_I^{jk} associated with the interval $\left[\underline{x}_I^{jk}, \overline{x}_I^{jk}\right]$ in the rule jk of the student model (4); $\mu^{a_{i,K}}(x_i)$ is the membership function of the variable x_i to the modified term $a_{i,K}$.

We use a bell-shaped membership function of the variable x to the term A [9]:

$$\mu^A(x) = \frac{1}{1 + \left(\frac{x-\beta}{\sigma}\right)^2},$$

where β is the coordinate of maximum; σ is the parameter of concentration.

If the value of the variable x in the SFRE (3) is given by the fuzzy term A, then the degree of membership $\mu^c(x = A)$ is defined using the *supremum* operator [6]:

$$\mu^c(x = A) = \sup_{x \in [\underline{x}, \overline{x}]}\left[\min\left(\mu^c(x), \mu^A(x)\right)\right].$$

The operation of defuzzification is defined in [9] as follows:

$$y = \frac{\sum_{j=1}^{m} \underline{y}_j \mu^{d_j}(y)}{\sum_{j=1}^{m} \mu^{d_j}(y)}.$$

Correlations (2), (3), (6), (7) define the hierarchical teacher-student architecture based on the granular solutions of the SFRE as follows:

$$y = f_T\left(\boldsymbol{X}, Z^*, \underline{\boldsymbol{G}}_T, \overline{\boldsymbol{G}}_T\right), \tag{8}$$

$$\mu^E(y) = f_R(\boldsymbol{X}, \boldsymbol{R}, \boldsymbol{P}_C, \boldsymbol{P}_E), \tag{9}$$

$$y = f_g\big(\boldsymbol{X}, f_R, Z, \underline{\boldsymbol{G}}_R, \overline{\boldsymbol{G}}_R\big), \tag{10}$$

$$y = f_c(\boldsymbol{X}, f_R, \boldsymbol{P}_a, Q, \boldsymbol{V}), \tag{11}$$

where $\underline{\boldsymbol{G}}_T = \left(\underline{\chi}_1^1, \ldots, \underline{\chi}_n^1, \ldots, \underline{\chi}_1^{Z^*}, \ldots, \underline{\chi}_n^{Z^*}\right)$, $\overline{\boldsymbol{G}}_T = \left(\overline{\chi}_1^1, \ldots, \overline{\chi}_n^1, \ldots, \overline{\chi}_1^{Z^*}, \ldots, \overline{\chi}_n^{Z^*}\right)$ are the vectors of lower and upper bounds of the interval rules (1); Z^* is the number of rules of the granular teacher model (1), $Z^* = z_1^* + \ldots + z_m^*$;

$\boldsymbol{P}_C = \left(\beta^{C_1}, \ldots, \beta^{C_N}, \sigma^{C_1}, \ldots, \sigma^{C_N}\right)$, $\boldsymbol{P}_E = \left(\beta^{E_1}, \ldots, \beta^{E_M}, \sigma^{E_1}, \ldots, \sigma^{E_M}\right)$ are the vectors of membership functions parameters for the primary terms C_I and E_J;

f_T, f_R are the operators of "inputs–output" connection, corresponding to the granular and relational teacher models (2), (3);

$\underline{\boldsymbol{G}}_R = \left(\underline{x}_1^1, \ldots, \underline{x}_N^1, \ldots, \underline{x}_1^Z, \ldots, \underline{x}_N^Z\right)$, $\overline{\boldsymbol{G}}_R = \left(\overline{x}_1^1, \ldots, \overline{x}_N^1, \ldots, \overline{x}_1^Z, \ldots, \overline{x}_N^Z\right)$ are the vectors of lower and upper bounds of the rule-based solutions (4); Z is the number of rules of the granular student model (4), $Z = z_1 + \ldots + z_m$;

$\boldsymbol{P}_a = \left(\beta^{A_1}, \ldots, \beta^{A_L}, \sigma^{A_1}, \ldots, \sigma^{A_L}\right)$ is the vector of membership functions parameters for the modified terms a_{IK}; Q is the number of rules of the linguistic student model (5), $Q = q_1 + \ldots + q_m$; $\boldsymbol{V} = \left(v_1^1, \ldots, v_L^1, \ldots, v_1^Q, \ldots, v_L^Q\right)$ is the vector of terms weights for the linguistic rules (5);

f_g, f_c are the operators of "inputs–output" connection, corresponding to the granular and linguistic student models (6), (7).

5 The Problem of Training the Teacher-Student Architecture

Before distillation, the granular teacher model (8) is trained using the set of experimental data $\left\langle \hat{\boldsymbol{X}}_l, \hat{y}_l \right\rangle$, $l = 1, \ldots, \Theta$, where $\hat{\boldsymbol{X}}_l = \left(\hat{x}_1^l, \ldots, \hat{x}_n^l\right)$, $\hat{x}_i^l \in \left[\underline{x}_i, \overline{x}_i\right]$, and $\hat{y}_l \in \left[\underline{y}, \overline{y}\right]$ are the vector of values of the input variables and the value of the output variable in the experiment number l [9].

At the first stage, knowledge is transferred from the granular teacher model (8) to the relational model (9). After compressing the training data, the relational model (9) is trained using the distilled expert dataset $\langle \underline{\boldsymbol{X}}_{js}, \overline{\boldsymbol{X}}_{js}, d_j \rangle$, $j = 1, \ldots, m$, $s = 1, \ldots, z_j^*$, where $\underline{\boldsymbol{X}}_{js} = \left(\underline{\chi}_1^{js}, \ldots, \underline{\chi}_n^{js}\right)$, $\overline{\boldsymbol{X}}_{js} = \left(\overline{\chi}_1^{js}, \ldots, \overline{\chi}_n^{js}\right)$ are the vectors of lower and upper values of the input variables in the rule js of the granular teacher model (8).

Given the pre-trained granular teacher model (8), the essence of training the relational model (9) is as follows. It is necessary to find the relation matrix **R** and the vectors of membership functions parameters $\boldsymbol{P}_C$ and $\boldsymbol{P}_E$, which provide the minimum distance between the model and experimental vectors of significance measures of the output primary terms:

$$\sum_{j=1}^{m}\sum_{s=1}^{z_j^*}\left[f_R\left(\underline{\boldsymbol{X}}_{js}, \overline{\boldsymbol{X}}_{js}, \boldsymbol{R}, \boldsymbol{P}_C\right) - \boldsymbol{\mu}^E\left(d_j, \boldsymbol{P}_E\right)\right]^2 = \min_{\boldsymbol{R}, \boldsymbol{P}_C, \boldsymbol{P}_E}. \quad (12)$$

At the second stage, knowledge is transferred from the relation-based teacher model (9) to the granular and linguistic student models (10), (11). Structure of the student model is defined by the set of interval or constrained solutions of the SFRE (3). Membership functions parameters for the rule-based solutions are trained using the set of experimental data.

Given the pre-trained relational teacher model (9), the essence of training the granular student model (10) is as follows. It is necessary to find the number of interval rules Z and the vectors of lower and upper bounds $\underline{\boldsymbol{G}}_R$, $\overline{\boldsymbol{G}}_R$, which provide the minimum distance between the model and experimental outputs of the object:

$$\sum_{l=1}^{\Theta}\left[f_g\left(\hat{\boldsymbol{X}}_l, f_R, Z, \underline{\boldsymbol{G}}_R, \overline{\boldsymbol{G}}_R\right) - \hat{y}_l\right]^2 = \min_{Z, \underline{\boldsymbol{G}}_R, \overline{\boldsymbol{G}}_R}. \quad (13)$$

For the predefined granularity level, the essence of training the linguistic student model (11) is as follows. It is necessary to find the vector of membership functions parameters $\boldsymbol{P}_a$, the number of linguistic rules Q and the vector of terms weights **V**, which provide the minimum distance between the model and experimental outputs of the object:

$$\sum_{l=1}^{\Theta}\left[f_c\left(\hat{\boldsymbol{X}}_l, f_R, \boldsymbol{P}_a, Q, \boldsymbol{V}\right) - \hat{y}_l\right]^2 = \min_{\boldsymbol{P}_a, Q, \boldsymbol{V}}. \quad (14)$$

The genetic-neural approach is used for solving the optimization problems (12)–(14).

6 Knowledge Distillation by Solving the SFRE

6.1 Structure of the Constrained Linguistic Solution

Let $\boldsymbol{\mu}^C = \left(\mu^{C_1}, \ldots, \mu^{C_N}\right)$ be the solution of the SFRE (3), $\mu^{C_I} \in \left[\underline{\mu}^{C_I}, \overline{\mu}^{C_I}\right]$, where $\underline{\mu}^{C_I}\left(\overline{\mu}^{C_I}\right)$ are the lower (upper) bounds of the significance measure μ^{C_I}.

Let us impose constraints on the significance measures of the primary terms μ^{C_I}, $I = 1, \ldots, N$, associated with the linguistic modifiers α_{IK} [44]:

$$\mu^{C_I}(\alpha_{IK}) = \left[\underline{\mu}_K^{C_I}, \overline{\mu}_K^{C_I}\right], I = 1, \ldots, N, K = 1, \ldots, g_I, \tag{15}$$

where $\underline{\mu}_K^{C_I}\left(\overline{\mu}_K^{C_I}\right)$ are the lower (upper) bounds of the significance measure μ^{C_I} associated with the linguistic modifier α_{IK}.

Given constraints (15), the structure of the linguistic solution is defined as follows. For each interval solution $\left[\underline{\mu}^{C_I}, \overline{\mu}^{C_I}\right]$, $I = 1, \ldots, N$, interconnection between the predefined terms a_{IK} and terms b_I associated with the intervals $\left[\underline{\mu}^{C_I}, \overline{\mu}^{C_I}\right]$ in the partition at the level of rules is modelled by the vector of weights $\mathbf{W}_I = (w_{I1}, \ldots, w_{Ig_I})$. An element of the vector $\mathbf{W}_I$ is the term weight w_{IK}, where $w_{IK} = 1(0)$ if the term a_{IK} is present (absent) in the linguistic description of the interval solution $\left[\underline{\mu}^{C_I}, \overline{\mu}^{C_I}\right]$.

The following *max–min* SFRE connects the significance measures of the primary terms C_I, $I = 1, \ldots, N$, for the interval and constrained solutions:

$$\mu^{C_I}(b_I) = \bigvee_{K=1}^{g_I} \left[\wedge\left(\mu^{C_I}(a_{IK}), w_{IK}\right)\right], I = 1, \ldots, N. \tag{16}$$

where $\mu^{C_I}(b_I)$ is the significance measure of the primary term C_I in partitioning at the level of rules by the terms b_I; $\mu^{C_I}(a_{IK})$ is the significance measure of the primary term C_I in partitioning by the predefined terms a_{IK}.

In the general case, the SFRE (16) has not a single, but a set of solutions for the vectors of weights $\mathbf{W}_I$, $I = 1, \ldots, N$, that is the interval partition (4) has a set of linguistic interpretations. The linguistic description of the interval solution follows from the properties of the solution set of the *max–min* SFRE [7, 8]. The unique maximum solution of the SFRE (16) completely covers the interval solution $\left[\underline{\mu}^{C_I}, \overline{\mu}^{C_I}\right]$, $I = 1, \ldots, N$, and the set of minimal solutions corresponds to subintervals which are subject to merging as redundant.

The unique maximum solution $\overline{V} = \left(\overline{\mathbf{W}}_1, \ldots, \overline{\mathbf{W}}_N\right) = \left(\overline{v}_1, \ldots, \overline{v}_L\right)$, where $\overline{\mathbf{W}}_I = \left(\overline{w}_{I1}, \ldots, \overline{w}_{Ig_I}\right)$ are the vectors of upper bounds of the terms weights w_{IK}, is called a complete crisp solution.

6.2 *The Problem of Structural Optimization of the Student Model*

For the given output classes d_j, $j = 1, \ldots, m$, the problem of transferring granular knowledge from the relation-based teacher model (9) to the student model (10) consists in finding the set of interval solutions of the SFRE (3). The solution elements are the coordinates of maximum of the membership functions of the fuzzy terms b_I^{jk} associated with the intervals in the partition at the level of rules (4).

We shall redenote:

$\boldsymbol{B}_j = \left(\beta_1^j, \ldots, \beta_N^j\right)$ is the vector of coordinates of maximum of the membership functions of the fuzzy terms b_I^j for the interval rule in the class d_j.

The problem of knowledge distillation based on solving the SFRE (3) is formulated as follows. For each distillation spot d_j, $j = 1, \ldots, m$, the vector of coordinates of maximum $\boldsymbol{B}_j = \left(\beta_1^j, \ldots, \beta_N^j\right)$, $\beta_I^j \in \left[\underline{x}_i, \overline{x}_i\right]$, $I = 1, \ldots, N$, should be found which provides the least distance between the teacher's and student's significance measures of the primary terms E_J:

$$F_g\left(\boldsymbol{B}_j\right) = \sum_{J=1}^{M}\left[\mu^{E_J}(d_j) - \mu^{E_J}\left(\mu^{C_1}\left(\beta_1^j\right), \ldots, \mu^{C_N}\left(\beta_N^j\right)\right)\right]^2 = \min_{\boldsymbol{B}_j}, \tag{17}$$

where $F_g\left(\boldsymbol{B}_j\right)$ is the distillation loss for the interval solution $\boldsymbol{B}_j$.

Properties of the set of solutions of the extended *max–min* SFRE (3) are proven for an arbitrary interval output value $d \in \left[\underline{y}_d, \overline{y}_d\right]$ for which $\mu^{E_J} \in \left[\underline{\mu}^{E_J}, \overline{\mu}^{E_J}\right]$.

Statement 1. The set S_g of interval solutions of the SFRE (3) is defined by the aggregating solution $\hat{\boldsymbol{\mu}}^C \in \left[\hat{\boldsymbol{\mu}}^{1C}, \hat{\boldsymbol{\mu}}^{2C}\right]$, for which there exist the set of minimal solutions $\underline{S} = \left\{\underline{\boldsymbol{\mu}}_l^C,\ l = 1, \ldots, z_1\right\}$ and the set of maximum solutions $\overline{S} = \left\{\overline{\boldsymbol{\mu}}_k^C,\ k = 1, \ldots, z_2\right\}$:

$$S_g = \bigcup_{\underline{\boldsymbol{\mu}}_l^C \in \underline{S}} \ \bigcup_{\overline{\boldsymbol{\mu}}_k^C \in \overline{S}} \left[\left[\underline{\boldsymbol{\mu}}_l^C,\ \hat{\boldsymbol{\mu}}^{2C}\right] \cup \left[\hat{\boldsymbol{\mu}}^{1C}, \hat{\boldsymbol{\mu}}^{2C}\right] \cup \left[\hat{\boldsymbol{\mu}}^{1C},\ \overline{\boldsymbol{\mu}}_k^C\right]\right]. \tag{18}$$

Here $\hat{\boldsymbol{\mu}}^{1C} = \left(\hat{\mu}^{1C_1}, \ldots, \hat{\mu}^{1C_N}\right)$ and $\hat{\boldsymbol{\mu}}^{2C} = \left(\hat{\mu}^{2C_1}, \ldots, \hat{\mu}^{2C_N}\right)$ are the vectors of lower and upper bounds of aggregating significance measures;

$\underline{\boldsymbol{\mu}}_l^C = \left(\underline{\mu}_l^{C_1}, \ldots, \underline{\mu}_l^{C_N}\right)$ and $\overline{\boldsymbol{\mu}}_k^C = \left(\overline{\mu}_k^{C_1}, \ldots, \overline{\mu}_k^{C_N}\right)$ are the vectors of lower and upper bounds of significance measures μ^{C_I}, where the union is taken over all $\underline{\boldsymbol{\mu}}_l^C \in \underline{S}$ and $\overline{\boldsymbol{\mu}}_k^C \in \overline{S}$.

Proof The formula (18) follows from the properties of the solution set of the *max–min* and dual *min–max* SFRE [7, 8].

Since the SFRE (3) contains subsystems with *max–min* composition, then for $\mu^{E_J} \in \left[\underline{\mu}^{E_J}, \overline{\mu}^{E_J}\right]$ the SFRE (3) has the lower solution subset $D_1 \subseteq S_g$, which is defined by the upper or aggregating solution $\hat{\boldsymbol{\mu}}^C \in \left[\hat{\boldsymbol{\mu}}^{1C}, \hat{\boldsymbol{\mu}}^{2C}\right]$ and the set of minimal solutions $\underline{S} = \left\{\underline{\boldsymbol{\mu}}_l^C,\ l = 1, \ldots, z_1\right\}$:

$$D_1 = \bigcup_{\underline{\boldsymbol{\mu}}_l^C \in \underline{S}} \left[\left[\underline{\boldsymbol{\mu}}_l^C, \hat{\boldsymbol{\mu}}^{2C}\right] \cup \left[\hat{\boldsymbol{\mu}}^{1C}, \hat{\boldsymbol{\mu}}^{2C}\right]\right]. \tag{19}$$

On the other hand, since aggregation of the *max–min* subsystems is carried out using the *min* operator, then for $\mu^{E_J} \in \left[\underline{\mu}^{E_J}, \overline{\mu}^{E_J}\right]$ the SFRE (3) with dual *min–max* composition has the upper solution subset $D_2 \subseteq S_g$, which is defined by the lower or aggregating solution $\hat{\boldsymbol{\mu}}^C \in \left[\hat{\boldsymbol{\mu}}^{1C}, \hat{\boldsymbol{\mu}}^{2C}\right]$ and the set of maximum solutions $\overline{S} = \{\overline{\boldsymbol{\mu}}_k^C, \ k = 1, \ldots, z_2\}$:

$$D_2 = \bigcup_{\overline{\boldsymbol{\mu}}_k^C \in \overline{S}} \left[\left[\hat{\boldsymbol{\mu}}^{1C}, \hat{\boldsymbol{\mu}}^{2C}\right] \cup \left[\hat{\boldsymbol{\mu}}^{1C}, \overline{\boldsymbol{\mu}}_k^C\right]\right]. \tag{20}$$

Then, by performing the union $D_1 \cup D_2$, we obtain the formula (18).

For the predefined granularity, the problem of transferring linguistic knowledge from the relation-based teacher model (9) to the student model (11) consists in finding the set of constrained solutions of the SFRE (3), (16). Given the coordinates of maximum of the membership functions of the fuzzy terms a_{IK}, the solution elements are the terms weights in the linguistic description (5).

We shall redenote:

$\boldsymbol{V}_j = \left(\boldsymbol{W}_1^j, \ldots, \boldsymbol{W}_N^j\right) = \left(v_1^j, \ldots, v_L^j\right)$, where $\boldsymbol{W}_I^j = \left(w_{I1}^j, \ldots, w_{Ig_I}^j\right)$ is the vector of the terms a_{IK} weights for the linguistic rule in the class d_j.

Given constraints (15), the problem of knowledge distillation based on solving the SFRE (3), (16) is formulated as follows. For each distillation spot d_j, $j = 1, \ldots, m$, the vector of terms weights $\boldsymbol{V}_j = \left(\boldsymbol{W}_1^j, \ldots, \boldsymbol{W}_N^j\right) = \left(v_1^j, \ldots, v_L^j\right)$, $w_{IK}^j \in \{0, 1\}$, $I = 1, \ldots, N$, $K = 1, \ldots, g_I$, should be found which provides the least distance between the teacher's and student's significance measures of the primary terms E_J:

$$F_c(\boldsymbol{V}_j) = \sum_{J=1}^{M} \left[\mu^{E_J}(d_j) - \mu^{E_J}\left(\mu^{C_1}\left(\boldsymbol{W}_1^j\right), \ldots, \mu^{C_N}\left(\boldsymbol{W}_N^j\right)\right)\right]^2 = \min_{\boldsymbol{V}_j}, \tag{21}$$

where $F_c(\boldsymbol{V}_j)$ is the distillation loss for the constrained solution $\boldsymbol{V}_j$.

Let us introduce the following concepts:

$\hat{\boldsymbol{V}} = \left(\hat{\boldsymbol{W}}_1, \ldots, \hat{\boldsymbol{W}}_N\right) = (\hat{v}_1, \ldots, \hat{v}_L)$ is the constrained aggregating solution, which is defined by the set of lower aggregating indices $\left\{\hat{X}_1^1, \ldots, \hat{X}_N^1\right\}$ and the set of upper aggregating indices $\left\{\hat{X}_1^2, \ldots, \hat{X}_N^2\right\}$, where $\hat{w}_{IK} = 1$, if $\hat{X}_I^1 \leq K \leq \hat{X}_I^2$;

$\overleftarrow{\boldsymbol{V}} = \left(\overleftarrow{\boldsymbol{W}}_1, \ldots, \overleftarrow{\boldsymbol{W}}_N\right) = (\overleftarrow{v}_1, \ldots, \overleftarrow{v}_L)$ is the upper bounded complete crisp solution, where for all $\overleftarrow{w}_{IK} = 1$, $K \leq \hat{X}_I^2$;

$\overrightarrow{\boldsymbol{V}} = \left(\overrightarrow{\boldsymbol{W}}_1, \ldots, \overrightarrow{\boldsymbol{W}}_N\right) = (\overrightarrow{v}_1, \ldots, \overrightarrow{v}_L)$ is the lower bounded complete crisp solution, where for all $\overrightarrow{w}_{IK} = 1$, $K \geq \hat{X}_I^1$.

Statement 2. The set S_c of constrained solutions of the SFRE (3), (16) is defined by the constrained aggregating solution $\hat{\boldsymbol{V}}$, for which there exist the lower subset of complete crisp solutions $U_1 = \left\{\overleftarrow{\boldsymbol{V}}_h,\ h = 1, \ldots, q_1\right\}$ bounded by the set of upper aggregating indices $\left\{\hat{X}_1^2, \ldots, \hat{X}_N^2\right\}$ and the upper subset of complete crisp solutions $U_2 = \left\{\overrightarrow{\boldsymbol{V}}_p,\ p = 1, \ldots, q_2\right\}$ bounded by the set of lower aggregating indices $\left\{\hat{X}_1^1, \ldots, \hat{X}_N^1\right\}$:

$$S_c = \bigcup_{\overleftarrow{\boldsymbol{V}}_h \in U_1} \bigcup_{\vec{\boldsymbol{V}}_p \in U_2} \left(\overleftarrow{\boldsymbol{V}}_h \cup \hat{\boldsymbol{V}} \cup \vec{\boldsymbol{V}}_p\right). \tag{22}$$

Here $\overleftarrow{\boldsymbol{V}}_h = \left(\overleftarrow{\boldsymbol{W}}_1^h, \ldots, \overleftarrow{\boldsymbol{W}}_N^h\right) = \left(\overleftarrow{v}_1^h, \ldots, \overleftarrow{v}_L^h\right)$ is the upper bounded complete crisp solution for the lower subset U_1, where for all $\overleftarrow{w}_{IK}^h = 1,\ K \le \hat{X}_I^2$, and the union is taken over all $\overleftarrow{\boldsymbol{V}}_h \in U_1$;

$\overrightarrow{\boldsymbol{V}}_p = \left(\overrightarrow{\boldsymbol{W}}_1^p, \ldots, \overrightarrow{\boldsymbol{W}}_N^p\right) = \left(\overrightarrow{v}_1^p, \ldots, \overrightarrow{v}_L^p\right)$ is the lower bounded complete crisp solution for the upper subset U_2, where for all $\overrightarrow{w}_{IK}^p = 1,\ K \ge \hat{X}_I^1$, and the union is taken over all $\vec{\boldsymbol{V}}_p \in U_2$.

Proof The formula (22) follows from the properties (18) of the solution set of the SFRE (3) with extended *max–min* composition.

6.3 *Genetic Algorithm for Parallel Deployment of the Student Model*

When deploying the interval rules, the genetic algorithm is performed in two stages: search for the lower and upper aggregating solutions $\hat{\mu}^{1C_I}$, $\hat{\mu}^{2C_I}$; parallel search for the lower and upper subsets D_1, D_2. When searching for the aggregating solution, the chromosome is defined as a string of binary codes of the solutions $\boldsymbol{\mu}^C = \left(\mu^{C_1}, \ldots, \mu^{C_N}\right)$. The cross-over operation is performed by exchanging parts of the chromosomes inside each solution μ^{C_I}, $I = 1, \ldots, N$. The fitness function is based on the criterion (17). The bounds of the aggregating solution are formed by a stepwise increment (decrement) until the widest interval is obtained.

As the null aggregating solution, we designate $\boldsymbol{\mu}_0^C = \left(\mu_0^{C_1}, \ldots, \mu_0^{C_N}\right)$. The lower (upper) aggregating solution is found in the range: if $\mu_0^{C_I} \in D_1$, then $\hat{\mu}^{2C_I} \in \left[\mu_0^{C_I},\ 1\right]$; if $\mu_0^{C_I} \in D_2$, then $\hat{\mu}^{1C_I} \in \left[0,\ \mu_0^{C_I}\right]$.

Let $\boldsymbol{\mu}^C(t) = \left(\mu^{C_1}(t), \ldots, \mu^{C_N}(t)\right)$ be some t-th aggregating solution, that is, $F_g\left(\boldsymbol{\mu}^C(t)\right) = F_g\left(\boldsymbol{\mu}_0^C\right)$ and $\mu^{E_J}(t) \in \left[\underline{\mu}^{E_J},\ \overline{\mu}^{E_J}\right]$ for all $\mu^{C_I}(t) \in D_1$ or $\mu^{C_I}(t) \in$

D_2. If $\mu^{C_I}(t) \in D_1$, then $\mu^{C_I}(t) \geq \mu^{C_I}(t-1)$ and $\mu_i^{E_J}\left(\mu^{C_I}(t)\right) = \mu_i^{E_J}\left(\mu^{C_I}(t-1)\right)$ while searching for the upper aggregating solution $\hat{\mu}^{2C_I}$; if $\mu^{C_I}(t) \in D_2$, then $\mu^{C_I}(t) \leq \mu^{C_I}(t-1)$ and $\mu_i^{E_J}\left(\mu^{C_I}(t)\right) \neq \mu_i^{E_J}\left(\mu^{C_I}(t-1)\right)$ while searching for the lower aggregating solution $\hat{\mu}^{1C_I}$.

The lower (upper) aggregating solution is defined as follows: if $\boldsymbol{\mu}^C(t) \neq \boldsymbol{\mu}^C(t-1)$, then $\hat{\mu}^{1C_I}$ $(\hat{\mu}^{2C_I}) = \mu^{C_I}(t)$. If $\boldsymbol{\mu}^C(t) = \boldsymbol{\mu}^C(t-1)$, then the search for the aggregating solution $\hat{\boldsymbol{\mu}}^C \in [\hat{\boldsymbol{\mu}}^{1C}, \hat{\boldsymbol{\mu}}^{2C}]$ is stopped.

When searching for the lower and upper subsets D_1 and D_2, the chromosome is distributed to solutions $\boldsymbol{\mu}^{1C} = \left(\mu^{1C_1}, \ldots, \mu^{1C_N}\right)$ and $\boldsymbol{\mu}^{2C} = \left(\mu^{2C_1}, \ldots, \mu^{2C_N}\right)$. Formation of the lower and upper subsets is accomplished by way of multiple solving the optimization problem (17). The interval solution is formed by a stepwise increment (decrement) until the widest interval is obtained. The set of intervals is formed by repeated runs of the genetic algorithm if new minimum (maximum) solutions are found. The criterion for stopping the genetic algorithm is the absence of new upper (lower) bounds within a given number of iterations.

As the null solutions we designate:

$$\boldsymbol{\mu}_0^{1C} = \left(\mu_0^{1C_1}, \ldots, \mu_0^{1C_N}\right) \text{ for the lower subset } D_1, \text{ where } \mu_0^{1C_I} \leq \hat{\mu}^{2C_I};$$
$$\boldsymbol{\mu}_0^{2C} = \left(\mu_0^{2C_1}, \ldots, \mu_0^{2C_N}\right) \text{ for the upper subset } D_2, \text{ where } \mu_0^{2C_I} \geq \hat{\mu}^{1C_I}.$$

The lower and upper bounds are found in the range: $\underline{\mu}_l^{C_I} \in \left[0, \mu_0^{1C_I}\right]$ for $l = 1$, and $\underline{\mu}_l^{C_I} \in \left[0, \hat{\mu}^{2C_I}\right]$ for $l > 1$, where the minimal solutions $\underline{\mu}_e^{1C_I}$, $e < l$, are excluded from the search space; $\overline{\mu}_k^{C_I} \in \left[\mu_0^{2C_I}, 1\right]$ for $k = 1$, and $\overline{\mu}_k^{C_I} \in \left[\hat{\mu}^{1C_I}, 1\right]$ for $k > 1$, where the maximum solutions $\overline{\mu}_e^{2C_I}$, $e < k$, are excluded from the search space.

Let $\boldsymbol{\mu}^{1C}(t) = \left(\mu^{1C_1}(t), \ldots, \mu^{1C_N}(t)\right)$ and $\boldsymbol{\mu}^{2C}(t) = \left(\mu^{2C_1}(t), \ldots, \mu^{2C_N}(t)\right)$ be some t-th lower and upper solutions, that is, $F_g\left(\boldsymbol{\mu}^{1C}(t)\right) = F_g\left(\boldsymbol{\mu}_0^{1C}\right)$ and $F_g\left(\boldsymbol{\mu}^{2C}(t)\right) = F_g\left(\boldsymbol{\mu}_0^{2C}\right)$ for all $\boldsymbol{\mu}^{1C} \in D_1$ and $\boldsymbol{\mu}^{2C} \in D_2$. While searching for the lower and upper bounds $\underline{\mu}_l^{C_I}$ and $\overline{\mu}_k^{C_I}$ it is suggested that $\mu^{1C_I}(t) \leq \mu^{1C_I}(t-1)$ and $\mu^{2C_I}(t) \geq \mu^{2C_I}(t-1)$ provided that $\mu^{E_J}(t) \in \left[\underline{\mu}^{E_J}, \overline{\mu}^{E_J}\right]$.

The lower (upper) bounds are defined as follows: if $\boldsymbol{\mu}^{1C}(t) \neq \boldsymbol{\mu}^{1C}(t-1)$, then $\underline{\mu}_l^{C_I} = \mu^{1C_I}(t)$ for the lower subset D_1; if $\boldsymbol{\mu}^{2C}(t) \neq \boldsymbol{\mu}^{2C}(t-1)$, then $\overline{\mu}_k^{C_I} = \mu^{2C_I}(t)$ for the upper subset D_2. If $\boldsymbol{\mu}^{1C}(t) = \boldsymbol{\mu}^{1C}(t-1)$ and $\boldsymbol{\mu}^{2C}(t) = \boldsymbol{\mu}^{2C}(t-1)$, then the search for the interval solutions $\left[\underline{\boldsymbol{\mu}}_l^C, \hat{\boldsymbol{\mu}}^{2C}\right]$ and $\left[\hat{\boldsymbol{\mu}}^{1C}, \overline{\boldsymbol{\mu}}_k^C\right]$ is stopped. The search will go on until the conditions $\underline{\boldsymbol{\mu}}_e^C \neq \underline{\boldsymbol{\mu}}_l^C$, $e < l$, and $\overline{\boldsymbol{\mu}}_e^C \neq \overline{\boldsymbol{\mu}}_k^C$, $e < k$, have been satisfied.

When deploying the linguistic rules, the genetic algorithm is also performed in two stages: search for the lower and upper aggregating indices $\hat{X}_I^1$, $\hat{X}_I^2$ for the constrained aggregating solution $\hat{\boldsymbol{V}}$; parallel search for the lower and upper subsets U_1, U_2.

When searching for the constrained aggregating solution, the chromosome is defined as a string of binary weights $\boldsymbol{V} = (\boldsymbol{W}_1, \ldots, \boldsymbol{W}_N)$. The cross-over operation is performed by exchanging parts of the chromosomes inside each solution $\boldsymbol{W}_I$,

$I = 1, \ldots, N$. The fitness function is based on the criterion (21). The weights are activated by a stepwise increment (decrement) until the widest interval is obtained.

As the null aggregating solution, we designate: $\boldsymbol{V}_0 = \left(\boldsymbol{W}_1^0, \ldots, \boldsymbol{W}_N^0\right)$, where $\left\{X_1^0, \ldots, X_N^0\right\}$ is the null set of aggregating indices, that is, $w_{IK}^0 = 1$, if $K = X_I^0$. The lower (upper) aggregating indices take values from the sets: if $\boldsymbol{W}_I^0 \in U_1$, then $\hat{X}_I^2 \in \left\{X_I^0, \ldots, g_I\right\}$; if $\boldsymbol{W}_I^0 \in U_2$, then $\hat{X}_I^1 \in \left\{1, \ldots, X_I^0\right\}$.

Let $\boldsymbol{V}(t) = (\boldsymbol{W}_1(t), \ldots, \boldsymbol{W}_N(t))$ be some t-th constrained aggregating solution with some set of aggregating indices $\{X_1(t), \ldots, X_N(t)\}$, where $w_{IK}(t) = 1$, if $K = X_I(t)$. In this case, $F_c(\boldsymbol{V}(t)) = F_c(\boldsymbol{V}_0)$ and $\mu^{E_J}(t) \in \left[\underline{\mu}^{E_J}, \overline{\mu}^{E_J}\right]$ for all $\boldsymbol{W}_I(t) \in U_1$ or $\boldsymbol{W}_I(t) \in U_2$. If $\boldsymbol{W}_I(t) \in U_1$, then $X_I(t) \geq X_I(t-1)$ and $\mu_i^{E_J}(\boldsymbol{W}_I(t)) = \mu_i^{E_J}(\boldsymbol{W}_I(t-1))$ while searching for the upper aggregating indices $\hat{X}_I^2$; if $\boldsymbol{W}_I(t) \in U_2$, then $X_I(t) \leq X_I(t-1)$ and $\mu_i^{E_J}(\boldsymbol{W}_I(t)) \neq \mu_i^{E_J}(\boldsymbol{W}_I(t-1))$ while searching for the lower aggregating indices $\hat{X}_I^1$.

The constrained aggregating solutions are defined as follows: if $\boldsymbol{V}(t) \neq \boldsymbol{V}(t-1)$, that is, $X_I(t) \neq X_I(t-1)$, then $\hat{X}_I^1\left(\hat{X}_I^2\right) = X_I(t)$. If $\boldsymbol{V}(t) = \boldsymbol{V}(t-1)$, that is, $X_I(t) = X_I(t-1)$, then the search for the lower and upper aggregating indices $\left\{\hat{X}_1^1, \ldots, \hat{X}_N^1\right\}$ and $\left\{\hat{X}_1^2, \ldots, \hat{X}_N^2\right\}$ is stopped.

When searching for the lower and upper subsets U_1 and U_2, the chromosome is distributed to solutions $\boldsymbol{V}^1 = \left(\boldsymbol{W}_1^1, \ldots, \boldsymbol{W}_N^1\right)$ and $\boldsymbol{V}^2 = \left(\boldsymbol{W}_1^2, \ldots, \boldsymbol{W}_N^2\right)$. Formation of the lower and upper subsets is accomplished by way of multiple solving the optimization problem (21). To cover the interval solution, the stepwise increment (decrement) is performed until the maximum number of weights is activated. To form the set of intervals under constraints, the genetic algorithm is repeatedly run if new complete crisp solutions are found. The criterion for stopping the genetic algorithm is the absence of new complete crisp solutions within a given number of iterations.

As the null solutions we designate:

$\boldsymbol{V}^{1,0} = \left(\boldsymbol{W}_1^{1,0}, \ldots, \boldsymbol{W}_N^{1,0}\right)$ for the lower subset U_1, where for all $w_{IK}^{1,0} = 1$, $K \leq \hat{X}_I^2$;

$\boldsymbol{V}^{2,0} = \left(\boldsymbol{W}_1^{2,0}, \ldots, \boldsymbol{W}_N^{2,0}\right)$ for the upper subset U_2, where for all $w_{IK}^{2,0} = 1$, $K \geq \hat{X}_I^1$.

The upper bounds $\overleftarrow{w}_{IK}^h$ and $\vec{w}_{IK}^p$ for $h = 1$ and $\mathrm{p} = 1$ take values from the sets $\left\{w_{IK}^{1,0}, 1\right\}$ and $\left\{w_{IK}^{2,0}, 1\right\}$, respectively; for $h > 1$ and $\mathrm{p} > 1$—from the set $\{0, 1\}$, where the maximum solutions $\overleftarrow{\boldsymbol{V}}_e$, $e < h$, and $\vec{\boldsymbol{V}}_e$, $e < p$, are excluded from the search space.

Let $\boldsymbol{V}^1(t) = \left(\boldsymbol{W}_1^1(t), \ldots, \boldsymbol{W}_N^1(t)\right)$ and $\boldsymbol{V}^2(t) = \left(\boldsymbol{W}_1^2(t), \ldots, \boldsymbol{W}_N^2(t)\right)$ be some t-th constrained solutions for the lower and upper subsets U_1 and U_2, that is, $F_c\left(\boldsymbol{V}^1(t)\right) = F_c\left(\boldsymbol{V}^{1,0}\right)$ and $F_c\left(\boldsymbol{V}^2(t)\right) = F_c\left(\boldsymbol{V}^{2,0}\right)$ for all $\boldsymbol{V}^1(t) \in U_1$ and $\boldsymbol{V}^2(t) \in U_2$. While searching for the upper bounds $\overleftarrow{w}_{IK}^h$ and $\vec{w}_{IK}^p$ it is suggested that $w_{IK}^1(t) \geq w_{IK}^1(t-1)$ and $w_{IK}^2(t) \geq w_{IK}^2(t-1)$, where for all $w_{IK}^1(t) = 1$, $K \leq \hat{X}_I^2$, and for all $w_{IK}^2(t) = 1$, $K \geq \hat{X}_I^1$, provided that $\mu^{E_J}(t) \in \left[\underline{\mu}^{E_J}, \overline{\mu}^{E_J}\right]$.

The upper bounds are defined as follows: if $V^1(t) \neq V^1(t-1)$, then $\overleftarrow{w}_{IK}^h = w_{IK}^1(t)$ for the lower subset U_1; if $V^2(t) \neq V^2(t-1)$, then $\vec{w}_{IK}^p = w_{IK}^2(t)$ for the upper subset U_2. If $V^1(t) = V^1(t-1)$ and $V^2(t) = V^2(t-1)$, then the search for the complete crisp solutions $\overleftarrow{V}_h$ and $\vec{V}_p$ is stopped. The search will go on until the conditions $\overleftarrow{V}_e \neq \overleftarrow{V}_h$, $e < h$, and $\vec{V}_e \neq \vec{V}_p$, $e < p$, have been satisfied.

7 Method for Training the Neuro-Fuzzy Relational System

The neuro-fuzzy model for transferring knowledge from the granular teacher model to the relation-based teacher model is shown in Fig. 2. The neuro-fuzzy model is obtained by embedding the expert rules (1) and the fuzzy relation matrix $\boldsymbol{R}$ into the neural network so that the weights of arcs subject to training are fuzzy relations and membership functions of the primary terms [9].

The neuro-fuzzy models for transferring knowledge from the relation-based teacher model to the granular and linguistic student models are shown in Fig. 3, 4. The relation matrix $\boldsymbol{R}$ and interval rules (4) are embedded into the fuzzy *min–max* neural network (Fig. 3) so that the self-organized set of rules based on solutions of the SFRE is autonomously identified using the incremental learning scheme. Switching between student models occurs after imposing constraints on the degree of knowledge granularity. In this case, the relation matrix $\boldsymbol{R}$ and linguistic rules (5) are embedded into the fuzzy relational neural network (Fig. 4).

For training the structure and parameters of the relation-based teacher model, the recurrent relations are used:

$$\begin{aligned} r_{IJ}(t+1) &= r_{IJ}(t) - \eta \frac{\partial \varepsilon_t^R}{\partial r_{IJ}(t)}; \\ \beta^{C_I}(t+1) &= \beta^{C_I}(t) - \eta \frac{\partial \varepsilon_t^R}{\partial \beta^{C_I}(t)}; \sigma^{C_I}(t+1) = \sigma^{C_I}(t) - \eta \frac{\partial \varepsilon_t^R}{\partial \sigma^{C_I}(t)}; \\ \beta^{E_J}(t+1) &= \beta^{E_J}(t) - \eta \frac{\partial \varepsilon_t^R}{\partial \beta^{E_J}(t)}; \sigma^{E_J}(t+1) = \sigma^{E_J}(t) - \eta \frac{\partial \varepsilon_t^R}{\partial \sigma^{E_J}(t)}, \end{aligned} \tag{23}$$

which minimize the criterion

$$\varepsilon_t^R = \frac{1}{2}(\hat{\mu}_t^{E_J} - \mu_t^{E_J})^2, \tag{24}$$

where $\hat{\mu}_t^{E_J}$ ($\mu_t^{E_J}$) are the expert (theoretical) significance measures of the output primary terms on the t-th training step; $r_{IJ}(t)$ are the fuzzy relations on the t-th training step; $\beta^{C_I}(t), \sigma^{C_I}(t), \beta^{E_J}(t), \sigma^{E_J}(t)$ are the membership functions parameters

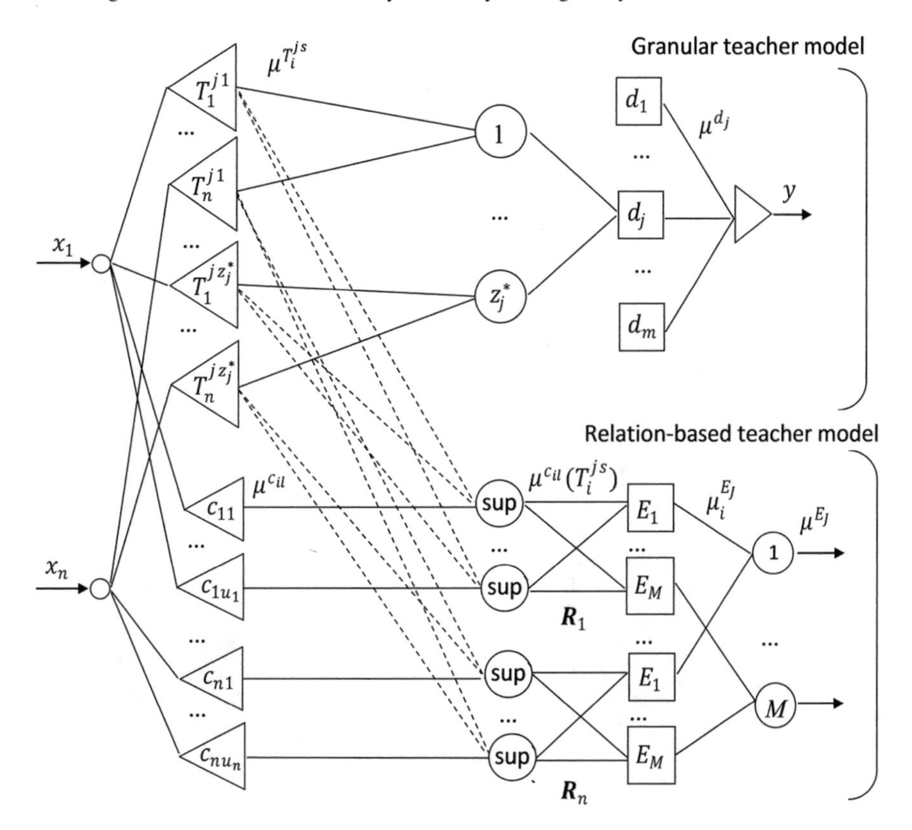

Fig. 2 Neuro-fuzzy model for transferring knowledge between teachers models

for the input and output primary terms on the t-th training step; η is the training parameter.

For training the structure of the student model based on the interval solutions of the SFRE, the recurrent relations are used:

$$\beta_I^{jk}(t+1) = \beta_I^{jk}(t) - \eta \frac{\partial \varepsilon_t^R}{\partial \beta_I^{jk}(t)}, \tag{25}$$

which minimize the criterion (24), where $\beta_I^{jk}(t)$ are the coordinates of the membership functions maximum in the partition at the level of rules on the t-th training step.

For training the structure of the student model based on the constrained solutions of the SFRE, the recurrent relations are used:

$$w_{IK}^{jp}(t+1) = w_{IK}^{jp}(t) - \eta \frac{\partial \varepsilon_t^R}{\partial w_{IK}^{jp}(t)}, \tag{26}$$

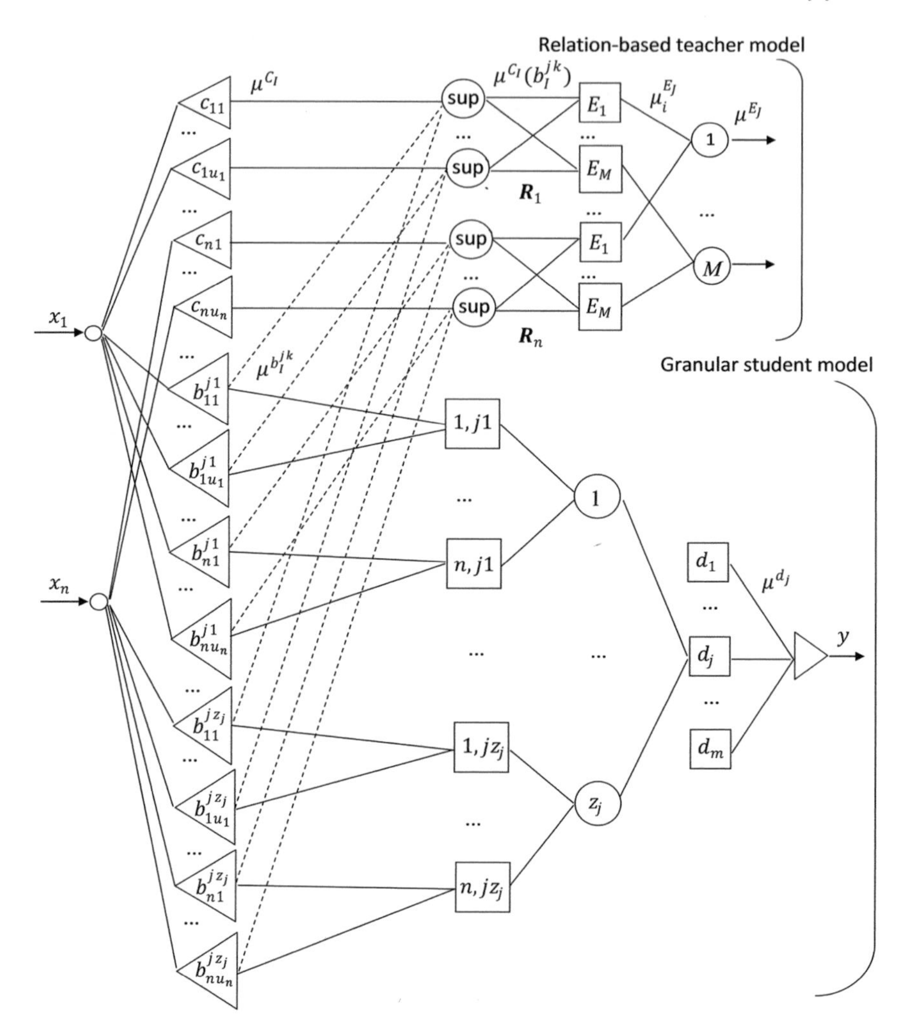

Fig. 3 Teacher-student neuro-fuzzy model based on the interval solutions of the SFRE

which minimize the criterion (24), where $w_{IK}^{jp}(t)$ are the weights of modified terms on the t-th training step.

For training the parameters of the granular student model, the recurrent relations are used:

$$\beta_I^{jk}(t+1) = \beta_I^{jk}(t) - \eta\frac{\partial\varepsilon_t^g}{\partial\beta_I^{jk}(t)}, \sigma_I^{jk}(t+1) = \sigma_I^{jk}(t) - \eta\frac{\partial\varepsilon_t^g}{\partial\sigma_I^{jk}(t)}, \quad (27)$$

which minimize the criterion

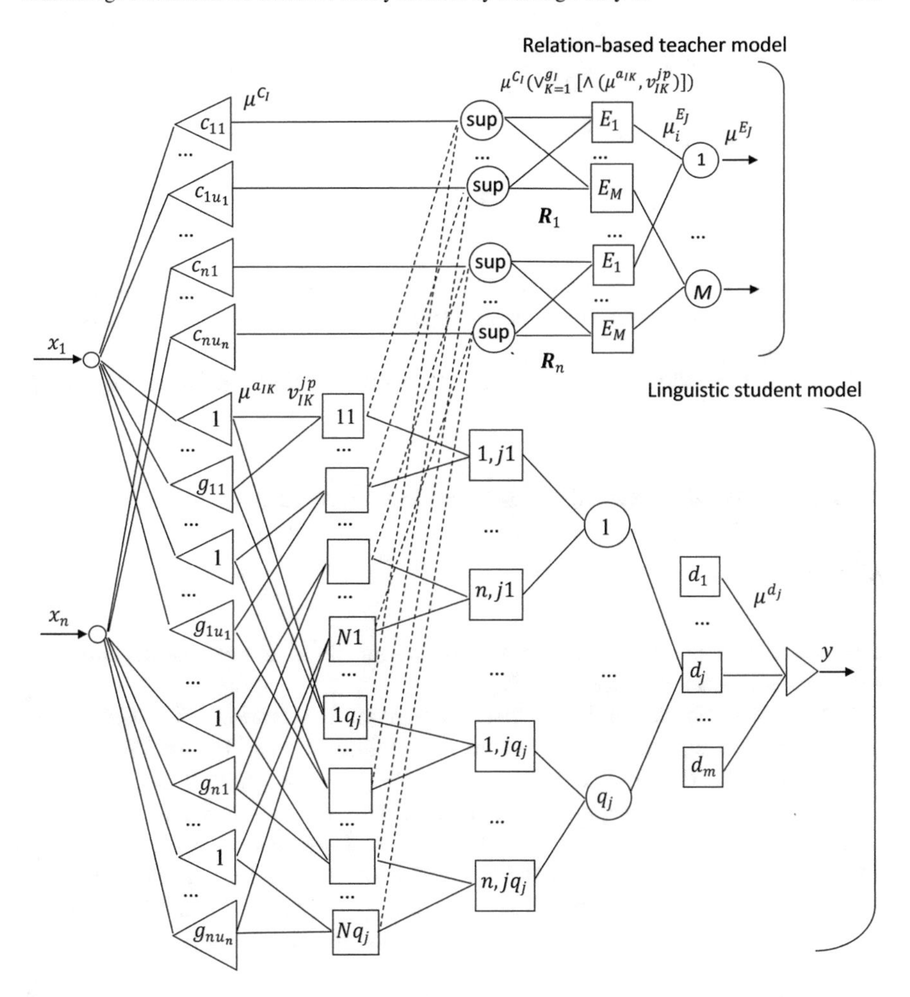

Fig. 4 Teacher-student neuro-fuzzy model based on the constrained solutions of the SFRE

$$\varepsilon_t^g = \frac{1}{2}(\hat{y}_t - y_t^g)^2,$$

where $\hat{y}_t\left(y_t^g\right)$ is the experimental (theoretical) output of the granular student model on the t-th training step; $\sigma_I^{jk}(t)$ are the concentration parameters of the membership functions in the partition at the level of rules on the t-th training step.

For training the parameters of the linguistic student model, the recurrent relations are used:

$$w_{IK}^{jp}(t+1) = w_{IK}^{jp}(t) - \eta \frac{\partial \varepsilon_t^c}{\partial w_{IK}^{jp}(t)},$$

$$\beta^{a_{IK}}(t+1) = \beta^{a_{IK}}(t) - \eta \frac{\partial \varepsilon_t^c}{\partial \beta^{a_{IK}}(t)}, \sigma^{a_{IK}}(t+1) = \sigma^{a_{IK}}(t) - \eta \frac{\partial \varepsilon_t^c}{\partial \sigma^{a_{IK}}(t)}, \quad (28)$$

which minimize the criterion

$$\varepsilon_t^c = \frac{1}{2}(\hat{y}_t - y_t^c)^2,$$

where y_t^c is the output of the linguistic student model on the t-th training step; $\beta^{a_{IK}}(t)$, $\sigma^{a_{IK}}(t)$ are the membership functions parameters for the modified input terms on the t-th training step.

The partial derivatives included in (23), (25)–(28) are calculated according to [9].

8 Experimental Results: Time Series Forecasting

8.1 Data Set Organization

The problem of predicting the number of comments after post publication over Facebook pages is considered [49, 50].

Let t_0 be some randomly selected moment of observation [50]. The essential features are: x_1, x_2—the time after post publication and the total number of comments at the time moment t_0, $x_1 \in [1, 72]$ h, $x_2 \in [0, 750]$; $x_3 \div x_5$—the number of comments for the last, the penultimate and the first day relative to the time moment t_0, respectively, $x_{3\div5} \in [0, 250]$. When aggregating by source, the five derived features are obtained for the essential features $x_2 \div x_5$ by calculating min, max, average, median and standard deviation [50]. Thus, the total number of features is 25. The output parameter is: y—the number of comments in the next day, $y \in [0, 250]$.

To study viewers' demand and TV programs ratings, the following categories of posts were chosen: public figure, politician, TV channel, telecommunication, media, news, publishing [51]. For this category, the dataset obtained from [49] is $\Theta = 19{,}500$ posts. Following [49, 50], the training and testing datasets were formed as follows. To refine the forecast while new experimental data is appearing, $k = 5$ observation moments $t_0^{p1}, \ldots, t_0^{p5}, t_0^{p,k-1} < t_0^{pk}, p = 1, \ldots, \Theta$, were randomly selected for each post. As a result, $k = 5$ datasets $\left\langle t_0^{pk}, \hat{\mathbf{X}}_{pk}, \hat{y}_{pk} \right\rangle$ of $k\Theta$ posts were formed (Table 2). To form the training and testing samples, $\Theta_{train} = 14{,}500$ posts (75%) and $\Theta_{test} = 5000$ posts (25%) were selected, respectively. The testing sample was formed from 10 subsets of 500 posts. The total number of observations was $19{,}500 \times 5 = 97{,}500$ including $14{,}500 \times 5 = 72{,}500$ for the training data and $5000 \times 5 = 25{,}000$ for the testing data.

Table 2 Data set organization

Dataset	Dataset 1	Dataset 2	Dataset 3	Dataset 4	Dataset 5
Training	14,500	29,000	43,500	58,000	72,500
Testing	10 × 500	10 × 1000	10 × 1500	10 × 2000	10 × 2500

To predict TV ratings, the indicator of the model accuracy is: P_{top20}—the average probability of a correct forecast of top 20 posts, which will receive the largest number of comments according to the test datasets [49, 50].

8.2 Training Results for the Teacher-Student Architecture

We shall describe the amplitude trends of the time series using the primary terms: c_{i1}, E_1—*Decreased (D)*, c_{i2}, E_2—*Sufficient (St)*, c_{i3}, E_3—*Increased (I)*. The amplitude trends for the essential features are described using the following gradual rules:

the more x_1, the less y; the more x_2, the more y;
the smaller x_3 and x_4 and x_5, the less y; the more x_3 and x_5, the more y;
the more $x_4 - x_3$, the less y; the less $x_3 - x_5$, the more y.

Bounds of the output classes for the rule-based teacher and student models were selected as follows: $d_1 = [0, 40]$; $d_2 = [40, 80]$; $d_3 = [80, 125]$; $d_4 = [125, 170]$; $d_5 = [170, 210]$; $d_6 = [210, 250]$.

After training the granular teacher model, the training dataset was compressed to $Z^* = 118$ interval rules. As a result of training the relational model using the distilled dataset, the teacher model can be represented in the form of the extended *max–min* SFRE:

$$\begin{aligned}
\mu^{E_1} &= \left[\left(\mu^{c_{11}} \wedge 0.39\right) \vee \left(\mu^{c_{12}} \wedge 0.47\right) \vee \left(\mu^{c_{13}} \wedge 0.85\right)\right] \\
&\wedge \left[\left(\mu^{c_{21}} \wedge 0.62\right) \vee \left(\mu^{c_{22}} \wedge 0.45\right)\right] \wedge \left[\left(\mu^{c_{31}} \wedge 0.86\right) \vee \left(\mu^{c_{32}} \wedge 0.53\right)\right] \\
&\wedge \left[\left(\mu^{c_{41}} \wedge 0.92\right) \vee \left(\mu^{c_{42}} \wedge 0.88\right) \vee \left(\mu^{c_{43}} \wedge 0.61\right)\right] \\
&\wedge \left[\left(\mu^{c_{51}} \wedge 0.79\right) \vee \left(\mu^{c_{52}} \wedge 0.46\right)\right]; \\
\mu^{E_2} &= \left[\left(\mu^{c_{11}} \wedge 0.65\right) \vee \left(\mu^{c_{12}} \wedge 0.81\right) \vee \left(\mu^{c_{13}} \wedge 0.43\right)\right] \\
&\wedge \left[\left(\mu^{c_{21}} \wedge 0.79\right) \vee \left(\mu^{c_{22}} \wedge 0.80\right) \vee \left(\mu^{c_{23}} \wedge 0.68\right)\right] \\
&\wedge \left[\left(\mu^{c_{32}} \wedge 0.80\right) \vee \left(\mu^{c_{33}} \wedge 0.57\right)\right] \wedge \left[\left(\mu^{c_{42}} \wedge 0.64\right) \vee \left(\mu^{c_{43}} \wedge 0.78\right)\right] \\
&\wedge \left[\left(\mu^{c_{52}} \wedge 0.85\right) \vee \left(\mu^{c_{53}} \wedge 0.54\right)\right]; \\
\mu^{E_3} &= \left[\left(\mu^{c_{11}} \wedge 0.77\right) \vee \left(\mu^{c_{12}} \wedge 0.60\right)\right] \wedge \left[\left(\mu^{c_{22}} \wedge 0.56\right) \vee \left(\mu^{c_{23}} \wedge 0.84\right)\right] \\
&\wedge \left[\mu^{c_{33}} \wedge 0.76\right] \wedge \left[\mu^{c_{43}} \wedge 0.45\right] \wedge \left[\left(\mu^{c_{52}} \wedge 0.60\right) \vee \left(\mu^{c_{53}} \wedge 0.82\right)\right].
\end{aligned} \tag{29}$$

Membership functions parameters for the primary terms are presented in Table 3.

For the given classes $d_{1\div 6}$, the granular knowledge is distilled by generating the set of interval solutions of the SFRE (29):

$$S_g(d_1) = \{\mu^{c_{13}}, \mu^{c_{31}} = [0.79,\ 1] \cup 0.79;\ \ \mu^{c_{41}}, \mu^{c_{42}} = [0.79,\ 1] \cup [0,\ 0.79];$$
$$\mu^{c_{51}} = [0.79,\ 1];\ \mu^{c_{21}} = [0.62,\ 1];\ \mu^{c_{43}} = [0,\ 0.62];\ \mu^{c_{32}} = [0,\ 0.43]\};$$

$$S_g(d_2) = \{\mu^{c_{21}} = [0.56,\ 1] \cup 0.56;\ \mu^{c_{13}}, \mu^{c_{51}} = [0.56,\ 0.79] \cup 0.56;$$
$$\mu^{c_{31}} = [0.56,\ 0.79] \cup [0,\ 0.56] \cup 0.56,$$
$$\mu^{c_{43}} = [0.56,\ 1] \cup [0.56,\ 0.79] \cup [0,\ 0.56];$$
$$\mu^{c_{41}}, \mu^{c_{42}} = [0.56,\ 0.79] \cup [0,\ 0.56];\ \mu^{c_{32}} = 0.56 \cup [0,\ 0.56];$$
$$\mu^{c_{12}}, \mu^{c_{22}}, \mu^{c_{52}} = [0,\ 0.43]\};$$

$$S_g(d_3) = \{\mu^{c_{11}} = [0.64,\ 1] \cup 0.64;\ \mu^{c_{21}}, \mu^{c_{22}} = [0.64,\ 0.80] \cup [0,\ 0.64];$$
$$\mu^{c_{42}} = [0.64,\ 1] \cup\ 0.64 \cup [0,\ 0.64];\ \mu^{c_{32}}, \mu^{c_{52}} = [0.64,\ 0.80] \cup 0.64;$$
$$\mu^{c_{23}}, \mu^{c_{53}} = 0.64 \cup [0,\ 0.64];\ \mu^{c_{43}} = [0.64,\ 0.80] \cup\ 0.64 \cup [0,\ 0.64];$$
$$\mu^{c_{31}} = [0,\ 0.53];\ \mu^{c_{12}} = [0,\ 0.39]\};$$

$$S_g(d_4) = \{\mu^{c_{12}}, \mu^{c_{22}}, \mu^{c_{43}} = [0.80,\ 1] \cup [0.73,\ 0.80] \cup [0,\ 0.73];$$
$$\mu^{c_{32}}, \mu^{c_{52}} = [0.80,\ 1] \cup [0.73,\ 0.80];\ \mu^{c_{21}} = [0.73,\ 0.80] \cup [0,\ 0.80];$$
$$\mu^{c_{11}}, \mu^{c_{42}}, \mu^{c_{53}} = 0.64 \cup [0,\ 0.64];\ \mu^{c_{23}}, \mu^{c_{31}} = [0,\ 0.56];\ \mu^{c_{13}}, \mu^{c_{51}} = [0,\ 0.43]\};$$

$$S_g(d_5) = \{\mu^{c_{11}}, \mu^{c_{12}}, \mu^{c_{53}} = [0.62,\ 0.76] \cup [0,\ 0.62];\ \mu^{c_{23}}, \mu^{c_{33}} = [0.62,\ 0.76];$$
$$\mu^{c_{52}} = [0.62,\ 1] \cup [0,\ 0.54];\ \mu^{c_{22}}, \mu^{c_{43}} = [0,\ 0.62];\ \mu^{c_{32}} = [0,\ 0.54]\};$$

$$S_g(d_6) = \{\mu^{c_{11}}, \mu^{c_{23}}, \mu^{c_{33}}, \mu^{c_{53}} = [0.76,\ 1];\ \mu^{c_{43}} = [0,\ 0.76];$$

Table 3 Membership functions parameters for the primary fuzzy terms

Input	Parameter	Primary fuzzy terms		
		D	*St*	*I*
x_1	β	5	35	70
	σ	12.81	9.54	21.28
x_2	β	25	357	722
	σ	128.51	79.14	186.11
$x_{3\div 5}$	β	4	121	247
	σ	45.28	26.34	54.86

$$\mu^{c_{12}}, \mu^{c_{22}} = [0,\ 0.65];\ \mu^{c_{32}}, \mu^{c_{52}} = [0,\ 0.54]\}.$$

For the essential features, the set of rules distilled to the granular student model is presented in Table 4. The subset of features in each rule is controlled by the minimal solutions. Given the derived features, the number of interval rules is $Z = 87$.

We shall describe the detailed patterns in the time series data using the modified terms: *weak, moderate, strong Decrease (Increase) (wD, mD, sD; wI, mI, sI)*. After training the relation-based teacher model for the predefined granularity, the constraints imposed on significance measures of the primary terms take the form:

$$\mu^{D(w)}(x_1) = [0, 0.43], \mu^{D(m)}(x_1) = [0.43, 0.64], \mu^{D(s)}(x_1) = [0.76, 1];$$
$$\mu^{I(w)}(x_1) = [0, 0.46], \mu^{I(m)}(x_1) = [0.56, 0.79], \mu^{I(s)}(x_1) = [0.79, 1];$$

$$\mu^{D(w)}(x_2) = [0, 0.39], \mu^{D(m)}(x_2) = [0.56, 0.73], \mu^{D(s)}(x_2) = [0.80, 1];$$
$$\mu^{I(w)}(x_2) = [0, 0.46], \mu^{I(m)}(x_2) = [0.53, 0.64], \mu^{I(s)}(x_2) = [0.76, 1];$$

$$\mu^{D(w)}(x_{3\div5}) = [0, 0.43], \mu^{D(m)}(x_{3\div5}) = [0.53, 0.79], \mu^{D(s)}(x_{3\div5}) = [0.79, 1];$$
$$\mu^{I(w)}(x_{3\div5}) = [0, 0.46], \mu^{I(m)}(x_{3\div5}) = [0.56, 0.73], \mu^{I(s)}(x_{3\div5}) = [0.76, 1]$$

Table 4 The set of interval rules for the essential features

	IF					THEN
	x_1	x_2	x_3	x_4	x_5	y
1	[23, 59]	[1, 127]	[28, 90]	[1, 28]	[1, 28]	$d_1 = [0, 40]$
2	[47, 72]	[1, 127]	[1, 28]	[28, 107]	[1, 28]	
3	[24, 51]	[91, 265]	[47, 101]	[27, 101]	[28, 44]	$d_2 = [40, 80]$
4	[24, 59]	[91, 265]	[44, 98]	[27, 98]	[27, 44]	
5	[46, 59]	[91, 265]	[27, 44]	[44, 121]	[27, 44]	
6	[46, 72]	[139, 357]	[44, 121]	[44, 98]	[44, 91]	
7	[15, 47]	[121, 298]	[108, 140]		[47, 108]	$d_3 = [80, 125]$
8	[24, 47]	[121, 265]	[101, 140]		[53, 121]	
9	[23, 72]	[265, 490]	[47, 101]	[101, 206]	[101, 134]	
10	[23, 72]	[265, 490]	[101, 134]	[140, 220]	[44, 108]	
11	[15, 41]	[203, 327]	[137, 163]		[107, 138]	$d_4 = [125, 170]$
12	[29, 41]	[310, 405]	[96, 145]	[140, 217]	[101, 142]	
13	[30, 72]	[310, 556]	[107, 137]	[101, 214]	[138, 206]	
14	[30, 72]	[316, 490]	[107, 134]	[142, 220]	[107, 134]	
15	[12, 35]	[295, 418]	[145, 205]		[145, 217]	$d_5 = [170, 210]$
16	[28, 72]	[418, 620]	[205, 217]	[140, 205]	[107, 205]	
17	[24, 72]	[415, 722]	[217, 250]	[140, 217]	[217, 250]	$d_6 = [210, 250]$

Given constraints, the linguistic knowledge is distilled by generating the set of constrained solutions of the SFRE (29):

$$S_c(d_1) = \{\mathbf{W}_1 = (0, 1, 1, 1, 1, 0) \cup (0, 0, 0, 1, 1, 1);\ \mathbf{W}_2 = (1, 1, 0, 0, 0, 0);$$
$$\mathbf{W}_3 = \mathbf{W}_4 = (1, 1, 1, 0, 0, 0) \cup (1, 0, 0, 0, 0, 0);\ \mathbf{W}_5 = (1, 0, 0, 0, 0, 0)\};$$

$$S_c(d_2) = \{\mathbf{W}_1 = (0, 1, 1, 1, 1, 0) \cup (0, 0, 0, 1, 1, 0) \cup (0, 0, 0, 1, 1, 1);$$
$$\mathbf{W}_2 = (1, 1, 1, 0, 0, 0) \cup (0, 1, 1, 0, 0, 0);\ \mathbf{W}_3 = (0, 1, 1, 0, 0, 0) \cup (1, 1, 0, 0, 0, 0) \cup (0, 1, 1, 1, 0, 0);$$
$$\mathbf{W}_4 = (1, 1, 1, 0, 0, 0) \cup (0, 1, 1, 1, 0, 0) \cup (0, 1, 1, 0, 0, 0);\ \mathbf{W}_5 = (1, 1, 0, 0, 0, 0) \cup (0, 1, 1, 0, 0, 0)\};$$

$$S_c(d_3) = \{\mathbf{W}_1 = (1, 1, 1, 1, 0, 0) \cup (0, 1, 1, 1, 1, 1);$$
$$\mathbf{W}_2 = (0, 1, 1, 0, 0, 0) \cup (0, 0, 1, 1, 0, 0);\ \mathbf{W}_3 = (0, 0, 1, 1, 0, 0) \cup (0, 1, 1, 0, 0, 0);$$
$$\mathbf{W}_4 = (0, 0, 1, 1, 1, 1) \cup (0, 0, 0, 1, 1, 1);\ \mathbf{W}_5 = (0, 0, 1, 1, 0, 0) \cup (0, 1, 1, 0, 0, 0)\};$$

$$S_c(d_4) = \{\mathbf{W}_1 = (1, 1, 1, 0, 0, 0) \cup (0, 0, 1, 1, 1, 1);$$
$$\mathbf{W}_2 = (0, 1, 1, 0, 0, 0) \cup (0, 0, 1, 1, 1, 0) \cup (0, 0, 1, 1, 0, 0);$$
$$\mathbf{W}_3 = (0, 0, 0, 1, 1, 0) \cup (0, 0, 1, 1, 0, 0);\ \mathbf{W}_4 = (0, 0, 1, 1, 1, 1) \cup (0, 0, 0, 1, 1, 1);$$
$$\mathbf{W}_5 = (0, 0, 1, 1, 0, 0) \cup (0, 0, 0, 1, 1, 1)\};$$

$$S_c(d_5) = \{\mathbf{W}_1 = (1, 1, 1, 0, 0, 0) \cup (0, 0, 1, 1, 1, 1);\ \mathbf{W}_2 = (0, 0, 1, 1, 0, 0) \cup (0, 0, 0, 1, 1, 0);$$
$$\mathbf{W}_3 = (0, 0, 0, 1, 1, 1) \cup (0, 0, 0, 0, 1, 1);\ \mathbf{W}_4 = (0, 0, 0, 1, 1, 1);$$
$$\mathbf{W}_5 = (0, 0, 0, 1, 1, 1) \cup (0, 0, 1, 1, 1, 1)\};$$

$$S_c(d_6) = \{\mathbf{W}_1 = (0, 1, 1, 1, 1, 1);\ \mathbf{W}_2 = \mathbf{W}_4 = (0, 0, 0, 1, 1, 1);\ \mathbf{W}_3 = \mathbf{W}_5 = (0, 0, 0, 0, 0, 1)\}.$$

For the essential features, the set of rules distilled to the linguistic student model is presented in Table 5. Given the derived features, the number of rules is $Q = 74$.

Membership functions parameters for the modified fuzzy terms are presented in Table 6. Membership functions of the primary and modified terms for the essential features are shown in Fig. 5.

8.3 A Comparison of KD Techniques in Time Series Forecasting

In time series forecasting, the model update time from the moment the new experimental data becomes available should not exceed the forecasting timeframe. To coordinate TV content with the surge of users' activity in the social networks, the maximum size of the time slot is 20–40 min [51]. To implement incremental updating of the prediction model, knowledge is transferred from the teacher model constructed using the (k-1)-th dataset to the student model constructed using the k-th dataset, $k = 2,\ldots,5$ [52]. After imposing constraints on the distillation loss, knowledge is transferred to a more compact student model to reduce the model update time.

Table 5 The set of linguistic rules for the essential features

Rule	IF					THEN
	x_1	x_2	x_3	x_4	x_5	y
1	*mD–mI*	*sD–mD*	*sD–wD*	*sD*	*sD*	*sD*
2	*wI–sI*	*sD–mD*	*sD*	*sD–wD*	*sD*	
3	*mD–mI*	*sD–wD*	*mD–wD*	*sD–wD*	*sD–mD*	*mD*
4	*wI–mI*	*sD–wD*	*sD–mD*	*mD-wI*	*sD–mD*	
5	*wI–sI*	*mD–wD*	*mD–wI*	*mD–wD*	*mD–wD*	
6	*sD–wI*	*mD–wD*	*wD–wI*		*mD–wD*	*wD*
7	*mD–sI*	*wD–wI*	*mD–wD*	*wD–sI*	*wD–wI*	
8	*mD–sI*	*wD–wI*	*wD–wI*	*wI–sI*	*mD–wD*	
9	*sD–wD*	*mD–wD*	*wI–mI*		*wD–wI*	*wI*
10	*wD–sI*	*wD–mI*	*wD–wI*	*wD–sI*	*wI–sI*	
11	*wD–sI*	*wD–wI*	*wD–wI*	*wI–sI*	*wD–wI*	
12	*sD–wD*	*wD–wI*	*wI–sI*		*wI–sI*	*mI*
13	*wD–sI*	*wD–mI*	*mI–sI*	*wI–sI*	*wD–sI*	
14	*mD–sI*	*wD–sI*	*sI*	*wI–sI*	*sI*	*sI*

Table 6 Membership functions parameters for the modified fuzzy terms

Input	Parameter	Modified fuzzy terms					
		sD	*mD*	*wD*	*wI*	*mI*	*sI*
x_1	β	10	18	28	45	56	64
	σ	6.27	6.85	5.12	8.15	5.39	6.12
x_2	β	65	162	279	457	562	680
	σ	49.42	53.75	74.83	45.21	59.17	75.08
$x_{3\div 5}$	β	21	61	92	142	186	225
	σ	19.84	18.70	16.03	22.61	19.55	21.04

The teacher model was constructed to provide $P_{\text{top20}} \longrightarrow \max$. To reduce the model update time for the given distillation loss, the student model was constructed to provide $P_{\text{top20}} \geq 0.7$ [46].

The papers [53–56] proposed methods of accelerated learning based on data and knowledge distillation. In [53], an expansion and pruning mechanism of gradient space based on integrity of hidden neurons is proposed for accelerated learning of the Radial Basis Function (RBF) neural network. In [54], transfer reinforcement learning via knowledge extraction from the pre-trained model is accelerated using auto-pruned Decision Trees (DT). In [55], an accelerator for Support Vector Machines (SVM) learning based on window technology is proposed, where the fixed-size training window is constructed to select the samples suitable for the hyperplane updating. In [56], to accelerate the training process of the Fuzzy *Min–Max* neural network

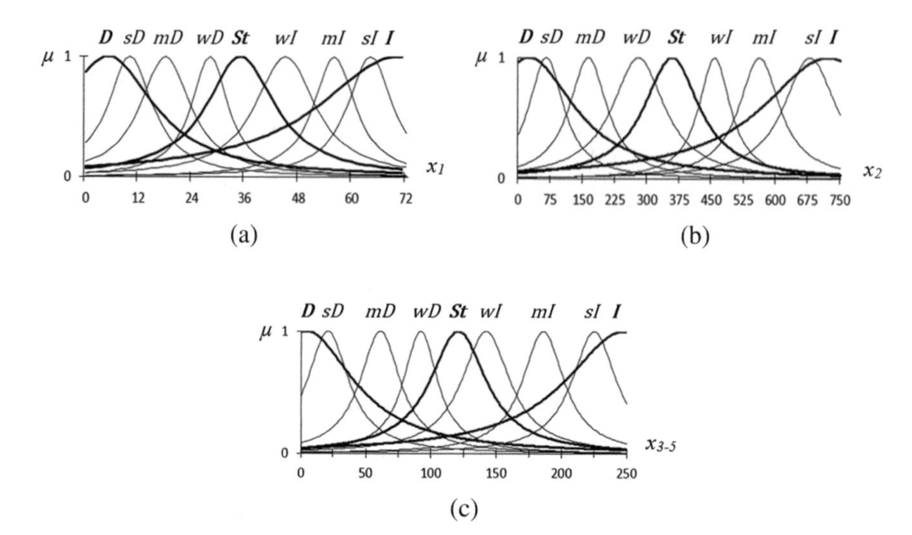

Fig. 5 Membership functions of the primary and modified fuzzy terms for the essential features x_1 (**a**); x_2 (**b**); $x_{3\div5}$ (**c**)

(FMM), a selection rule is proposed which allows removing unsuitable hyperboxes when covering a new input pattern at the step of expansion or aggregation.

To provide equal conditions for comparison, the KD algorithms were implemented in MATLAB. The structure of the teacher-student model was determined as follows:

- RBF neural network design with iterative addition of neurons to reach the required accuracy (distillation loss), where the cluster number is set 100 for the teacher (90 for the student) [53];
- deep tree for the teacher with default parameters (82 rules), post-pruned tree for the student that ensures splitting with less entropy calculated using the teacher soft labels (69 rules) [54];
- Gaussian kernel SVM classifier (box_constraint = 1000 (Inf penalty), kernel_ scale = 10) for the teacher, compact SVM classifier with data distillation using pre-fixed windows (box_constraint = 100, kernel_scale = 0.1) for the student [55];
- fine-grained FMM with minimum hyperbox_size = $0.02 \times (y_{\max} - y_{\min})$ for 110 hyperboxes of the teacher, coarse-grained FMM with data distillation based on the selection rule covering the maximum hyperbox_size = $0.05 \times (y_{\max} - y_{\min})$ (83 hyperboxes for the student) [56].

The results of training the teacher and student models (average P_{top20} and training time) for the bagging meta learning algorithm applied to various training and testing datasets are presented in Tables 7, 8 and Figs. 6, 7. Table 7 shows the total training time of the granular and relational teacher models, and Table 8 shows the total training time of the granular and linguistic student models.

Table 7 Training results for the teacher model

Teacher model	Indicator	Dataset 1	Dataset 2	Dataset 3	Dataset 4	Dataset 5
RBF	P_{top20}	0.53 ± 0.12	0.61 ± 0.10	0.65 ± 0.12	0.71 ± 0.09	0.73 ± 0.08
	Time	1224 s	2914 s	4432 s	5323 s	6384 s
DT	P_{top20}	0.57 ± 0.14	0.63 ± 0.11	0.71 ± 0.09	0.75 ± 0.11	0.76 ± 0.10
	Time	474 s	780 s	1210 s	2362 s	3897 s
SVM	P_{top20}	0.61 ± 0.12	0.65 ± 0.10	0.74 ± 0.10	0.78 ± 0.09	0.80 ± 0.11
	Time	562 s	875 s	1186 s	1776 s	2419 s
FMM	P_{top20}	0.64 ± 0.11	0.69 ± 0.13	0.75 ± 0.12	0.82 ± 0.10	0.84 ± 0.10
	Time	757 s	1053 s	1516 s	2187 s	2940 s
SFRE	P_{top20}	0.62 ± 0.12	0.67 ± 0.11	0.78 ± 0.09	0.81 ± 0.10	0.83 ± 0.09
	Time	345 + 63 = 408 s	531 + 112 = 643 s	810 + 185 = 995 s	1168 + 300 = 1468 s	1538 + 432 = 1970 s

Table 8 Training results for the student model

Student model	Indicator	Dataset 1	Dataset 2	Dataset 3	Dataset 4	Dataset 5
RBF	P_{top20}	0.49 ± 0.10	0.58 ± 0.11	0.62 ± 0.10	0.67 ± 0.08	0.70 ± 0.10
	Time	1020 s	2490 s	3821 s	4670 s	5700 s
DT	P_{top20}	0.54 ± 0.13	0.59 ± 0.12	0.67 ± 0.11	0.71 ± 0.10	0.73 ± 0.12
	Time	389 s	650 s	1035 s	2054s	3480 s
SVM	P_{top20}	0.58 ± 0.10	0.63 ± 0.09	0.70 ± 0.12	0.74 ± 0.11	0.76 ± 0.12
	Time	446 s	712 s	980 s	1480 s	2068 s
FMM	P_{top20}	0.62 ± 0.10	0.65 ± 0.12	0.71 ± 0.11	0.79 ± 0.09	0.81 ± 0.08
	Time	574 s	823 s	1193 s	1750 s	2390 s
SFRE	P_{top20}	0.59 ± 0.13	0.65 ± 0.10	0.74 ± 0.10	0.78 ± 0.11	0.80 ± 0.09
	Time	237 + 45 = 282 s	380 + 70 = 450 s	590 + 125 = 715 s	852 + 212 = 1064 s	1145 + 325 = 1470 s

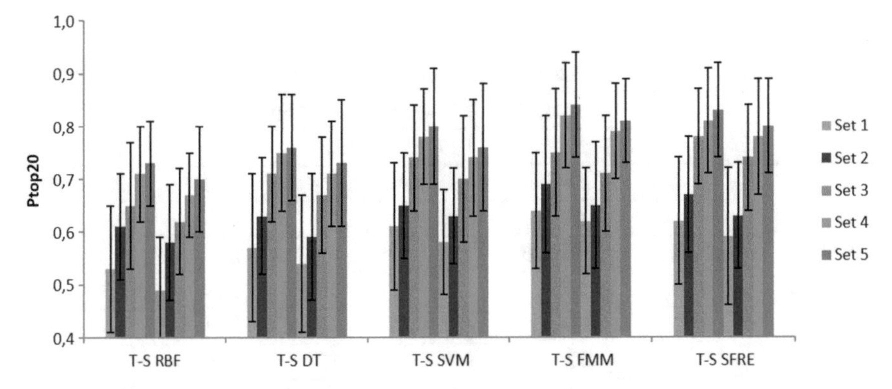

Fig. 6 Average P_{top20} for the Teacher (T) and Student (S) models

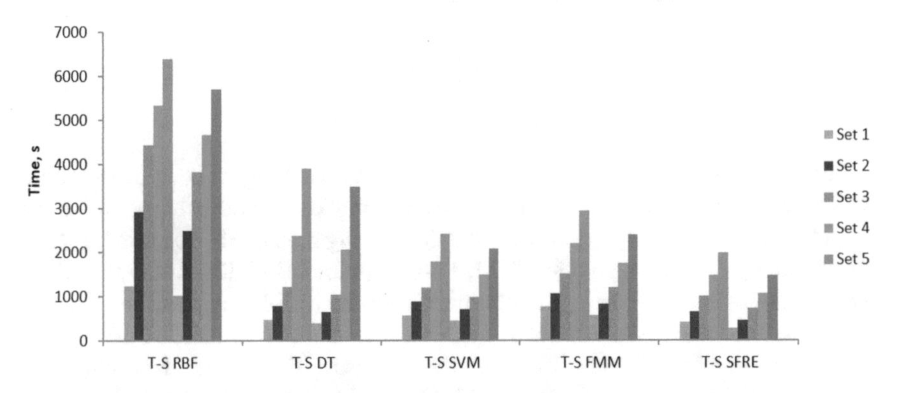

Fig. 7 Training time for the Teacher (T) and Student (S) models

We evaluate the distillation loss and compression ratio by comparing the values of P_{top20} and training time for the teacher and student models (Table 9). The teacher-student architecture was trained locally (Intel Core i5-7400 4.2 Ghz Processor) to demonstrate the dynamics of changes in the distillation loss and compression degree with the growth of the dataset. When using MATLAB mobile, a desktop session is used for training the granular and relational teacher models, and the multitask distillation scheme can be distributed across two or more mobile devices. The number of published comments is stored on MATLAB Drive, allowing us to access the dataset from any device running MATLAB Mobile. MATLAB Drive storage is limited to 5 GB while supporting full functionality [57].

The distillation loss λ is estimated as a ratio of the difference between the teacher and student correctness P_{top20}^{T} and P_{top20}^{St} to the teacher correctness P_{top20}^{T}; the compression ratio γ is estimated as a ratio of the teacher training time τ_T to the student training time τ_{St}:

Table 9 Distillation loss and compression degree for the student model

Student model	Indicator	Dataset 1	Dataset 2	Dataset 3	Dataset 4	Dataset 5
RBF	λ, %	3.77–11.32	3.28–6.56	1.54–7.69	4.22–7.04	1.37–6.85
	γ	1.20	1.17	1.16	1.14	1.12
DT	λ, %	3.51–7.02	4.76–7.94	2.82–8.45	4.00–6.67	1.32–6.58
	γ	1.22	1.20	1.17	1.15	1.12
SVM	λ, %	1.64–8.20	1.54–4.61	2.70–8.11	2.56–7.69	3.75–6.25
	γ	1.26	1.23	1.21	1.20	1.17
FMM	λ, %	1.56–4.69	4.35–7.25	4.00–6.67	2.44–4.88	1.19–5.95
	γ	1.32	1.28	1.27	1.25	1.23
SFRE	λ, %	3.22–6.45	4.48–7.46	3.85–6.41	2.47–4.94	2.41–4.82
	γ	1.45	1.43	1.39	1.38	1.34

$$\lambda = 1 - \frac{P_{top20}^{St}}{P_{\text{top }20}^{T}}; \gamma = \frac{\tau_T}{\tau_{St}}.$$

Table 9 shows that the proposed technique ensures the higher compression ratio with the minimal distillation loss. For the teacher model with the average prediction accuracy at the level of $P_{top20} = 0.7 - 0.8$ the training time is: 73–106 min for RBF; 20–65 min for DT; 15–40 min for SVM; 17–49 min for FMM; 12–35 min for SFRE, which exceeds the forecasting timeframe. Given the distillation loss, i.e., average prediction accuracy at the level of $P_{top20} \geq 0.7$, the training time is: 64–95 min for RBF; 17–58 min for DT, that still exceeds the forecasting timeframe. The deployment time that satisfies time constraints is: 12–35 min for SVM; 14–40 min for FMM; 8–25 min when solving the SFRE. However, the SVM student model has a lower accuracy, and the FMM student model has a longer update time.

9 Model Compression Estimations

The use of the proposed KD technique is limited to fuzzy classifiers based on relations and rules [9, 10]. The assumption of monotonicity in the semantic trends of the teacher model underlies the student learning through hedge-based distillation. The principal difference of the hierarchical distillation scheme is:

- compression of the training dataset to the set of interval rules when training the granular teacher model;
- compression of the granular teacher model when transferring expert knowledge in the form of intervals into the relational teacher model;
-

quantization of the student model when generating interval (constrained) solutions of the SFRE.

Following [9], training the granular teacher model requires solving the optimization problem with $2nZ$ parameters for the interval partition at the level of rules; linguistic interpretation requires solving the optimization problem with LQ parameters for the predefined fuzzy partition and terms weights. Training the relation-based teacher model is the optimization problem with $NM + 2(N + M)$ parameters for the fuzzy relation matrix and membership functions parameters of the primary terms. Following [4], parameters of the hierarchical fuzzy partition are sequentially determined for each class of the modified rules. When training parameters for the certain class, the remaining parameters are fixed. Generation of the interval solutions of the SFRE is reduced to solving z_j optimization problems with $2N$ parameters for each interval rule in the class d_j, $j = 1, \ldots, m$. Generation of the constrained solutions of the SFRE is reduced to solving q_j optimization problems with L parameters for each linguistic rule in the class d_j, $j = 1, \ldots, m$.

For predicting the number of comments in social networks, the model updating time determines the next observation moment t_0. Deployment of the SFRE solutions using principles of parallel computing allows to reduce the waiting time, i.e., increase the frequency of observations that guarantees the accuracy of predictions based on highly dynamic behavior of time series describing the social network users' activity.

10 Conclusion and Future Work

The hierarchical teacher-student architecture based on granular solutions of the SFRE is proposed to provide the interpretable KD mechanism under uncertainty. The relation-based model serves as a carrier of expert knowledge transferred from the granular teacher model for significance measures of the primary terms. From the perspective of human beings' learning, the abductive reasoning underlies the response-based process of generating the teacher's statements IF which correspond to the student's query THEN. In this case, expert knowledge is transferred from the relation-based teacher model to the granular student model by solving the SFRE. To control the level of granularity, knowledge is distilled in the form of interval or constrained solutions. Thus, given the relation-based teacher model, knowledge transfer occurs simultaneously with the search for the optimal structure of the student model.

The model compression method based on data and knowledge distillation is proposed to deploy the granular fuzzy classifier on devices with limited computing resources. The hierarchical distillation scheme that uses a relational model for transferring knowledge between teacher and student granular models provides consistent model compression. Incorporation of expert knowledge in the form of interval rules makes it possible to replace the granular teacher model with a compact relational model. The multitask distillation scheme ensures transferring distributed knowledge into the granular student model. Deployment of the student model can be parallelized when solving the SFRE for each output class. The self-organized set of rules based on solutions of the SFRE is autonomously identified using the incremental learning scheme. Compression of the student model is achieved through quantization at the level of granular or constrained solutions of the SFRE. The quality of distillation is defined by the fuzzy relation matrix. The completeness and accuracy of the rule set is ensured by the genetic-neural technology for solving the SFRE.

A further area of research is the development of a deep neuro-fuzzy network based on solutions of the SFRE. The problem is to ensure the interpretable inner structure of deep rule-based classifiers for large-scale complex classification problems [37, 38]. A feasible solution is to construct a hierarchical classification model by aggregating the detailed prototypes in the form of solutions of the SFRE into abstract feature maps in the form of fuzzy relation matrices. The hierarchical system of granular and relational fuzzy models is integrated into the multilayer modular structure. The training process evolves to the generalized relational structure by stacking modules of rule-based solutions of the SFRE. When transferring knowledge from the deep rule-based classifier, the degree of compression depends on the number of output terms and the granularity of solutions in the stacked modules.

References

1. Hinton, G., Vinyals, O., Dean, J.: Distilling the knowledge in a neural network. ArXiv, abs/1503.02531 (2015). https://doi.org/10.48550/arXiv.1503.02531
2. Gou, J., Yu, B., Maybank, S., Tao, D.: Knowledge distillation: a survey. Int. J. Comput. Vision **129**, 1789–1819 (2021). https://doi.org/10.1007/s11263-021-01453-z
3. Cheng, Y., Wang, D., Zhou, P., Zhang, T.: Model compression and acceleration for deep neural networks: the principles, progress, and challenges. IEEE Signal Process. Mag. **35**(1), 126–136 (2018). https://doi.org/10.1109/MSP.2017.2765695
4. Pedrycz, W.: Granular Computing: Analysis and Design of Intelligent Systems. CRC Press, Bosa Roca (2018) https://doi.org/10.1201/9781315216737
5. Pedrycz, W., Chen, S.-M. (Eds.): Interpretable Artificial Intelligence: A Perspective of Granular Computing. Studies in Computational Intelligence. Springer Cham (2021) https://doi.org/10.1007/978-3-030-64949-4
6. Yager, R., Filev, D.: Essentials of Fuzzy Modeling and Control. Willey, New York (1994)
7. Di Nola, A., Sessa, S., Pedrycz, W., Sancez, E.: Fuzzy Relational Equations and Their Applications to Knowledge Engineering. Kluwer, Dordrecht (1989)
8. Peeva, K., Kyosev, Y.: Fuzzy Relational Calculus. Theory, Applications and Software. World Scientific, New York (2004)
9. Rotshtein, A., Rakytyanska, H.: Fuzzy Evidence in Identification, Forecasting and Diagnosis, vol. 275. Studies in Fuzziness and Soft Computing. Springer, Heidelberg (2012). https://doi.org/10.1007/978-3-642-25786-5
10. Rotshtein, A., Rakytyanska, H.: Optimal design of rule-based systems by solving fuzzy relational equations. In: Hippe, Z., Kulikowski, L., Mroczek, T., Wtorek, J. (Eds.), Issues and Challenges in Artificial Intelligence, vol. 559, Studies in Computational Intelligence, pp. 167–178. Springer, Cham (2014). https://doi.org/10.1007/978-3-319-06883-1_14
11. Shao, B., Chen, Y.: Multi-granularity for knowledge distillation. Image Vis. Comput. **115**, 104286 (2021). https://doi.org/10.1016/j.imavis.2021.104286

12. Chen, H., Wang, Y., Xu, C., Xu, C., Tao, D.: Learning student networks via feature embedding. IEEE Trans. Neural Netw. Learn. Syst. **32**(1), 25–35 (2021). https://doi.org/10.1109/TNNLS.2020.2970494
13. Samek, W., Montavon, G., Lapuschkin, S., Anders, C.J., Müller, K.: Explaining deep neural networks and beyond: A review of methods and applications. Proc. IEEE **109**, 247–278 (2021). https://doi.org/10.1109/JPROC.2021.3060483
14. Bao, H., Wang, G., Li, S., Liu, Q.: Multi-granularity visual explanations for CNN. Knowl.-Based Syst. 109474 (2022). https://doi.org/10.1016/j.knosys.2022
15. Li, Y., Liu, L., Wang, G., Du, Y., Chen, P.: EGNN: Constructing explainable graph neural networks via knowledge distillation. Knowl.-Based Syst. **241**, 108345 (2022). https://doi.org/10.1016/j.knosys.2022.108345
16. Chavan, T., Nandedkar, A.: A convolutional fuzzy min-max neural network. Neurocomputing **405**, 62–71 (2020). https://doi.org/10.1016/j.neucom.2020.04.003
17. Guo, T., Xu, C., He, S., Shi, B., Xu, C., Tao, D.: Robust student network learning. IEEE Trans. Neural Netw. Learn. Syst. **31**(7), 2455–2468 (2020). https://doi.org/10.1109/TNNLS.2019.2929114
18. Phuong, M., Lampert, C.: Towards understanding knowledge distillation. In: Proceedings of the 36th International Conference on Machine Learning, vol. 97, pp. 5142–5151 (2019)
19. Roychowdhury, S., Diligenti, M., Gori, M.: Regularizing deep networks with prior knowledge: a constraint-based approach. Knowl.-Based Syst. **222**, 106989 (2021). https://doi.org/10.1016/j.knosys.2021.106989
20. Li, J., et al.: Explainable CNN with fuzzy tree regularization for respiratory sound analysis. IEEE Trans. Fuzzy Syst. **30**(6), 1516–1528 (2022). https://doi.org/10.1109/TFUZZ.2022.3144448
21. Liu, X., Wang, X., Matwin, S.: Improving the interpretability of deep neural networks with knowledge distillation. In: 2018 IEEE International Conference on Data Mining Workshops (ICDMW), pp. 905–912 (2018)
22. Sujatha, K., Nalinashini, G., Ponmagal, R.S., Ganesan, A., Kalaivani, A., Hari, R.: Hybrid deep learning neuro-fuzzy networks for industrial parameters estimation. In: Pandey, R., Khatri, S.K., Singh, N.K., Verma, P. (Eds.), Artificial Intelligence and Machine Learning for EDGE Computing, pp. 325–341. Academic Press (2022). https://doi.org/10.1016/B978-0-12-824054-0.00028-9
23. Yeganejou, M., Kluzinski, R., Dick, S., Miller, J.: An end-to-end trainable deep convolutional neuro-fuzzy classifier. In: Proceedings of the 2022 IEEE International Conference on Fuzzy Systems (FUZZ-IEEE), pp. 1–7 (2022). https://doi.org/10.1109/FUZZ-IEEE55066.2022.9882723
24. Wen, Y.-W., Peng, S.-H., Ting, C.-K.: Two-stage evolutionary neural architecture search for transfer learning. IEEE Trans. Evol. Comput. **25**(5), 928–940 (2021). https://doi.org/10.1109/TEVC.2021.3097937
25. Zhang, X., Gong, Y.-J., Xiao, X.: Adaptively transferring deep neural networks with a hybrid evolution strategy. In: Proceedings of the 2020 IEEE International Conference on Systems, Man, and Cybernetics (SMC), pp. 1068–1074 (2020). https://doi.org/10.1109/SMC42975.2020.9283487
26. Fernandes, F.E., Jr., Yen, G.G.: Pruning deep convolutional neural networks architectures with evolution strategy. Inf. Sci. **552**, 29–47 (2021). https://doi.org/10.1016/j.ins.2020.11.009
27. Zhou, Y., Yen, G.G., Yi, Z.: Evolutionary shallowing deep neural networks at block levels. IEEE Trans. Neural Netw. Learn. Syst. **33**(9), 4635–4647 (2022). https://doi.org/10.1109/TNNLS.2021.3059529
28. Stanley, K.O., Miikkulainen, R.: Evolving neural networks through augmenting topologies. Evol. Comput. **10**(2), 99–127 (2002). https://doi.org/10.1162/106365602320169811
29. He, C., Tan, H., Huang, S., Cheng, R.: Efficient evolutionary neural architecture search by modular inheritable crossover. Swarm Evol. Comput. **64**, 100894 (2021). https://doi.org/10.1016/j.swevo.2021.100894

30. Hassanzadeh, T., Essam, D., Sarker, R.: EvoDCNN: an evolutionary deep convolutional neural network for image classification. Neurocomputing **488**, 271–283 (2022). https://doi.org/10.1016/j.neucom.2022.02.003
31. Wen, L., Gao, L., Li, X., Li, H.: A new genetic algorithm based evolutionary neural architecture search for image classification. Swarm Evol. Comput. **75**, 101191 (2022). https://doi.org/10.1016/j.swevo.2022.101191
32. Wang, Z., Li, F., Shi, G., Xie, X., Wang, F.: Network pruning using sparse learning and genetic algorithm. Neurocomputing **404**, 247–256 (2020). https://doi.org/10.1016/j.neucom.2020.03.082
33. Yang, C., An, Z., Li, C., Diao, B., Xu, Y.: Multi-objective pruning for CNNs using genetic algorithm. In: Tetko, I., Kurkova, V., Karpov, P., Theis, F. (eds.) Artificial Neural Networks and Machine Learning—ICANN 2019: Deep Learning. ICANN 2019. Lecture Notes in Computer Science, vol. 11728, pp. 299–305. Springer, Cham (2019). https://doi.org/10.1007/978-3-030-30484-3_25
34. Xu, K., Zhang, D., An, J., Liu, L., Liu, L., Wang, D.: GenExp: multi-objective pruning for deep neural network based on genetic algorithm. Neurocomputing **451**, 81–94 (2021). https://doi.org/10.1016/j.neucom.2021.04.022
35. Nagae, S., Kawai, S., Nobuhara, H.: Transfer learning layer selection using genetic algorithm. In: Proceedings of the 2020 IEEE Congress on Evolutionary Computation (CEC), pp. 1–6 (2020). https://doi.org/10.1109/CEC48606.2020.9185501
36. Poyatos, J., Molina, D., Martinez, A.D., Del Ser, J., Herrera, F.: EvoPruneDeepTL: an evolutionary pruning model for transfer learning based deep neural networks. Neural Netw. **158**, 59–82 (2023). https://doi.org/10.1016/j.neunet.2022.10.011
37. Angelov, P., Gu, X.: Deep rule-based classifier with human-level performance and characteristics. Inf. Sci. **463–464**, 196–213 (2018). https://doi.org/10.1016/j.ins.018.06.048
38. Gu, X., Angelov, P.: Highly interpretable hierarchical deep rule-based classifier. Appl. Soft Comput. **92**, 106310 (2020). https://doi.org/10.1016/j.asoc.2020.106310
39. Pratama, M., Pedrycz, W., Webb, G.I.: An incremental construction of deep neuro fuzzy system for continual learning of nonstationary data streams. IEEE Trans. Fuzzy Syst. **28**(7), 1315–1328 (2020). https://doi.org/10.1109/TFUZZ.2019.2939993
40. Wang, Z., Pan, X., Wei, G., Fei, J., Lu, X.: A faster convergence and concise interpretability TSK fuzzy classifier deep-wide-based integrated learning. Appl. Soft Comput. **85**, 105825 (2019). https://doi.org/10.1016/j.asoc.2019.105825
41. Xie, C., Rajan, D., Prasad, D.K., Quek, C.: An embedded deep fuzzy association model for learning and explanation. Appl. Soft Comput. **131**, 109738 (2022). https://doi.org/10.1016/j.asoc.2022.109738
42. Liu, Y., Lu, X., Peng, W., Li, C., Wang, H.: Compression and regularized optimization of modules stacked residual deep fuzzy system with application to time series prediction. Inform. Sci. **608**, 551–577 (2022). https://doi.org/10.1016/j.ins.2022.06.088
43. Talpur, N., Abdulkadir, S.J., Alhussian, H., Hasan, M.H., Abdullah, M.A.: Optimizing deep neuro-fuzzy classifier with a novel evolutionary arithmetic optimization algorithm. J. Comput. Sci. **64**, 101867 (2022). https://doi.org/10.1016/j.jocs.2022.101867
44. Bartl, E., Belohlavek, R., Vychodil, V.: Bivalent and other solutions of fuzzy relational equations via linguistic hedges. Fuzzy Sets Syst. **187**(1), 103–112 (2012). https://doi.org/10.1016/j.fss.2011.05.020
45. Rakytyanska, H.: Classification rule hierarchical tuning with linguistic modification based on solving fuzzy relational equations. East.-Eur. J. Enterp. Technol. **1**(4), 50–58 (2018). https://doi.org/10.15587/1729-4061.2018.123567
46. Rakytyanska, H.: Optimization of fuzzy classification knowledge bases using improving transformations. East.-Eur. J. Enterp. Technol. **5**(2), 33–41 (2017). https://doi.org/10.15587/1729-4061.2017.110261
47. Rakytyanska H.: Solving systems of fuzzy logic equations in inverse inference problems. Herald of the National University "Lviv Polytechnic". Comput. Sci. Inform. Technol. **826**, 248–259 (2015)

48. Rotshtein, A., Rakytyanska, H.: Fuzzy logic and the least squares method in diagnosis problem solving. In: Sarma, R.D. (ed.) Genetic Diagnoses, pp. 53–97. Nova Science Publishers, New York (2011)
49. Singh, K., Sandhu, R., Kumar, D.: Machine Learning Repository: Facebook Comment Volume Dataset (2016). https://archive.ics.uci.edu/ml/datasets/Facebook+Comment+Volume+Dataset
50. Singh, K.: Facebook comment volume prediction. Int. J. Simul. Syst. Sci. Technol. **16**(5) 16.1–16.9 (2015). https://doi.org/10.5013/IJSSST.a.16.05.16
51. Azarov, O., Krupelnitsky, L., Rakytyanska, H.: Television rating control in the multichannel environment using trend fuzzy knowledge bases and monitoring results. Data **3**(4), 57 (2018). https://doi.org/10.3390/data3040057
52. Wang, H., Li, M., Yue, X.: IncLSTM: incremental ensemble LSTM model towards time series data. Comput. Electr. Eng. **92**, 107156 (2021). https://doi.org/10.1016/j.compeleceng.2021.107156
53. Han, H.-G., Ma, M.-L., Yang, H.-Y., Qiao, J.-F.: Self-organizing radial basis function neural network using accelerated second-order learning algorithm. Neurocomputing **469**, 1–12 (2022). https://doi.org/10.1016/j.neucom.2021.10.065
54. Lan, Y., Xu, X., Fang, Q., Zeng, Y., Liu, X., Zhang, X.: Transfer reinforcement learning via meta-knowledge extraction using auto-pruned decision trees. Knowl.-Based Syst. **242**, 108221 (2022). https://doi.org/10.1016/j.knosys.2022.108221
55. Guo, H., Zhang, A., Wang, W.: An accelerator for online SVM based on the fixed-size KKT window. Eng. Appl. Artif. Intell. **92**, 103637 (2020). https://doi.org/10.1016/j.engappai.2020.103637
56. Khuat, T., Gabrys, B.: Accelerated learning algorithms of general fuzzy min-max neural network using a novel hyperbox selection rule. Inf. Sci. **547**, 887–909 (2021). https://doi.org/10.1016/j.ins.2020.08.046
57. MATLAB mobile (Accessed 2023). Available at: https://www.mathworks.com/products/matlab-mobile.html

Ensemble Knowledge Distillation for Edge Intelligence in Medical Applications

Yuri Gordienko, Maksym Shulha, Yuriy Kochura, Oleksandr Rokovyi, Oleg Alienin, Vladyslav Taran, and Sergii Stirenko

Abstract The "bucket of models" ensemble technique for search of the best-per-family model and then the best-among-families model of some DNN architecture was applied to choose the absolutely best model for each separate medical problem. It is based on usage of the previously obtained knowledge distillation (KD) approaches where an ensemble of "student" DNNs can be trained with regard to set of various flavours of teacher family with the same architecture. The bucket of student-teacher models is applied to various medical data from quite different medical domains with the purpose to check feasibility of such approaches for the practical medical datasets for Edge Intelligence devices with the limited computational abilities. The training and validation runs were performed on the standard datasets like CIFAR10/CIFAR100 in comparison to the specific medical datasets like MedMNIST including medical data of the quite various types and complexity. Several families of ResNet DNN architecture families were used and they demonstrated various performance on the standard CIFAR10/CIFAR100 and specific medical MedMNIST datasets. As a result, no relationship was found between CIFAR10/CIFAR100 performance and MedMNIST performance, the choice of model family effects performance more than some model within a family, the significant boost in performance can be obtained by smaller models, some powerful CIFAR10/CIFAR100 architectures were unnecessarily large for MedMNIST and can be made more parameter-efficient (smaller) without a significant drop in performance. In the future research the more specific combination of distillation approaches is planned to be used with taking into account the finer hyperparameter tuning, data augmentation procedures, and their additional influences on the original data of MedMNIST datasets.

Keywords Deep neural network · Convolutional neural network · Knowledge distillation · Classification · CIFAR · MedMNIST · Healthcare · Edge computing

---

Y. Gordienko (✉) · M. Shulha · Y. Kochura · O. Rokovyi · O. Alienin · V. Taran · S. Stirenko
National Technical University of Ukraine, Igor Sikorsky Kyiv Polytechnic Institute, Kyiv, Ukraine
e-mail: yuri.gordienko@gmail.com
URL: https://comsys.kpi.ua/en

W. Pedrycz and S.-M. Chen (eds.), *Advancements in Knowledge Distillation: Towards New Horizons of Intelligent Systems*, Studies in Computational Intelligence 1100,
https://doi.org/10.1007/978-3-031-32095-8_5

1 Introduction

Various artificial intelligence (AI) approaches, for example deep learning (DL) methods including usage of deep neural networks (DNN), have been used successfully to process data in different domains beginning from the initial attempts [9, 19, 21, 26, 33, 52] to the recent advances [30, 41–43]. These activities have become very intensive for the data processing in medical applications [5, 8].

They were thoroughly investigated in their solitary usage for the standard datasets like CIFAR10, CIFAR100 [28], ImageNet [29], and others. But the current necessity and problem is to port these approaches in the health care domain for classification tasks where datasets were usually quite different from the aforementioned standard datasets and have the smaller number of images. It is especially important due to the known problem of porting the best DNNs trained on ImageNet dataset to Edge Computing devices with low computational resources.

The main objective of this work is to investigate the "bucket of models" ensemble technique in which a model selection algorithm is used to choose the best model for each separate medical problem. It is based on usage of the previously obtained knowledge distillation (KD) approaches where an ensemble of "student" DNNs can be trained with regard to set of various flavours of teacher family with the same architecture. The bucket of student-teacher models is applied to various medical data from quite different medical domains with the purpose to check feasibility of such approaches for the practical medical datasets for Edge Intelligence devices with the limited computational abilities.

The structure of this article is as follows: Sect. 2 gives a brief overview of some similar attempts to study KD for medical purposes, Sect. 3 presents the standard and specific (medical) datasets, DNN types and metrics used, the Sect. 4 describes the results on different DNNs, the Sect. 5 contains a summary of all results with the comparative analysis of the applied DNNs, and the Sect. 6 gives an overview of the potential directions for further studies.

2 Background and Related Work

DL-based approaches play an important role at the intersection of AI and various applications, including healthcare [5, 8]. Usually, selection of the best model among many models is motivated by their performance on the standard datasets like CIFAR10, CIFAR100, or ImageNet. But because of the inherent complexity of medical data (classes) and their quite different structure in comparison to common objects (classes) from the standard datasets, the best models for the standard datasets can be not so efficient for the specific medical datasets. Moreover, it takes huge amounts of efforts to tune the DL models for usage in real practical applications, especially for the devices with the limited computing resources used for Edge Intelligence. On the other hand, it is also very hard to identify whether the specific selected model could be generalised if it is only evaluated on a few datasets, especially on the standard ones only. That is why the specific and diverse datasets should be used by the research communities to fairly evaluate generalization performance of the selected models.

This problem is especially important in medical applications where artificial intelligence (AI) can automate and dramatically accelerate Computer-Aided Detection (CADe) and Computer-Aided Diagnosis (CADx) by automatically processing medical data without the involvement of medical personnel on the screening stage and making it available on a regular basis as AI-platform of CADe/CADx services [2]. The additional promising way of the further development of CADe/CADx should include AI-based personalized medicine (AIPM) that can allow practitioners to find cures tuned for patients at Edge Computing level [53]. The specialized neural network accelerators become widely used for AIPM with Edge Intelligence, but the main limitation of these accelerators is related with the necessity to adapt and convert the neural network model to the limits imposed by the accelerators [45]. Modern DNNs were characterized by very high demands to resources that can limit their usage under conditions of limited bandwidth and low power consumption. The devices for Edge Intelligence with the limited GPU capabilities and scarce battery power were hardly ready for the huge size of the current state-of-the-art (SOTA) very powerful DNNs. That is why in recent years various model compression techniques (like pruning, knowledge distillation, etc.) were of great importance to decrease the size of DNNs of interest with a purpose to retain the desirable accuracy.

For example, knowledge distillation (KD), introduced by Hinton et al [17], is widely known as an approach where a "student" deep neural network (DNN) uses "soft" labels coming from a strong "teacher" DNN. It is usually believed that such a training improves the performance of the "student" DNN. Recently, several improved KD techniques were proposed: activation-boundary distillation (ABD) [16], overhaul feature distillation (OFD) [15], relational knowledge distillation (RKD) [37], teacher assistant distillation (TAD) [34], patient knowledge distillation (PKD) [44] adapted for vision tasks as a simple feature distillation (SFD) [39], and some others like unsupervised distillation (UD) with data augmentation and various ensemble approaches [3, 35, 39, 57]. In contrast to the traditional KD, recently, self-distillation (SD) was proposed that distills knowledge within network itself, i.e. in SD the same networks are used for the teacher and the student models [13, 56]. Despite the vast number of methods and applications of KD in general, and SD in part, understanding of their theoretical principles and empirical results remains insufficient.

Some of these KD-related approaches were successfully used for medical applications, for example, for efficient medical image segmentation based on knowledge distillation [22, 38, 58] with ensembles of convolutional neural networks for medical image segmentation [35], for classification of chest X-ray abnormalities [18], efficient histology image classification [7], imbalanced medical image distillation [32], melanoma detection [27], breast cancer classification [31], chest X-ray diagnosis [36], skin lesion classification using dermoscopic images [47] among many others.

In the context of the medical application several other aspects were of great interest that were found recently for chest X-ray interpretation by application of the DL models developed and pretrained for ImageNet [25] for the large chest X-ray dataset

(CheXpert). First, no relationship was found between ImageNet performance and CheXpert performance for both models without pretraining and models with pre-training. Second, the choice of model family effects performance more than some model within a family. Third, the statistically significant boost in performance can be obtained by smaller models. Fourth, some powerful ImageNet architectures were unnecessarily large for CheXpert and can be made more parameter-efficient (smaller) without a statistically significant drop in performance.

In relation to the above mentioned results, the aim of this work is to investigate and compare KD effect on performance of the "bucket of models" ensemble technique in which a model selection algorithm is used to choose the best model for each separate medical problem. The main idea is that the best model for some medical domain cannot be so highly efficient for other medical domain despite its high performance on several standard datasets. It is especially important for the practical medical datasets used by Edge Intelligence devices with the limited computational abilities.

3 Methodology

In this section some experimental aspects were described, namely: the standard and specific (medical) datasets with different types of visual data, some types of DNN models, and metrics. They were used to investigate the effect of KD on the performance of DNNs applied to the standard and specific (medical) datasets to solve the relevant classification problems.

3.1 Datasets

3.1.1 Standard Datasets CIFAR10 and CIFAR100

The standard CIFAR10 and CIFAR100 were labeled subsets (named after Canadian Institute For Advanced Research) of the many small images dataset [28]. The CIFAR10 dataset is a set of images that were widely used to test various computer vision (CV), machine learning (ML) and deep learning (DL) algorithms. It contains 60,000 color images in 10 different classes with image size 32×32 pixels (Fig. 1a). The 10 different classes includes airplanes, cars, birds, cats, deer, dogs, frogs, horses, ships, and trucks with 6,000 images per each class. The CIFAR100 dataset is similar to the CIFAR10 dataset, but it contains images of 100 classes with 600 images per class (500 training images and 100 testing images) (Fig. 1b). The 100 classes in the CIFAR100 were grouped into 20 so-called superclasses where each image has a "fine" label (for the class to which it belongs) and a "coarse" label (for the superclass to which it belongs).

Fig. 1 The examples of images from CIFAR10 (a) and CIFAR100 (b) datasets

3.1.2 Specific Medical Datasets

The MedMNIST v2 dataset is a large-scale MNIST-like super-set of datasets with 2D and 3D standardized biomedical data [54, 55]. All images were pre-processed into a small size of 28×28 (2D) or 28×28×28 (3D) with the relevant classification labels. They include some basic data modalities in biomedical images and designed to perform classification on lightweight 2D and 3D images with dataset of various volumes (from 100 to 100,000 images) and different tasks (for example, for binary, multi-class, and multi-label classification, etc.). These datasets were used in various medical and general-purpose research and educational purposes in biomedical CV, ML and DL applications.

The following datasets from MedMNIST collection were used here:

- BloodMNIST dataset includes images of blood analysis images (Fig. 2a),
- DermaMNIST dataset contains dermatoscopic images (Fig. 2b),
- PathMNIST dataset has histological images (Fig. 2c),
- RetinaMNIST dataset contains retina fundus images (Fig. 2d).

3.1.3 BloodMNIST

The BloodMNIST dataset is based on the original dataset [1] that contains images of individual normal cells, obtained from individuals without infection, hematologic or oncologic disease and free of any pharmacologic treatment at the moment of blood collection. It has more than 17,000 images of 8 classes. The source images with sizes of 3×360×363 pixels were center-cropped into 3×200×200 pixels, and then resized into images with the sizes of 3×28×28 pixels (Fig. 2a).

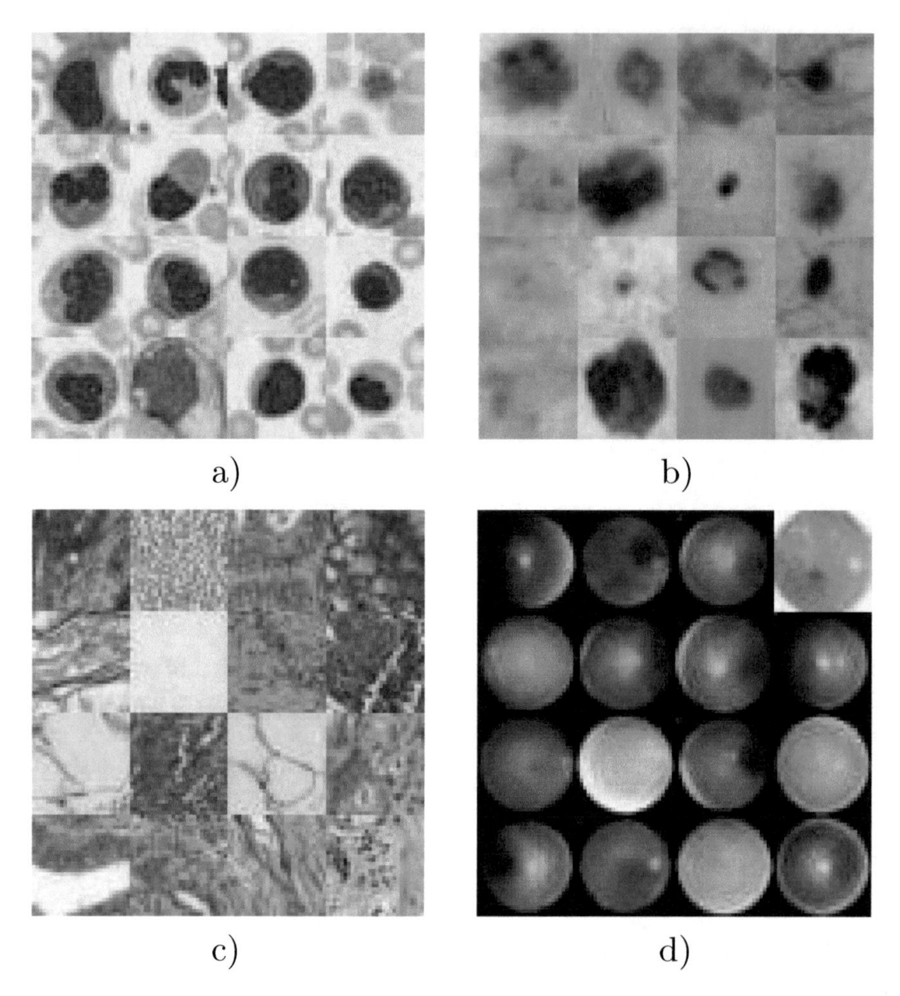

Fig. 2 The examples of images from datasets in MedMNIST collection that were used in this work: BloodMNIST (**a**), DermaMNIST (**b**), PathMNIST (**c**), RetinaMNIST (**d**)

3.1.4 DermaMNIST

The DermaMNIST dataset is based on the HAM10000 [6, 46] that contains dermatoscopic images of pigmented skin lesions. It has more than 10,000 dermatoscopic images of 7 different classes. The source images with sizes of $3\times600\times450$ were resized into images with the sizes of $3\times28\times28$ pixels (Fig. 2b).

3.1.5 PathMNIST

The PathMNIST dataset is based on the dataset (NCT-CRC-HE-100K) obtained after the study for predicting survival from colorectal cancer histology slides for more than 100,000 image patches from stained histological images, and a test dataset (CRC-VAL-HE-7K) with more than 7,000 image patches from a different clinical center [23, 24]. The dataset has images of 9 types of classes. The source images with sizes of $3 \times 224 \times 224$ were resized into images with the sizes of $3 \times 28 \times 28$ pixels (Fig. 2c).

3.1.6 RetinaMNIST

The RetinaMNIST dataset is based on the original dataset used in the DeepDRiD [20] challenge. It has more than 1,500 retina fundus images of 5 classes for 5-level grading of diabetic retinopathy severity. The source images with sizes of $3 \times 1{,}736 \times 1{,}824$ pixels were center-cropped with a window size of length of the short edge, and then resized into images with the sizes of $3 \times 28 \times 28$ pixels (Fig. 2d).

3.2 Models

The DNN models were limited to the widely popular ResNet architecture [14] which demonstrated its high performance not only on ImageNet dataset [29], but also on many other practical datasets with the similar objects (classes). Several different sources of the model various representations for the PyTorch framework were used due to their open source availability and detailed descriptions elsewhere [40]. Several widely known ResNet model families were used, namely, 3 layer weak, 3 layer, 4 layer, wide, and torchvision, with their correspondent model names and indicative number of parameters (P_t) for training on the CIFAR10 datasets mentioned in Tables 1, 2, 3, 4, 5, 6 below. Also the analysis of the model performance is given below as a dependence on the increase of the ratio P_t / P_s, where P_t—is the number of parameters in the teacher model, and P_s—is the number of parameters in the student model (Fig. 17).

For KD students the stripped versions, namely, ResNet8, ResNet8_sm, ResNet8_v, WRN10_1 that were mentioned in the first rows in Tables 1, 2, 3, 4, 5, 6 below, were used. In general, they has substantially less parameters and memory footprint. But as it is shown below the smaller student models can achieve the significant accuracy depending on the dataset and teacher flavour.

Table 1 Teacher (A_t) and student (A_s) validation accuracy values and the number of model parameters P_t for the 3 layer weak ResNet models used for CIFAR10 and CIFAR100 datasets (here and below the highest values per dataset-model pairs were emphasized by a **bold** font, and the lowest values—by a *italic* font)

	CIFAR10		CIFAR100		
Teacher model	A_s	A_t	A_s	A_t	P_t
ResNet8_sm	0.701	0.711	0.279	0.291	78042
ResNet14_sm	0.740	0.801	0.303	0.370	175258
ResNet20_sm	0.733	0.812	0.316	0.411	272474
ResNet32_sm	0.738	**0.825**	0.314	0.439	466906
ResNet44_sm	**0.755**	0.823	0.319	0.445	661338
ResNet56_sm	0.736	**0.825**	**0.320**	**0.453**	855770
ResNet110_sm	0.715	0.804	0.300	0.423	1730714
ResNet164_sm	*0.696*	*0.721*	*0.286*	*0.343*	1704154

Table 2 Teacher (A_t) and student (A_s) validation accuracy values and model parameters P_t for the 3 layer ResNet models used for CIFAR10 and CIFAR100 datasets

	CIFAR10		CIFAR100		
Teacher model	A_s	A_t	A_s	A_t	P_t
ResNet8	0.803	0.805	0.474	0.475	89322
ResNet14	*0.816*	*0.841*	*0.501*	*0.535*	186538
ResNet20	0.824	0.854	0.506	0.553	283754
ResNet26	0.823	0.859	0.503	0.562	380970
ResNet32	**0.830**	0.865	0.507	0.567	478186
ResNet44	0.821	0.867	0.499	0.578	672618
ResNet56	0.825	**0.872**	**0.510**	**0.584**	867050

Table 3 Teacher (A_t) and student (A_s) validation accuracy values and model parameters P_t for the 4 layer ResNet models used for CIFAR10 and CIFAR100 datasets

	CIFAR10		CIFAR100		
Teacher model	A_s	A_t	A_s	A_t	P_t
ResNet10	0.882	0.877	0.625	0.618	4903242
ResNet18	0.895	*0.900*	*0.647*	*0.661*	11173962
ResNet34	*0.894*	0.902	0.649	0.664	21282122
ResNet50	0.895	0.902	0.653	0.689	23520842
ResNet101	0.896	**0.911**	0.656	0.694	42512970
ResNet152	**0.897**	0.909	**0.660**	**0.696**	58156618

Table 4 Teacher (A_t) and student (A_s) validation accuracy values and model parameters P_t for the torchvision ResNet models used for CIFAR10 and CIFAR100 datasets

	CIFAR10		CIFAR100		
Teacher model	A_s	A_t	A_s	A_t	P_t
ResNet8_v	0.689	0.703	0.313	0.328	95082
ResNet10_v	**0.718**	0.771	0.342	0.471	4910922
ResNet14_v	0.691	0.696	0.318	0.336	192298
ResNet18_v	**0.718**	**0.773**	**0.344**	**0.456**	11181642
ResNet20_v	0.691	0.694	0.306	0.313	289514
ResNet34_v	0.706	0.759	0.332	0.395	21289802
ResNet50_v	0.620	0.534	0.259	0.237	23528522
ResNet101_v	0.550	0.471	0.195	0.196	42520650
ResNet152_v	*0.548*	*0.349*	*0.167*	*0.094*	58164298
wrn50_2_v	0.617	0.534	0.220	0.288	66854730
wrn101_2_v	0.581	0.502	0.218	0.194	124858186

Table 5 Teacher (A_t) and student (A_s) validation accuracy values and model parameters P_t for the wide ResNet models used for CIFAR10 and CIFAR100 datasets

	CIFAR10		CIFAR100		
Teacher model	A_s	A_t	A_s	A_t	P_t
WRN10_1	0.703	0.707	0.245	0.268	77850
WRN16_1	*0.736*	*0.786*	0.281	*0.322*	175066
WRN28_1	0.743	0.811	*0.278*	0.346	369498
WRN40_1	0.735	0.816	0.283	0.371	563930
WRN16_2	0.747	0.833	0.305	0.431	691674
WRN28_2	0.760	0.854	0.305	0.455	1467610
WRN10_4	0.752	0.812	0.302	0.426	1198810
WRN16_4	0.767	0.867	0.303	0.518	2748890
WRN22_4	0.756	0.880	0.306	0.545	4298970
WRN40_4	0.754	0.885	0.309	0.566	8949210
WRN16_8	0.757	0.885	0.310	0.595	10961370
WRN22_8	**0.764**	**0.899**	**0.320**	**0.626**	17158106

Table 6 Teacher (A_t) and student (A_s) validation accuracy values and model parameters P_t for VGG models used for CIFAR10 datasets

Teacher model	A_s	A_t	P_t
VGG11	0.869	0.871	9231114
VGG13	*0.876*	**0.892**	9416010
VGG16	**0.881**	0.891	14728266
VGG19	*0.876*	*0.882*	20040522

3.3 Metrics

Validation accuracy for teacher (A_t) and student (A_s) were used as metrics to evaluate and compare the model performance. In general, the area under curve (AUC) [4] should be measured for receiver operating characteristic (ROC) with their micro and macro versions, and their mean and standard deviation values. In fact, for a given threshold, accuracy measures the percentage of correctly classified objects, regardless of the class they belong to. AUC is threshold-invariant and can measure the quality of the models used here independently of the chosen classification threshold. As far as accuracy depends on the threshold chosen, AUC accounts for all possible thresholds. Because of this, it can sometimes provide a broader view of classifier performance. AUC is less sensitive to class imbalance than accuracy, but as far as there were no significant class imbalance in the datasets used, accuracy was selected as a main metric only for intuitive simplicity and brevity of the numerous results obtained for all datasets.

3.4 Workflow

The general workflow was organized for the aforementioned datasets and models where standard CIFAR10 and CIFAR100 datasets were selected because of their wide usage for the model comparison and computational feasibility. The specific medical MedMNIST datasets were selected as similar representatives of the typical medical data of the same size, but possibly different complexity.

The main focus was not to reach the maximal possible performance for each pair model-dataset, but to provide even conditions for all KD experiments in the training/testing procedure (protocol) proposed, developed, and explained in details elsewhere [39]. All runs were performed with the same optimizer, stochastic gradient descent with Nesterov enabled, a momentum of 0.9 and an initial learning rate of 0.1 that is gradually decreased by 0.1.

For the standard CIFAR10 and CIFAR100 datasets, the training data subset was augmented by using traditional CIFAR augmentation methods (namely, normalized, randomly horizontally flipped, and randomly cropped to a size of 32×32 with a padding of 4) and the validation data subset was only normalized. For the specific medical MedMNIST datasets the training data subset and the validation data subset were normalized only. The training and validation (optimization) curves for teacher and student demonstrated the standard gradual decrease of loss and increase of accuracy without overfitting (they are similar to [39] and not shown here to save space in favor of more meaningful results).

The main objective of this work is to investigate the "bucket of models" ensemble technique in which a model selection algorithm is used to choose the best model for each separate medical problem. It is based on usage of the previously obtained knowledge distillation (KD) approaches where an ensemble of "student" DNNs can be trained with regard to set of various flavours of teacher family with the same architecture. When tested with only one problem, a bucket of models can produce no better results than the best model in the set, but when evaluated across many problems, it will typically produce much better results, on average, than any model in the set.

4 Experiment

The numerous results below are presented in tables (to provide the absolute values that could be of interest for practitioners working with the specific medical datasets who would like to reproduce and compare results) and in the form of bar charts (to compare them with each other in the different settings). Each table contains the first row with the performance metric, validation accuracy, for a smallest scale teacher model (A_t) and for the same smallest scale student model (A_s) trained on the correspondent dataset.

In addition, some comparisons of different values that are scattered across different tables are presented in the additional summary figures (Fig. 17) that illustrate the final conclusions.

4.1 Standard CIFAR10 and CIFAR100 Datasets

4.1.1 3 Layer Weak Models

The teacher validation accuracy values (A_t in Table 1) and the student validation accuracy values (A_s in Table 1) demonstrate the gradual increase to the maximal value and then decrease (Fig. 3). For CIFAR10 dataset the maximal student validation accuracy value $A_s = 0.755$ was observed for ResNet44_sm teacher model. It should be noted that the maximal teacher validation accuracy values $A_t = 0.825$ were observed for the other ResNet32 and ResNet56 teacher models. For CIFAR100 dataset the maximal student validation accuracy value $A_s = 0.320$ was observed for ResNet56_sm teacher model that also demonstrated the maximal teacher validation accuracy value $A_t = 0.453$. The lowest validation values A_s and A_t for both CIFAR10 and CIFAR100 were observed for ResNet164_sm teacher model. Here in Table 1 and in other tables below the highest values per dataset-model pairs were emphasized by a **bold** font, and the lowest values—by a *italic* font.

4.1.2 3 Layer Models

Despite the gradual increase of the teacher validation accuracy values (A_t in Table 2) for all models, the student validation accuracy values (A_s in Table 2) have non-

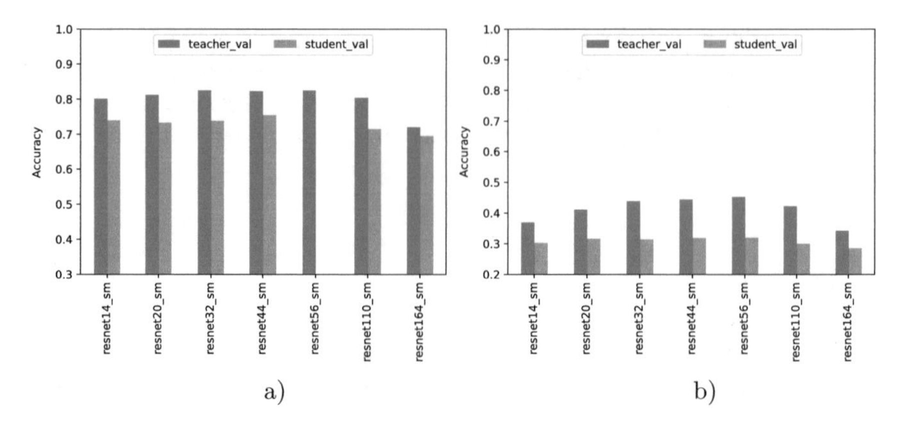

Fig. 3 The validation accuracy values for the teacher and student models from the 3 layer weak ResNet model family (details in Table 1) for CIFAR10 (**a**) CIFAR100 (**b**) datasets

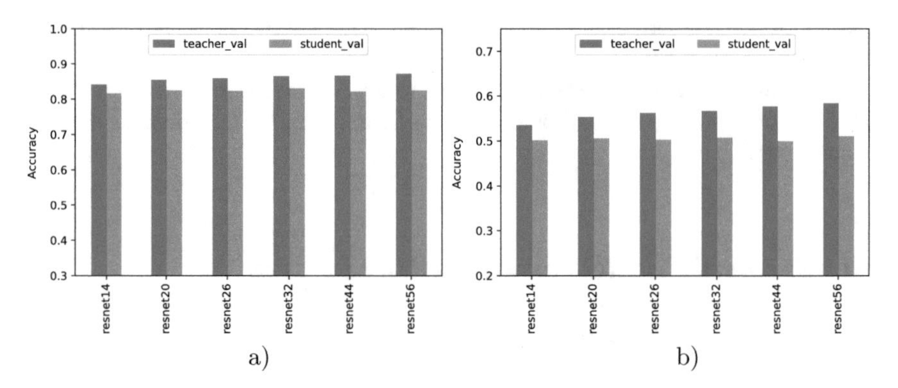

Fig. 4 The validation accuracy values for the teacher and student models from the 3 layer ResNet model family (details in Table 2) for CIFAR10 (**a**) CIFAR100 (**b**) datasets

monotonous dependence on the correspondent teacher model complexity (Fig. 4). For CIFAR10 the maximal student validation accuracy value $A_s = 0.830$ is observed for ResNet32 teacher model, and it is observed in contrast to the maximal teacher validation accuracy value $A_t = 0.872$ for ResNet56 teacher model. For CIFAR100 the maximal student validation accuracy value $A_s = 0.510$ was observed for ResNet56 teacher model that also demonstrated the maximal teacher validation accuracy value $A_t = 0.584$. The lowest validation values A_s and A_t for both CIFAR10 and CIFAR100 were observed for the smallest ResNet14 teacher model.

4.1.3 4 Layer Models

The teacher validation accuracy values (A_t in Table 3) and the student validation accuracy values (A_s in Table 3) demonstrate the gradual increase (Fig. 5). For CIFAR10

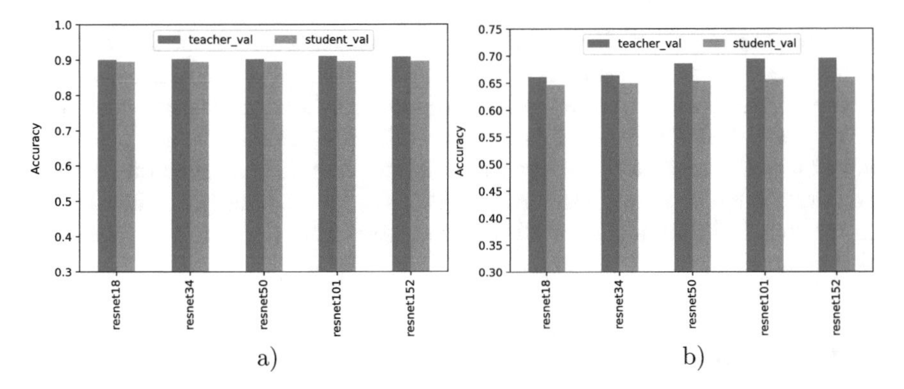

Fig. 5 The validation accuracy values for the teacher and student models from the 4 layer ResNet model family (details in Table 3) for CIFAR10 (**a**) CIFAR100 (**b**) datasets

dataset the maximal student validation accuracy value $A_s = 0.897$ is observed for ResNet152 teacher model which is quite close to ResNet101 teacher model with the maximal teacher validation accuracy value $A_t = 0.911$. For CIFAR100 the maximal student validation accuracy value $A_s = 0.660$ was observed for ResNet152 teacher model that also demonstrated the maximal teacher validation accuracy value $A_t = 0.696$. The lowest validation values A_s and A_t for both CIFAR10 and CIFAR100 were observed for the smallest ResNet18 and ResNet34 teacher models.

4.1.4 Torchvision Models

The teacher validation accuracy values (A_t in Table 4) and the student validation accuracy values (A_s in Table 4) demonstrate the gradual increase (Fig. 6) with the

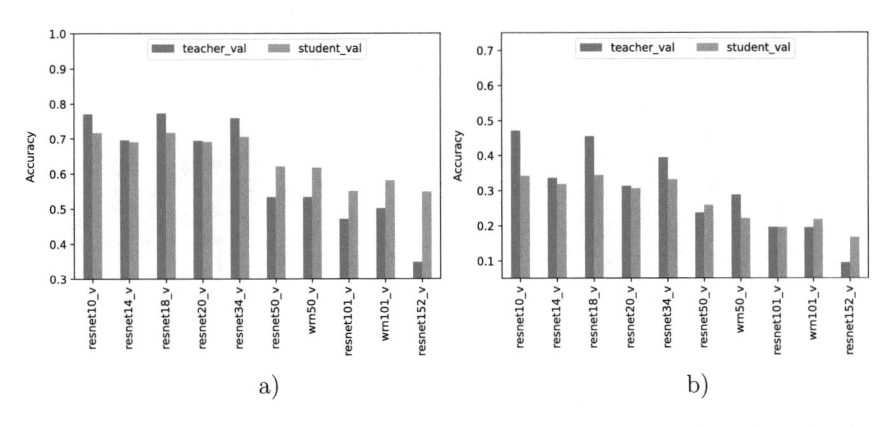

Fig. 6 The validation accuracy values for the teacher and student models from the torchvision ResNet model family (details in Table 4) for CIFAR10 (**a**) CIFAR100 (**b**) datasets

following decrease. For CIFAR10 dataset the maximal student validation accuracy value $A_s = 0.718$ is observed for ResNet18_v and ResNet10_v teacher models where ResNet18_v also demonstrated the maximal teacher validation accuracy value $A_t = 0.773$. For CIFAR100 the maximal student validation accuracy value $A_s = 0.344$ was observed for ResNet18_v teacher model that also demonstrated the maximal teacher validation accuracy value $A_t = 0.456$. The lowest validation values A_s and A_t for both CIFAR10 and CIFAR100 were observed for the largest ResNet152_v teacher models.

4.1.5 Wide Models

For CIFAR10 dataset the maximal student validation accuracy value $A_s = 0.764$ was observed for WRN22_8 teacher model that also demonstrated the maximal teacher validation accuracy value $A_t = 0.899$ (Fig. 7, Table 5). For CIFAR100 the maximal student validation accuracy value $A_s = 0.320$ was observed for the same WRN22_8 teacher model that also demonstrated the maximal teacher validation accuracy value $A_t = 0.626$. The lowest validation values A_s and A_t for both CIFAR10 and CIFAR100 were observed for the smallest WRN16_1 (and partially for WRN28_1) teacher model.

4.1.6 VGG Models

The VGG family of models was used here for the general comparison of the "bucket of models" ensemble technique for the quite other standard DNN architecture (Fig. 8) but for CIFAR10 dataset only.

For CIFAR10 dataset the maximal student validation accuracy value $A_s = 0.881$ was observed for VGG16 teacher model which is quite close to VGG13 teacher model

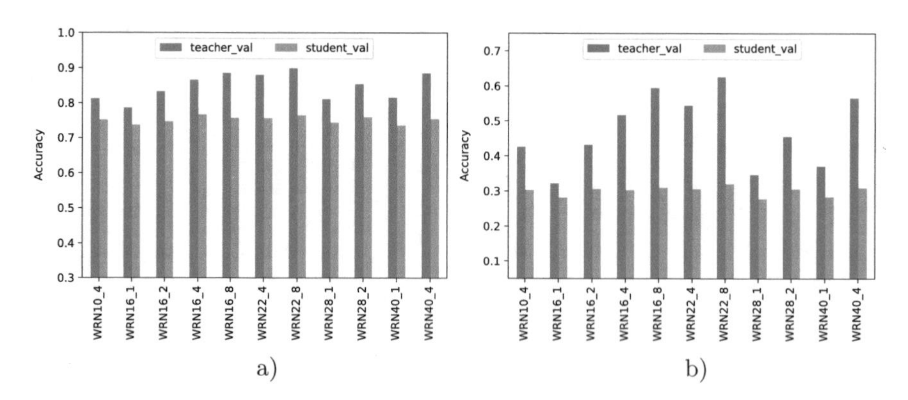

Fig. 7 The validation accuracy values for the teacher and student models from the wide ResNet model family (details in Table 5) for CIFAR10 (**a**) CIFAR100 (**b**) datasets

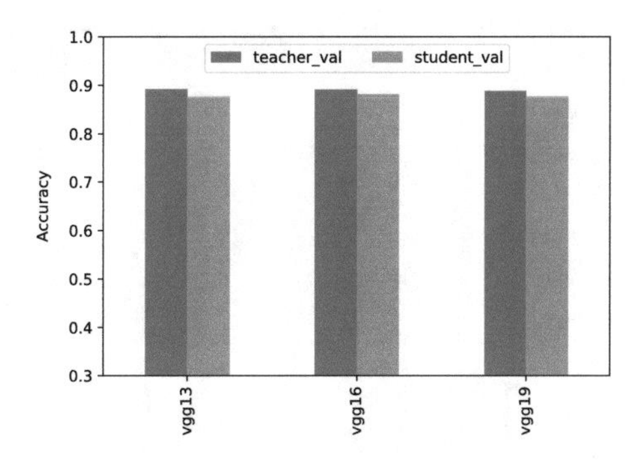

Fig. 8 The validation accuracy values for the VGG models used (details in Table 6)

with the maximal teacher validation accuracy value $A_t = 0.892$ (Fig. 8, Table 6). The lowest validation values A_s and A_t for CIFAR10 dataset were observed for the biggest VGG19 teacher model.

4.2 Specific MedMNIST Medical Datasets

4.2.1 3 Layer Weak Models

The 3 layer weak ResNet models demonstrated the quite different behavior as to the teacher validation accuracy values (A_t in Table 7) and the student validation accuracy values (A_s in Table 7) for various specific MedMNIST medical datasets (Fig. 9). For the BloodMNIST dataset the ResNet32_sm demonstrated the maximal $A_s = 0.938$ and $A_t = 0.948$ values. For the DermaMNIST dataset the other ResNet14_sm teacher model demonstrated the maximal $A_s = 0.732$ and $A_t = 0.744$ values. For the PathMNIST dataset also the other ResNet20_sm teacher model demonstrated the maximal $A_s = 0.898$ and $A_t = 0.893$ values. For the RetinaMNIST dataset the different models demonstrated the maximal $A_s = 0.558$ for ResNet110_sm and the maximal $A_t = 0.548$ for the other ResNet32_sm teacher model. The lowest validation values A_s and A_t were observed for the biggest ResNet164_sm teacher model.

4.2.2 3 Layer Models

The 3 layer ResNet models demonstrated the similar (to the previous model family) behavior as to the teacher validation accuracy values (A_t in Table 8) and the student validation accuracy values (A_s in Table 8) for various specific MedMNIST medical datasets (Fig. 10). For the BloodMNIST dataset the ResNet32 teacher model demonstrated the maximal $A_s = 0.952$ value and the ResNet26 teacher model did the maximal $A_t = 0.961$ value. For the DermaMNIST dataset the same ResNet32 teacher model demonstrated the maximal $A_s = 0.776$ and $A_t = 0.775$ values. For

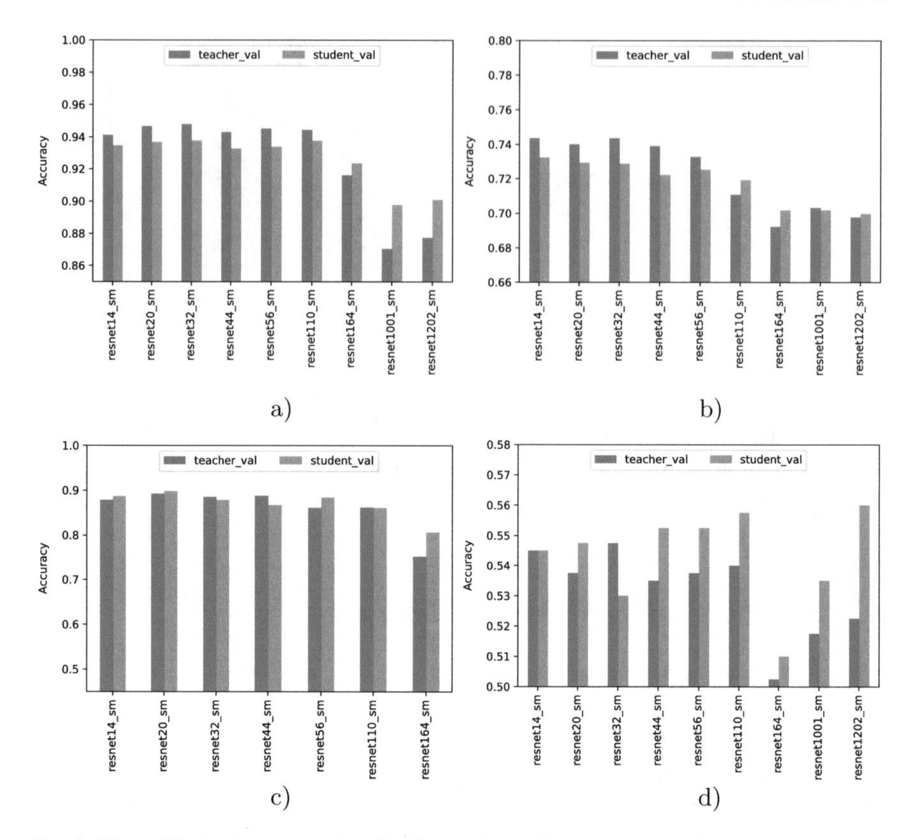

Fig. 9 The validation accuracy values for the teacher and student models from the 3 layer weak ResNet model family for BloodMNIST (**a**), DermaMNIST (**b**), PathMNIST (**c**), and RetinaMNIST (**d**) datasets (details in Table 7)

Table 7 Teacher (A_t) and student (A_s) validation accuracy values for the 3 layer weak ResNet models used for MedMNIST datasets

	Blood		Derma		Path		Retina	
Teacher model	A_s	A_t	A_s	A_t	A_s	A_t	A_s	A_t
ResNet8_sm	0.918	0.918	0.721	0.724	0.891	0.889	0.535	0.550
ResNet14_sm	0.935	0.941	**0.732**	**0.744**	0.887	0.878	0.545	0.545
ResNet20_sm	0.937	0.947	0.729	0.740	**0.898**	**0.893**	0.548	0.538
ResNet32_sm	**0.938**	**0.948**	0.729	**0.744**	0.878	0.885	0.530	**0.548**
ResNet44_sm	0.933	0.943	0.722	0.739	0.867	0.888	0.553	0.535
ResNet56_sm	0.934	0.945	0.725	0.733	0.884	0.861	0.553	0.538
ResNet110_sm	**0.938**	0.945	0.719	0.711	0.861	0.862	**0.558**	0.540
ResNet164_sm	*0.923*	*0.916*	*0.702*	*0.692*	*0.806*	*0.752*	*0.510*	*0.503*

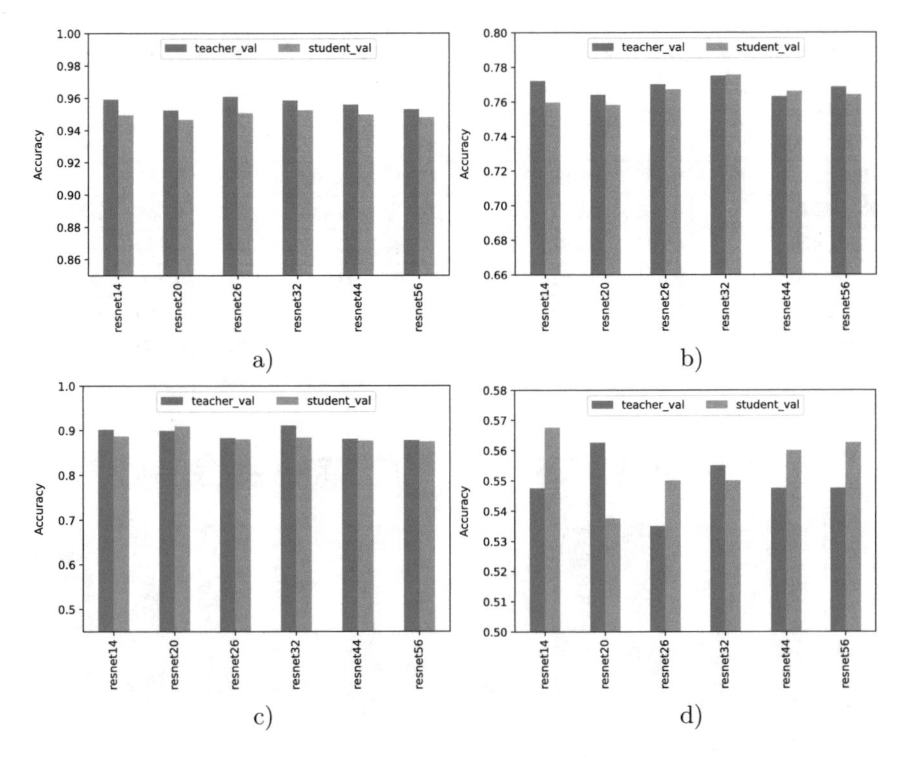

Fig. 10 The validation accuracy values for the teacher and student models from the 3 layer ResNet model family for BloodMNIST (**a**), DermaMNIST (**b**), PathMNIST (**c**), and RetinaMNIST (**d**) datasets (details in Table 8)

the PathMNIST dataset the other ResNet20 teacher model demonstrated the maximal $A_s = 0.909$ and $A_t = 0.911$ values. For the RetinaMNIST dataset the different models demonstrated the maximal $A_s = 0.568$ for ResNet14 and the maximal $A_t = 0.563$ for the other ResNet20 teacher model. The lowest validation values A_s and A_t were observed for the various teacher models: ResNet20 (BloodMNIST, DermaMNIST, RetinaMNIST), ResNet44 (PathMNIST).

4.2.3 4 Layer Models

For the BloodMNIST dataset the ResNet50 teacher model demonstrated the maximal $A_s = 0.961$ value and the ResNet101 teacher model did the maximal $A_t = 0.958$ value. For the DermaMNIST dataset the same ResNet18 teacher model demonstrated the maximal $A_s = 0.765$ and $A_t = 0.763$ values. For the PathMNIST dataset the other ResNet34 teacher model demonstrated the maximal $A_s = 0.915$ and ResNet18 teacher model—for $A_t = 0.911$ values. For the RetinaMNIST dataset also the different models demonstrated the maximal $A_s = 0.568$ for ResNet18 and the maximal

Table 8 Teacher (A_t) and student (A_s) validation accuracy values for the 3 layer ResNet models used for MedMNIST datasets

3 layer	Blood		Derma		Path		Retina	
Teacher model	A_s	A_t	A_s	A_t	A_s	A_t	A_s	A_t
ResNet8	0.946	0.950	0.756	0.754	0.898	0.896	0.545	0.543
ResNet14	0.949	0.959	0.760	0.772	0.886	0.902	**0.568**	0.548
ResNet20	*0.947*	*0.952*	*0.758*	0.764	**0.909**	0.899	*0.538*	**0.563**
ResNet26	0.951	**0.961**	0.767	0.770	0.880	0.883	0.550	*0.535*
ResNet32	**0.952**	0.959	**0.776**	**0.775**	0.883	**0.911**	0.550	0.555
ResNet44	0.950	0.956	0.766	*0.763*	0.877	0.881	0.560	0.548
ResNet56	0.948	0.953	0.764	0.769	*0.875*	*0.878*	0.563	0.548

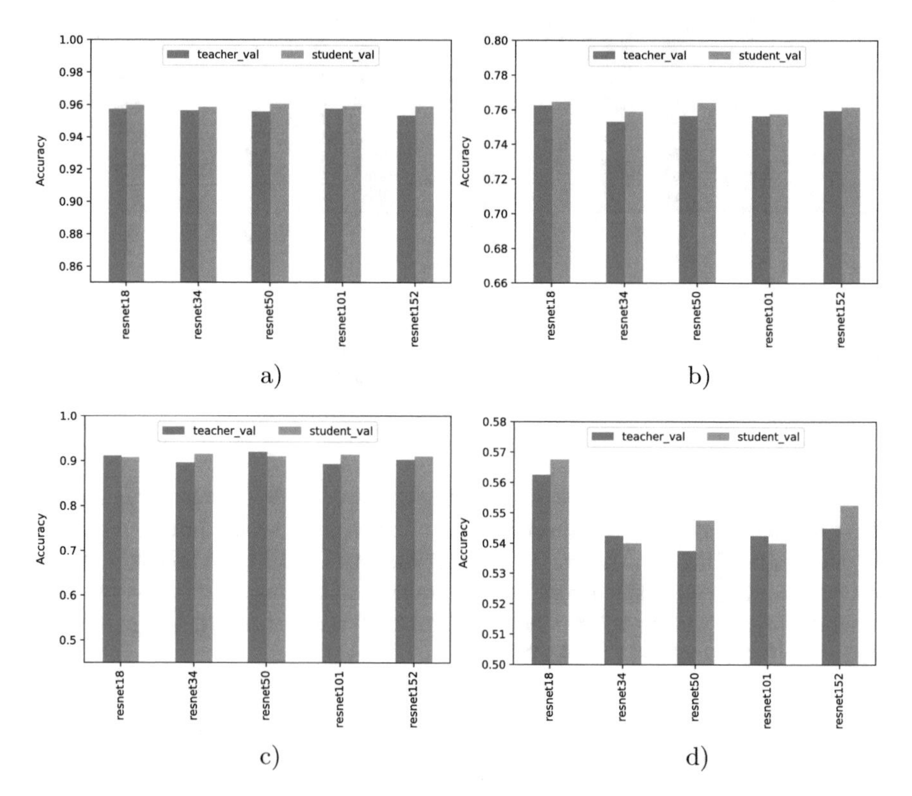

Fig. 11 The validation accuracy values for the teacher and student models from the 4 layer ResNet model family for BloodMNIST (**a**), DermaMNIST (**b**), PathMNIST (**c**), and RetinaMNIST (**d**) datasets (details in Table 9)

Table 9 Teacher (A_t) and student (A_s) validation accuracy values for the 4 layer ResNet models used for MedMNIST datasets

4 layer	Blood		Derma		Path		Retina	
Teacher model	A_s	A_t	A_s	A_t	A_s	A_t	A_s	A_t
ResNet10	0.961	0.958	0.766	0.766	0.898	0.896	0.545	0.543
ResNet18	0.960	0.957	**0.765**	**0.763**	*0.901*	**0.911**	**0.568**	0.548
ResNet34	*0.959*	0.956	0.759	*0.753*	**0.915**	0.896	*0.538*	**0.563**
ResNet50	**0.961**	0.956	0.764	0.757	0.910	0.920	0.550	*0.535*
ResNet101	*0.959*	**0.958**	*0.758*	0.757	0.914	*0.893*	0.550	0.555
ResNet152	*0.959*	*0.953*	0.762	*0.760*	0.910	0.902	0.560	0.548

Table 10 Teacher (A_t) and student (A_s) validation accuracy values for the torchvision ResNet models used for MedMNIST datasets

torchvision	Blood		Derma		Path		Retina	
Teacher model	A_s	A_t	A_s	A_t	A_s	A_t	A_s	A_t
ResNet8_v	0.928	0.928	0.750	0.752	0.829	0.821	0.575	0.548
ResNet10_v	**0.931**	**0.949**	0.745	**0.766**	**0.825**	**0.845**	**0.553**	0.525
ResNet14_v	0.923	0.924	**0.755**	0.747	0.813	0.811	**0.553**	**0.550**
ResNet18_v	0.927	0.937	0.751	0.760	0.812	0.815	0.533	0.523
ResNet20_v	0.928	0.918	0.753	0.748	0.811	0.786	0.540	0.530
ResNet34_v	0.926	0.929	0.737	0.730	0.818	0.807	0.548	0.540
ResNet50_v	0.896	0.864	0.723	0.718	0.801	0.721	0.545	0.500
ResNet101_v	0.867	0.811	0.719	0.690	*0.765*	*0.486*	0.515	0.493
ResNet152_v	*0.848*	*0.627*	*0.698*	*0.671*	0.789	0.591	*0.435*	*0.435*

$A_t = 0.563$ for the other ResNet34 teacher model. The lowest validation values A_s and A_t were scattered non-systematically among all models: ResNet18 (PathMNIST), ResNet34 (BloodMNIST, DermaMNIST, RetinaMNIST), ResNet50 (RetinaMNIST), ResNet101 (RetinaMNIST), ResNet152 (BloodMNIST, DermaMNIST) (Fig. 11).

4.2.4 Torchvision Models

For all MedMNIST datasets the smallest ResNet10_v and ResNet14_v teacher models demonstrated the maximal A_s and A_t values (Table 10). In the similar way, for all MedMNIST datasets the largest ResNet101_v and ResNet152_v teacher models demonstrated the lowest A_s and A_t values (Table 10, Fig. 12).

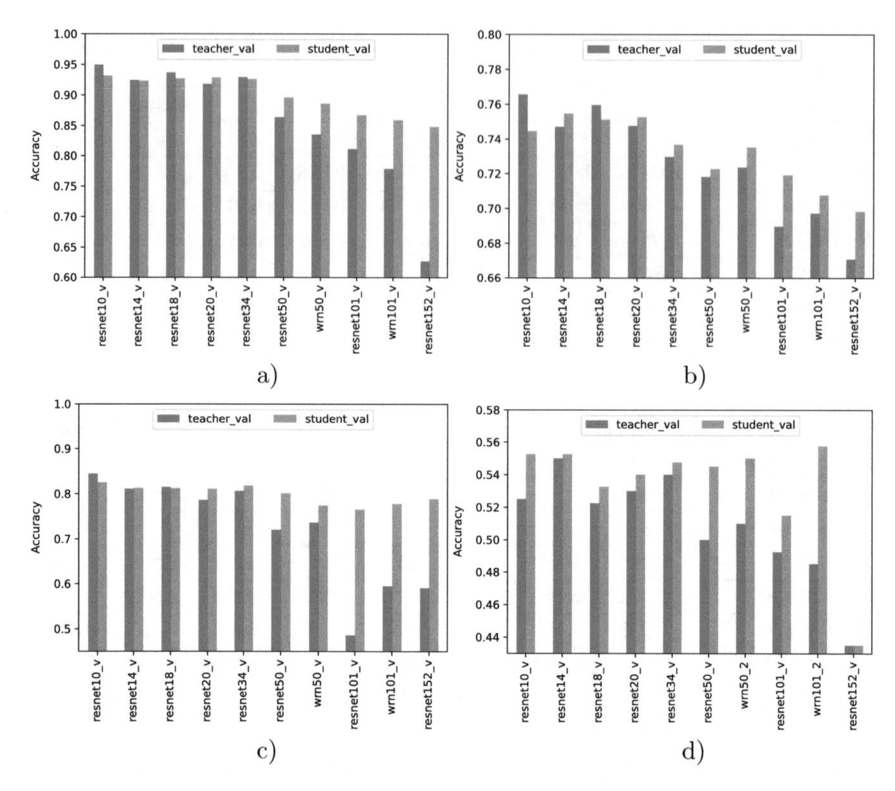

Fig. 12 The validation accuracy values for the teacher and student models from the torchvision ResNet model family for BloodMNIST (**a**), DermaMNIST (**b**), PathMNIST (**c**), and RetinaMNIST (**d**) datasets (details in Table 10)

4.2.5 Wide Models

For all of MedMNIST datasets (except of partially DermaMNIST and RetinaMNIST) the biggest WRN22_8 teacher model demonstrated the maximal A_s and A_t values (Table 11). The lowest validation values A_s and A_t were scattered mainly among the smaller models: WRN16_1 (BloodMNIST, DermaMNIST), WRN40_1 (DermaMNIST), WRN16_2 (PathMNIST), WRN10_4 (DermaMNIST, RetinaMNIST) (Fig. 13).

4.2.6 VGG Models

The best and worst teacher validation accuracy values (A_t in Table 12) and the student validation accuracy values (A_s in Table 12) were scattered non-systematically among all models (Fig. 14, Table 12).

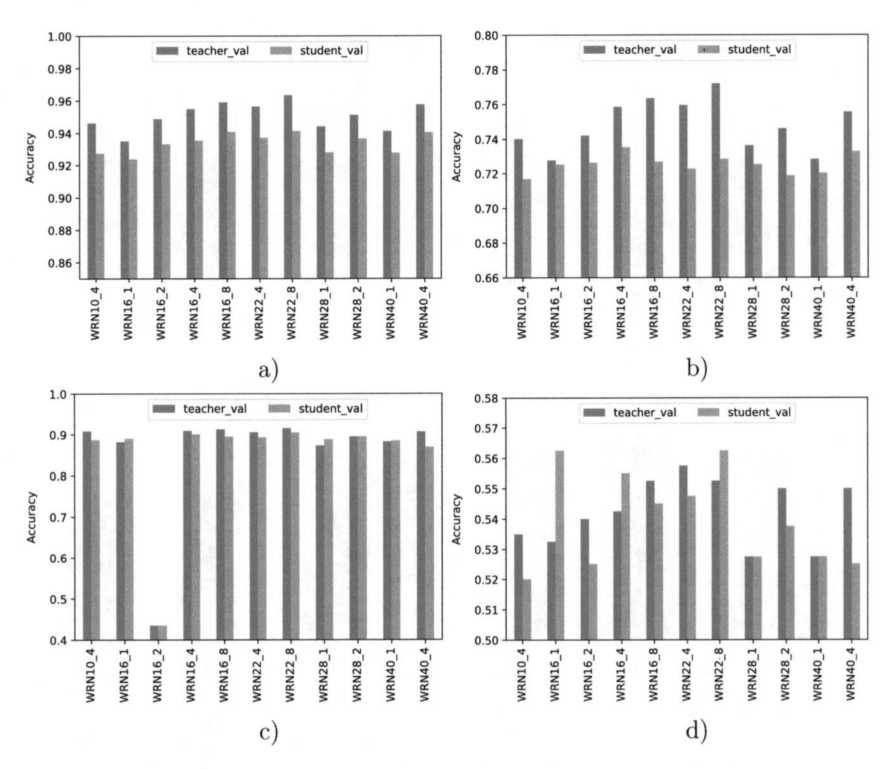

Fig. 13 The validation accuracy values for the teacher and student models from the wide ResNet model family for BloodMNIST (**a**), DermaMNIST (**b**), PathMNIST (**c**), and RetinaMNIST (**d**) datasets (details in Table 11)

Table 11 Teacher (A_t) and student (A_s) validation accuracy values for the wide ResNet models used for MedMNIST datasets

Wide	Blood		Derma		Path		Retina	
Teacher model	A_s	A_t	A_s	A_t	A_s	A_t	A_s	A_t
WRN10_1	0.914	0.918	0.711	0.716	0.435	0.435	0.538	0.543
WRN16_1	*0.924*	*0.935*	0.725	*0.728*	0.890	0.882	**0.563**	0.533
WRN28_1	0.928	0.944	0.725	0.736	0.888	0.873	0.523	*0.523*
WRN40_1	0.928	0.941	0.720	*0.728*	0.885	0.883	0.528	0.528
WRN16_2	0.933	0.949	0.726	0.742	*0.435*	*0.435*	0.525	0.540
WRN28_2	0.937	0.951	0.719	0.746	0.895	0.895	0.538	0.550
WRN10_4	0.927	0.946	*0.717*	0.740	0.887	0.908	*0.520*	0.535
WRN16_4	0.935	0.955	**0.735**	0.759	0.900	0.909	0.555	0.543
WRN22_4	0.937	0.956	0.723	0.760	0.893	0.905	0.548	**0.558**
WRN40_4	0.940	0.958	0.733	0.756	0.869	0.907	0.525	0.550
WRN16_8	**0.941**	0.959	0.727	0.764	0.895	0.913	0.545	0.553
WRN22_8	**0.941**	**0.963**	0.728	**0.772**	**0.905**	**0.916**	**0.563**	0.553

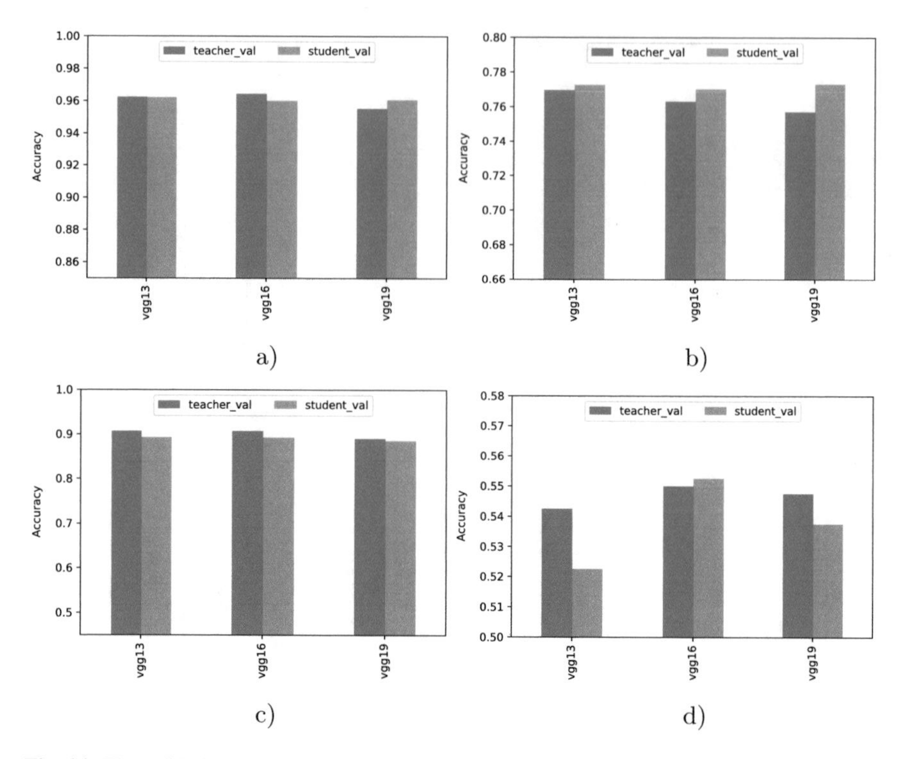

Fig. 14 The validation accuracy values for the teacher and student models from the VGG model family for BloodMNIST (**a**), DermaMNIST (**b**), PathMNIST (**c**), and RetinaMNIST (**d**) datasets (details in Table 12)

Table 12 Teacher (A_t) and student (A_s) validation accuracy values for the VGG models used for MedMNIST datasets

VGG	Blood		Derma		Path		Retina	
Teacher model	A_s	A_t	A_s	A_t	A_s	A_t	A_s	A_t
vgg11	0.962	0.964	0.778	0.769	0.894	0.889	0.538	0.540
vgg13	**0.963**	*0.962*	**0.893**	**0.907**	**0.890**	0.882	*0.522*	*0.543*
vgg16	*0.960*	**0.964**	0.892	**0.907**	0.888	*0.873*	**0.553**	**0.550**
vgg19	0.961	0.955	*0.773*	*0.757*	*0.885*	**0.890**	0.538	0.548

5 Discussion

5.1 Models-in-Family Comparison

To summarize the results obtained the maximal and minimal values of the teacher (A_t) and student (A_s) validation accuracy values were determined for each model family for all of the models used for CIFAR10, CIFAR100, and MedMNIST datasets.

It should be noted that for the family of 3 layer weak ResNet models trained on CIFAR10/CIFAR100 datasets the best teacher models (that allow to get the highest student validation accuracy values) were ResNet32_sm, ResNet44_sm, ResNet56_sm and they were not the biggest ones, and the worst teacher model is the biggest and more complicated ResNet164_sm (Table 1).

For the family of 3 layer ResNet models trained on CIFAR10/CIFAR100 datasets the best teacher models were ResNet32_sm, ResNet44_sm, ResNet56_sm and they were not the biggest ones, and the worst teacher model is the biggest and more complicated ResNet164_sm (Table 2).

For the family of 4 layer ResNet models trained on CIFAR10/CIFAR100 datasets the best teacher models were ResNet101, ResNet152 and they were the biggest ones, and the worst teacher model is the smallest and simplest ResNet18 and ResNet34 (Table 3).

For the family of torchvision ResNet models trained on CIFAR10/CIFAR100 datasets the best teacher models were ResNet18_v (and partially ResNet18_v), and the worst teacher model is the largest and most complex ResNet152_v (Table 4).

For the family of wide ResNet models trained on CIFAR10/CIFAR100 datasets the best teacher models was WRN22_8 and it was the biggest one, and the worst teacher model was the smallest and simplest WRN16_1 (and partially for WRN28_1) (Table 5).

For the family of VGG models trained on CIFAR10 dataset the best teacher model were VGG13, VGG16 and they were not the biggest ones, and the worst teacher model is the biggest and more complicated VGG19 (Table 6).

As to the trials on MedMNIST datasets it should be noted that for the all families (except for wide) of ResNet models trained on MedMNIST datasets the best teacher models (that allow to get the highest student validation accuracy values) were different for different MedMNIST datasets in contrast to the best models for CIAF10/CIFAR100 datasets where they were the same inside family. For the family of wide ResNet models trained on MedMNIST datasets the best teacher models was WRN22_8 and it was the biggest one, and the worst teacher models were mainly among the smallest and simplest models that is similar to the best models for CIAF10/CIFAR100 datasets where they were the same inside family (Table 5).

5.2 Family-to-Family Comparison

For the general comparison of the "bucket of models" ensemble technique the best models for the same datasets from different families were collected and presented on the same plots and tables for teachers (Fig. 15, Table 13) and students (Fig. 16, Table 14).

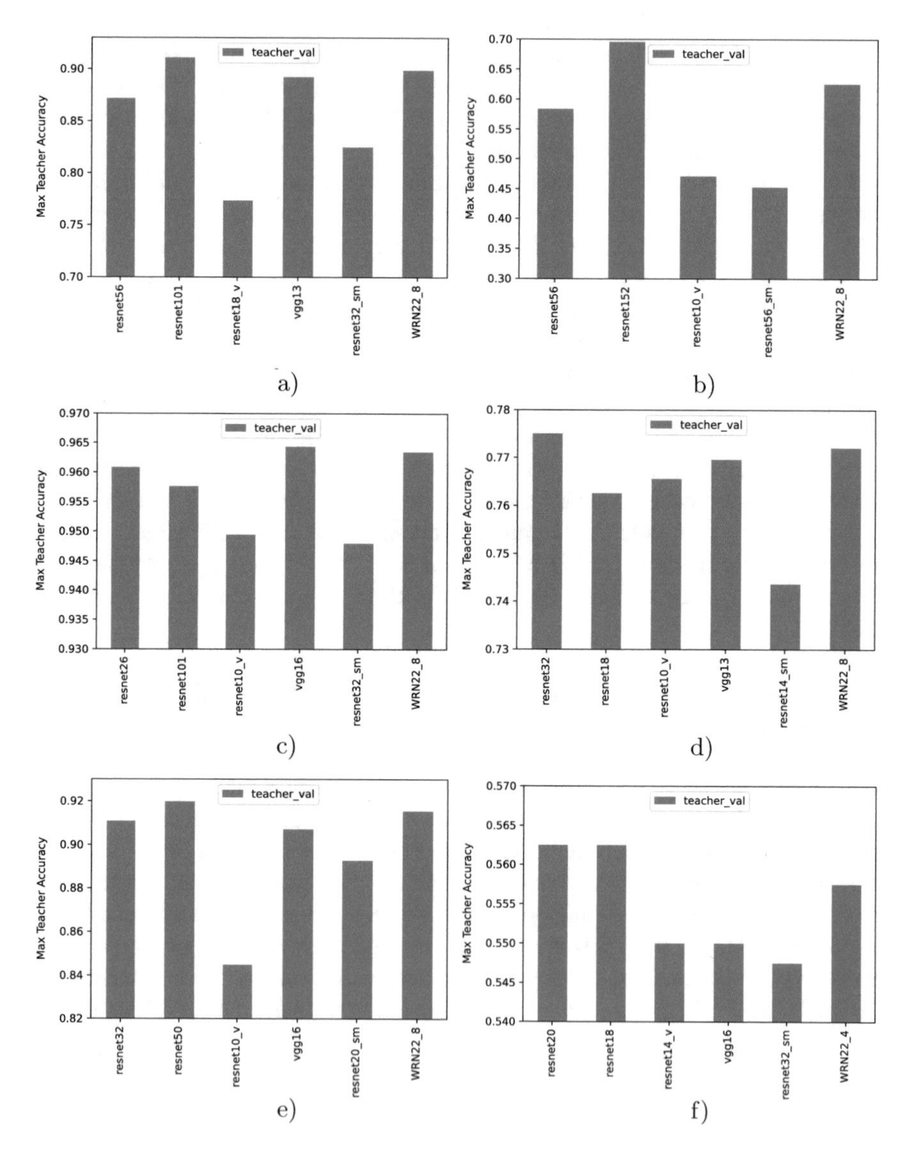

Fig. 15 The max-per-family teacher (A_t) validation accuracy values for the best ensemble teacher models for CIFAR10 (**a**), CIFAR100 (**b**), BloodMNIST (**c**), DermaMNIST (**d**), PathMNIST (**e**), and RetinaMNIST (**f**) datasets (details in Table 13)

Table 13 The max-per-family teacher (A_t) validation accuracy values for all model sets (the absolutely highest values per dataset were emphasized by a **bold** font)

Teacher model	CIFAR10	CIFAR100	Blood	Derma	Path	Retina
3 layer						
resnet56	0.872	0.584	–	–	–	–
resnet26	–	–	0.961	–	–	–
resnet32	–	–	–	**0.775**	0.911	–
resnet20	–	–	–	–	–	0.563
4 layer						
resnet101	**0.911**	–	0.958	–	–	–
resnet152	–	**0.696**	–	–	–	–
resnet18	–	–	–	0.763	–	0.563
resnet50	–	–	–	–	**0.920**	–
Torchvision						
resnet18_v	0.773	–	–	–	–	–
resnet10_v	–	0.471	0.949	0.766	0.845	–
resnet14_v	–	–	–	–	–	0.550
3 layer weak						
resnet32_sm	0.825	–	0.948	–	–	0.548
resnet56_sm	–	0.453	–	–	–	–
resnet14_sm	–	–	–	0.744	–	–
resnet20_sm	–	–	–	–	0.893	–
Wide						
WRN22_8	0.899	0.626	**0.963**	0.772	0.916	–
WRN22_4	–	–	–	–	–	**0.558**
VGG (for comparison)						
vgg13	0.892	na	–	0.770	–	–
vgg16	–	na	0.964	–	0.907	0.550

The max-per-family teacher (A_t) validation accuracy values for the best per family teacher models (Fig. 15, Table 13) were scattered among all model families considered here. The best performance teacher models for CIFAR10/CIFAR100 datasets were not the best performance teacher models for MedMNIST datasets. It should be noted that 4 layer family demonstrated the absolutely best performance models for the standard CIFAR10/CIFAR100 datasets, but they were not so good for any other specific medical MedMNIST dataset. Moreover, ResNet50 was max-per-family teacher for PathMNIST dataset, but not the max-per-family teacher for CIFAR10/CIFAR100 datasets.

The similar situation was observed for the max-per-family student (A_s) validation accuracy values for the best ensemble teacher models (Fig. 16, Table 14) that were scattered among all model families considered here. The best performance teacher

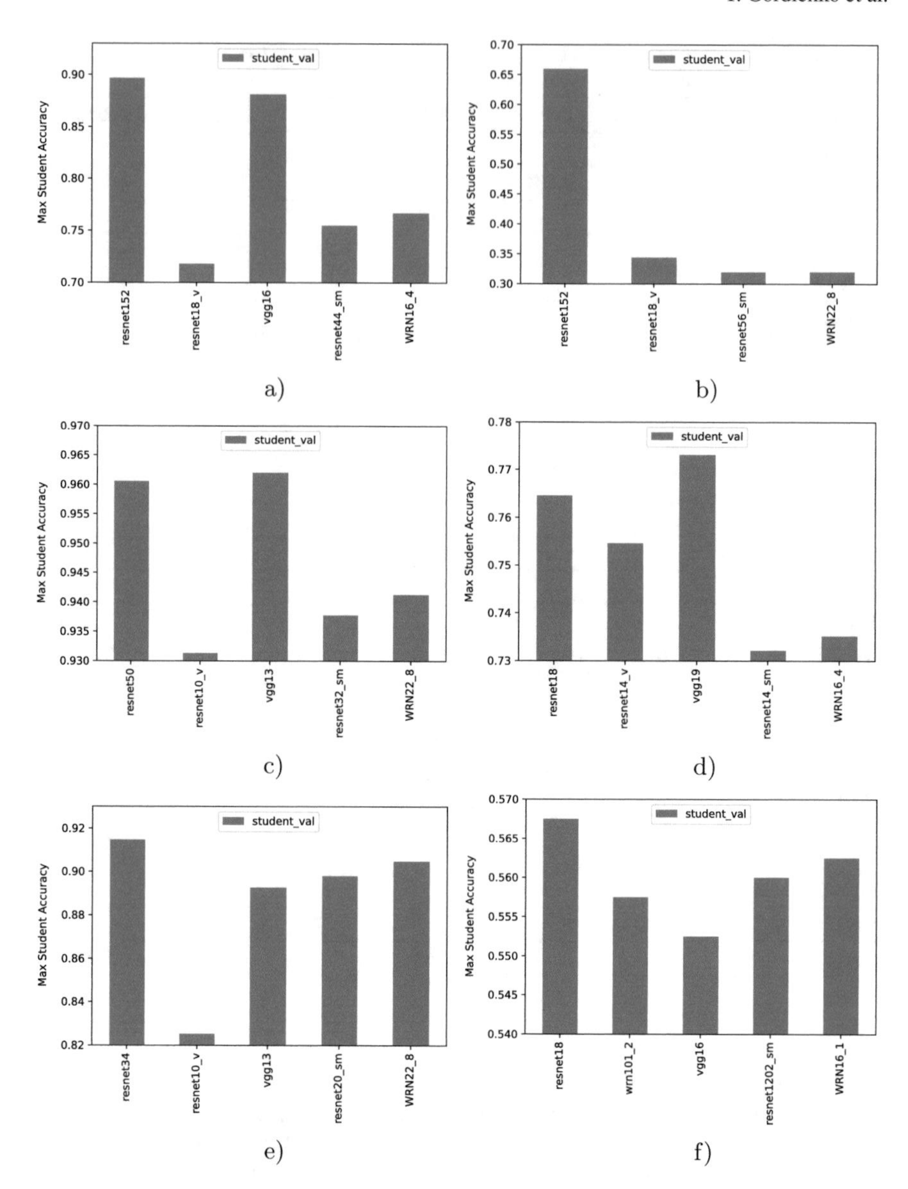

Fig. 16 The max-per-family student (A_s) validation accuracy values for the best ensemble teacher models for CIFAR10 (**a**), CIFAR100 (**b**), BloodMNIST (**c**), DermaMNIST (**d**), PathMNIST (**e**), and RetinaMNIST (**f**) datasets (details in Table 14)

Table 14 The max-per-family student (A_s) validation accuracy values for all model sets (the absolutely highest values per dataset were emphasized by a **bold** font)

Teacher model	CIFAR10	CIFAR100	Blood	Derma	Path	Retina
3 layer						
resnet32	0.830	–	0.952	**0.776**	–	–
resnet56	–	0.510	–	–	–	–
resnet20	–	–	–	–	0.909	–
resnet14	–	–	–	–	–	**0.568**
4 layer						
resnet152	**0.897**	**0.660**	–	–	–	–
resnet50	–	–	**0.961**	–	–	–
resnet18	–	–	–	0.765	–	**0.568**
resnet34	–	–	–	–	**0.915**	–
Torchvision						
resnet18_v	0.718	0.344	–	–	–	–
resnet10_v	–	–	0.931	–	0.825	–
resnet14_v	–	–	–	0.755	–	–
wrn101_2	–	–	–	–	–	0.558
3 layer weak						
resnet44_sm	0.755	–	–	–	–	–
resnet56_sm	–	0.320	–	–	–	–
resnet32_sm	–	–	0.938	–	–	–
resnet14_sm	–	–	–	0.732	–	–
resnet20_sm	–	–	–	–	0.898	–
resnet1202_sm	–	–	–	–	–	0.560
Wide						
WRN16_4	0.767	–	–	0.735	–	–
WRN22_8	–	0.320	0.941	–	0.905	–
WRN16_1	–	–	–	–	–	0.563
VGG (for comparison)						
vgg16	0.881	na	–	–	–	0.553
vgg13	–	na	0.962	–	0.893	–
vgg19	–	na	–	0.773	–	–

models for the standard CIFAR10/CIFAR100 datasets were not the best performance teacher models for MedMNIST datasets. It should be noted that 4 layer family contained the largest number of the absolutely best performance models the standard CIFAR10/CIFAR100 datasets and for the specific medical MedMNIST datasets. The other two absolutely best performance models were obtained in 3 layer family: ResNet32 for DermaMNIST and ResNet14 for RetinaMNIST.

The originality and importance of these results for the practical medical applications on Edge Intelligence devices with the limited computational abilities can be summarized in the following concluding observations as to the best teacher and student models trained and tested on CIFAR10/CIFAR100 and MedMNIST datasets (Table 13):

1. no relationship was found between performance obtained on CIFAR10/CIFAR100 datasets and performance obtained on MedMNIST datasets,
2. the choice of model family effects performance more than some model within a family,
3. the significant boost in performance can be obtained by smaller models,
4. some architectures can be efficient for one dataset, but can be inefficient for the others,
5. some architectures with the high performance obtained on CIFAR10/CIFAR100 datasets were unnecessarily large for MedMNIST and can be made more parameter-efficient (smaller) without a significant drop in the performance.

The formulations of these observations were prepared in the manner of the similar results and conclusions that were obtained for the quite different practical medical use case, namely, for chest X-ray interpretation by application of the DL models developed and pretrained for ImageNet [25] for the large chest X-ray dataset (CheXpert).

These concluding observations are especially evident after comparing not absolute, but relative values, namely, accuracy changes (increases and decreases) for various models and datasets versus the increase of the number of parameters P_t/P_s, where P_t—is the number of parameters in the teacher model, and P_s—is the number of parameters in the student model (Fig. 17).

1. For example, the increase of performance obtained on CIFAR10/CIFAR100 datasets is not always followed by the increase of performance obtained on MedMNIST datasets, moreover it can be followed by the decrease of performance like for PathMNIST dataset (green color in Fig. 17a).
2. Also the choice of model family like wide ResNet for PathMNIST dataset (green color in Fig. 17c) results in the much higher performance than the choice of some model within other families.
3. The significant increase of performance can be obtained by the smaller models like for the small teacher models even in comparison to the bigger teacher models, for example in the wide ResNet family for PathMNIST dataset (green color in Fig. 17c).
4. Some architectures, for example, wide ResNet family, can be efficient for one dataset like PathMNIST dataset (green color in Fig. 17c), but can be inefficient for the others like RetinaMNIST dataset (red color in Fig. 17c). And in reverse, the other architecture, for example, 3 layer ResNet family (Fig. 17a), can be inefficient for one dataset like PathMNIST dataset (green color in Fig. 17a), but can be efficient for the others like RetinaMNIST dataset (red color in Fig. 17a).
5. Some architectures allowed us to get the higher performance on CIFAR10 and CIFAR100 datasets for the larger models (Fig. 17), but these models were unnecessarily large for MedMNIST datasets and can be made more parameter-efficient

(smaller) without a significant decrease of the performance values, for example, resnet20 gives higher A_s values than bigger models like resnet32, resnet56 from 3 layer ResNet family (Table 14, Fig. 17a).

The possible reasons for such results can be explained by the relatively low complexity of the standard datasets like CIFAR10/CIFAR100 in comparison to specific medical datasets like MedMNIST and their original source images. The advantages of distillation in terms of computing overhead are especially crucial for usage of small models (with decay of the model performance with decrease of model sizes) on Edge Computing devices in various applications described in our previous publications [10, 11, 45]. In this connection, the question about data complexity estimation should be arisen and researched by taking into account some complexity metrics. For example, new or standard metrics can be applied like some image structural similarity index measure (SSIM) [50], multiscale SSIM (MS-SSIM) [51], complex wavelet transform variant of the SSIM (CW-SSIM) [49], and other measures used in other DL-related fields like image super resolution, image compression, image restoration, etc. [48]. In fact, it means that these findings can have the more general original and importance with regard to other and not only medical applications. In this context, the future more detailed and thorough similar investigations should be carried out on the original data of MedMNIST datasets.

6 Conclusions

The "bucket of models" ensemble technique for search of the best-per-family model and then the best-among-families model of some DNN architecture was applied to choose the absolutely best model for each separate medical problem. It is based on usage of the previously obtained knowledge distillation (KD) approaches where an ensemble of "student" DNNs can be trained with regard to set of various flavours of teacher family with the same architecture.

The bucket of teacher and small-scale student models was applied to various medical data from quite different medical domains. The main purpose was to check feasibility of such approaches for the practical medical datasets for Edge Intelligence devices with the limited computational abilities. From our experience the reliable small-scale models are crucial for porting these models to Edge Computing devices with low computing resources [10, 12, 45].

The training and validation runs were performed on the standard datasets like CIFAR10/CIFAR100 in comparison to the specific medical datasets like MedMNIST including medical data of the quite various types and complexity including BloodMNIST dataset with images of blood anaysis, DermaMNIST dataset with dermatoscopic images, PathMNIST dataset with histological images, and RetinaMNIST dataset with retina fundus images. They were selected because of their standard nature and possibility to reproduce and compare the results obtained on image classification problem.

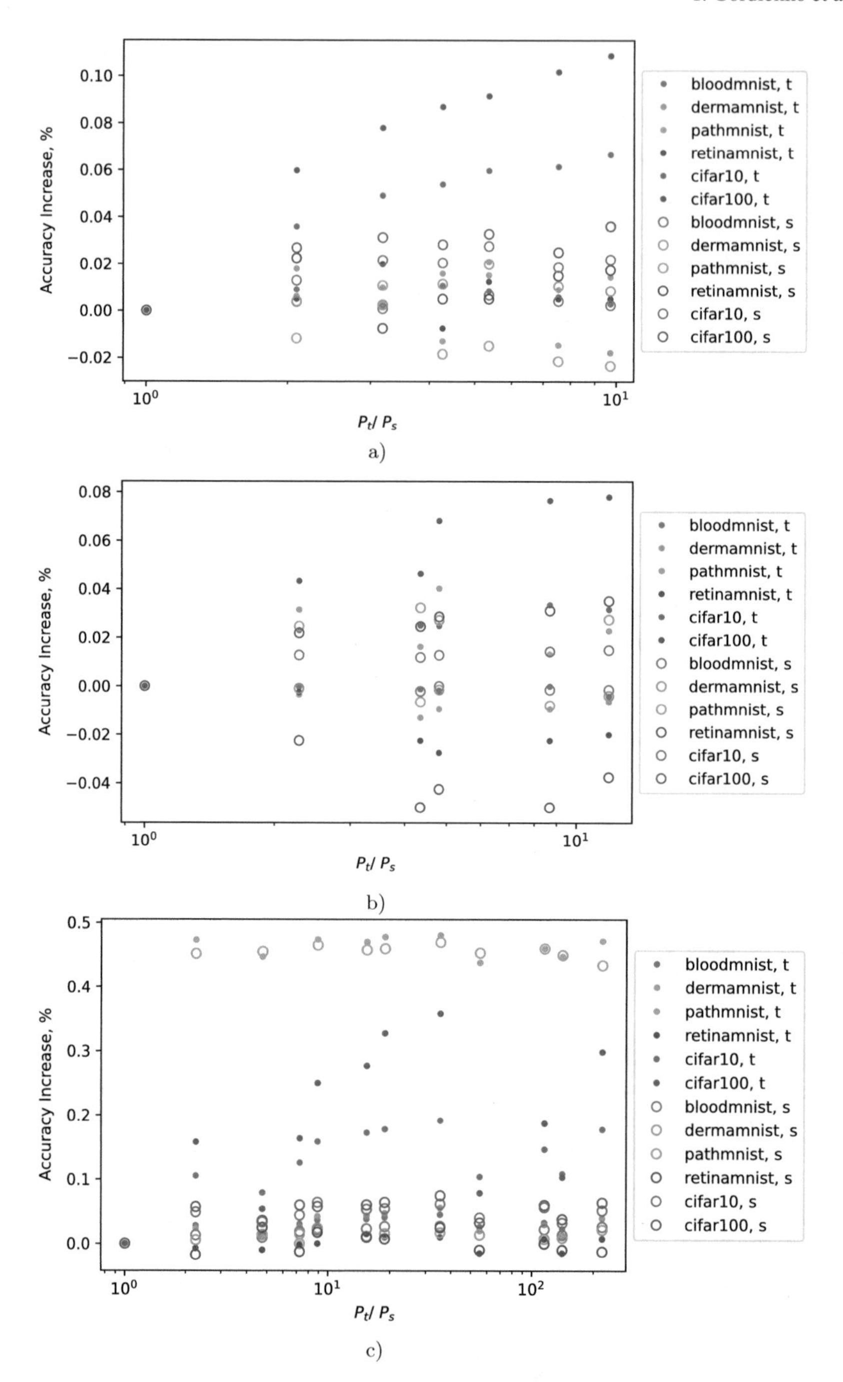

Fig. 17 The increase of student (A_s) and teacher (A_t) validation accuracy values for the following families of models: 3 layer ResNet (**a**), 4 layer ResNet (**b**), and wide ResNet. Legend contains the notation "*dataset, model*", where *model* = "s" is for a student model, and *model* = "t" is for a teacher model

Several families of ResNet DNN architecture families were used and they demonstrated various performance on the standard CIFAR10/CIFAR100 and specific medical MedMNIST datasets.

As a result, no relationship was found between CIFAR10/CIFAR100 performance and MedMNIST performance, the choice of model family effects performance more than some model within a family, the significant boost in performance can be obtained by smaller models, some powerful CIFAR10/CIFAR100 architectures were unnecessarily large for MedMNIST and can be made more parameter-efficient (smaller) without a significant drop in performance.

In the future research the more specific combination of distillation approaches is planned to be used with taking into account the finer hyperparameter tuning, data augmentation procedures, and their additional influences on the original data of MedMNIST datasets.

That is why these results should be verified for larger datasets and with more attention to the contribution of other architectures and original medical data (and not their small versions used here). Also, the next step of this work should be dedicated to investigation of other DNN families in order to obtain more data and increase the robustness of the measures obtained. In general, such studies can potentially hold promise for classification of various medical data The proposed methods can be used in the development of new Edge Intelligence devices and applications for various medical purposes under field conditions with the limited computing resources.

Acknowledgements This work was partially supported by "Knowledge At the Tip of Your fingers: Clinical Knowledge for Humanity" (KATY) project funded from the European Union's Horizon 2020 research and innovation program under grant agreement No. 101017453.

References

1. Acevedo, A., Merino, A., Alférez, S., Molina, Á., Boldú, L., Rodellar, J.: A dataset of microscopic peripheral blood cell images for development of automatic recognition systems. Data Brief **30** (2020)
2. Alienin, O., Rokovyi, O., Gordienko, Y., Kochura, Y., Taran, V., Stirenko, S.: Artificial intelligence platform for distant computer-aided detection (cade) and computer-aided diagnosis (cadx) of human diseases. In: The International Conference on Artificial Intelligence and Logistics Engineering, pp. 91–100. Springer (2022)
3. Asif, U., Tang, J., Harrer, S.: Ensemble knowledge distillation for learning improved and efficient networks (2019). arXiv:1909.08097
4. Bradley, A.P.: The use of the area under the roc curve in the evaluation of machine learning algorithms. Pattern Recogn. **30**(7), 1145–1159 (1997)
5. Chen, Y.W., Jain, L.C.: Deep Learning in Healthcare. Springer (2020)
6. Codella, N., Rotemberg, V., Tschandl, P., Celebi, M.E., Dusza, S., Gutman, D., Helba, B., Kalloo, A., Liopyris, K., Marchetti, M., et al.: Skin lesion analysis toward melanoma detection 2018: a challenge hosted by the international skin imaging collaboration (isic) (2019). arXiv:1902.03368
7. DiPalma, J., Suriawinata, A.A., Tafe, L.J., Torresani, L., Hassanpour, S.: Resolution-based distillation for efficient histology image classification. Artif. Intell. Med. **119**, 102136 (2021)

8. Esteva, A., Robicquet, A., Ramsundar, B., Kuleshov, V., DePristo, M., Chou, K., Cui, C., Corrado, G., Thrun, S., Dean, J.: A guide to deep learning in healthcare. Nat. Med. **25**(1), 24–29 (2019)
9. Fukushima, K.: Neural network model for a mechanism of pattern recognition unaffected by shift in position-neocognitron. IEICE Tech. Rep., A **62**(10), 658–665 (1979)
10. Gordienko, Y., Kochura, Y., Taran, V., Gordienko, N., Rokovyi, A., Alienin, O., Stirenko, S.: Scaling analysis of specialized tensor processing architectures for deep learning models. In: Deep Learning: Concepts and Architectures, pp. 65–99. Springer (2020)
11. Gordienko, Y., Kochura, Y., Taran, V., Gordienko, N., Rokovyi, O., Alienin, O., Stirenko, S.: "last mile" optimization of edge computing ecosystem with deep learning models and specialized tensor processing architectures. In: Advances in Computers, vol. 122, pp. 303–341. Elsevier (2021)
12. Gordienko, Y., Kochura, Y., Taran, V., Gordienko, N., Rokovyi, O., Alienin, O., Stirenko, S.: "last mile" optimization of edge computing ecosystem with deep learning models and specialized tensor processing architectures. In: Advances in Computers, vol. 122, pp. 303–341. Elsevier (2021). https://doi.org/10.1016/bs.adcom.2020.10.003
13. Gou, J., Yu, B., Maybank, S.J., Tao, D.: Knowledge distillation: a survey. Int. J. Comput. Vis. **129**(6), 1789–1819 (2021)
14. He, K., Zhang, X., Ren, S., Sun, J.: Deep residual learning for image recognition. In: Proceedings of the IEEE Conference on Computer Vision and Pattern Recognition, pp. 770–778 (2016)
15. Heo, B., Kim, J., Yun, S., Park, H., Kwak, N., Choi, J.Y.: A comprehensive overhaul of feature distillation. In: Proceedings of the IEEE/CVF International Conference on Computer Vision, pp. 1921–1930 (2019)
16. Heo, B., Lee, M., Yun, S., Choi, J.Y.: Knowledge transfer via distillation of activation boundaries formed by hidden neurons. In: Proceedings of the AAAI Conference on Artificial Intelligence, vol. 33, pp. 3779–3787 (2019)
17. Hinton, G., Vinyals, O., Dean, J., et al.: Distilling the knowledge in a neural network. **2**(7) (2015). arXiv:1503.02531
18. Ho, T.K.K., Gwak, J.: Utilizing knowledge distillation in deep learning for classification of chest x-ray abnormalities. IEEE Access **8**, 160749–160761 (2020)
19. Hochreiter, S., Schmidhuber, J.: Long short-term memory. Neural Comput. **9**(8), 1735–1780 (1997)
20. IEEE: the 2nd diabetic retinopathy—grading and image quality estimation, Challenge (2020). https://isbi.deepdr.org/data.html, Accessed 30 July 2022
21. Ivakhnenko, A., Lapa, V.: Cybernetic predicting devices (1966). https://apps.dtic.mil/sti/citations/AD0654237, Accessed 24 Oct. 2022
22. Kang, J., Gwak, J.: Kd-resunet++: automatic polyp segmentation via self-knowledge distillation. In: MediaEval (2020)
23. Kather, J.N., Halama, N., Marx, A.: 100,000 histological images of human colorectal cancer and healthy tissue. Zenodo10 **5281** (2018)
24. Kather, J.N., Krisam, J., Charoentong, P., Luedde, T., Herpel, E., Weis, C.A., Gaiser, T., Marx, A., Valous, N.A., Ferber, D., et al.: Predicting survival from colorectal cancer histology slides using deep learning: a retrospective multicenter study. PLoS Med. **16**(1), e1002730 (2019)
25. Ke, A., Ellsworth, W., Banerjee, O., Ng, A.Y., Rajpurkar, P.: Chextransfer: performance and parameter efficiency of imagenet models for chest x-ray interpretation. In: Proceedings of the Conference on Health, Inference, and Learning, pp. 116–124 (2021)
26. Kelley, H.J.: Gradient theory of optimal flight paths. Ars J. **30**(10), 947–954 (1960)
27. Khan, M.S., Alam, K.N., Dhruba, A.R., Zunair, H., Mohammed, N.: Knowledge distillation approach towards melanoma detection. Comput. Biol. Med. 105581 (2022)
28. Krizhevsky, A., Hinton, G., et al.: Learning multiple layers of features from tiny images (2009)
29. Krizhevsky, A., Sutskever, I., Hinton, G.E.: Imagenet classification with deep convolutional neural networks. Adv. Neural Inf. Process. Syst. **25**, 1097–1105 (2012)
30. LeCun, Y., Bengio, Y., Hinton, G.: Deep learning. Nature **521**(7553), 436–444 (2015)

31. Li, G., Li, C., Wu, G., Xu, G., Zhou, Y., Zhang, H.: Mf-omkt: Model fusion based on online mutual knowledge transfer for breast cancer histopathological image classification. Artif. Intell. Med. 102433 (2022)
32. Li, J., Chen, G., Mao, H., Deng, D., Li, D., Hao, J., Dou, Q., Heng, P.A.: Flat-aware cross-stage distilled framework for imbalanced medical image classification. In: International Conference on Medical Image Computing and Computer-Assisted Intervention, pp. 217–226. Springer (2022)
33. Linnainmaa, S.: Taylor expansion of the accumulated rounding error. BIT Numer. Math. **16**(2), 146–160 (1976)
34. Mirzadeh, S.I., Farajtabar, M., Li, A., Levine, N., Matsukawa, A., Ghasemzadeh, H.: Improved knowledge distillation via teacher assistant. In: Proceedings of the AAAI Conference on Artificial Intelligence, vol. 34, pp. 5191–5198 (2020)
35. Noothout, J.M., Lessmann, N., van Eede, M.C., van Harten, L.D., Sogancioglu, E., Heslinga, F.G., Veta, M., van Ginneken, B., Išgum, I.: Knowledge distillation with ensembles of convolutional neural networks for medical image segmentation. J. Med. Imaging **9**(5), 052407 (2022)
36. Park, S., Kim, G., Oh, Y., Seo, J.B., Lee, S.M., Kim, J.H., Moon, S., Lim, J.K., Park, C.M., Ye, J.C.: Ai can evolve without labels: self-evolving vision transformer for chest x-ray diagnosis through knowledge distillation (2022). arXiv:2202.06431
37. Park, W., Kim, D., Lu, Y., Cho, M.: Relational knowledge distillation. In: Proceedings of the IEEE/CVF Conference on Computer Vision and Pattern Recognition, pp. 3967–3976 (2019)
38. Qin, D., Bu, J.J., Liu, Z., Shen, X., Zhou, S., Gu, J.J., Wang, Z.H., Wu, L., Dai, H.F.: Efficient medical image segmentation based on knowledge distillation. IEEE Trans. Med. Imaging **40**(12), 3820–3831 (2021)
39. Ruffy, F., Chahal, K.: The state of knowledge distillation for classification (2019). arXiv:1912.10850
40. Ruffy, F., Chahal, K., Mirzadeh, I.: Distiller, pytorch implementation of resnet model families
41. Schmidhuber, J.: Deep learning: our miraculous year 1990–1991 (2020). arXiv:2005.05744
42. Schmidhuber, J.: Deep learning in neural networks: an overview. Neural Netw. **61**, 85–117 (2015)
43. Schmidhuber, J., Blog, A.: The 2010s: our decade of deep learning/outlook on the 2020s. The recent decade's most important developments and industrial applications based on our AI, with an outlook on the 2020s, also addressing privacy and data markets (2020)
44. Sun, S., Cheng, Y., Gan, Z., Liu, J.: Patient knowledge distillation for bert model compression (2019). arXiv:1908.09355
45. Taran, V., Gordienko, Y., Rokovyi, O., Alienin, O., Kochura, Y., Stirenko, S.: Edge intelligence for medical applications under field conditions. In: The International Conference on Artificial Intelligence and Logistics Engineering, pp. 71–80. Springer (2022)
46. Tschandl, P., Rosendahl, C., Kittler, H.: The ham10000 dataset, a large collection of multi-source dermatoscopic images of common pigmented skin lesions. Sci. Data **5**(1), 1–9 (2018)
47. Wang, Y., Wang, Y., Lee, T.K., Miao, C., Wang, Z.J.: Ssd-kd: a self-supervised diverse knowledge distillation method for lightweight skin lesion classification using dermoscopic images. arXiv e-prints pp. arXiv–2203 (2022)
48. Wang, Z., Chen, J., Hoi, S.C.: Deep learning for image super-resolution: a survey. IEEE Trans. Pattern Anal. Mach. Intell. **43**(10), 3365–3387 (2020)
49. Wang, Z., Bovik, A.C.: Mean squared error: love it or leave it? a new look at signal fidelity measures. IEEE Signal Process. Mag. **26**(1), 98–117 (2009)
50. Wang, Z., Bovik, A.C., Sheikh, H.R., Simoncelli, E.P.: Image quality assessment: from error visibility to structural similarity. IEEE Trans. Image Process. **13**(4), 600–612 (2004)
51. Wang, Z., Simoncelli, E.P., Bovik, A.C.: Multiscale structural similarity for image quality assessment. In: The Thrity-Seventh Asilomar Conference on Signals, Systems & Computers, 2003, vol. 2, pp. 1398–1402. IEEE (2003)
52. Williams, R.: Complexity of exact gradient computation algorithms for recurrent neural networks (technical report nu-ccs-89-27). Northeastern University, College of Computer Science, Boston (1989)

53. Yakimenko, Y., Stirenko, S., Koroliouk, D., Gordienko, Y., Zanzotto, F.M.: Implementation of personalized medicine by artificial intelligence platform. In: Soft Computing for Security Applications, pp. 597–611. Springer (2023)
54. Yang, J., Shi, R., Ni, B.: Medmnist classification decathlon: A lightweight automl benchmark for medical image analysis. In: IEEE 18th International Symposium on Biomedical Imaging (ISBI), pp. 191–195 (2021)
55. Yang, J., Shi, R., Wei, D., Liu, Z., Zhao, L., Ke, B., Pfister, H., Ni, B.: Medmnist v2: a large-scale lightweight benchmark for 2d and 3d biomedical image classification (2021). arXiv:2110.14795
56. Zhang, L., Song, J., Gao, A., Chen, J., Bao, C., Ma, K.: Be your own teacher: improve the performance of convolutional neural networks via self distillation. In: Proceedings of the IEEE/CVF International Conference on Computer Vision, pp. 3713–3722 (2019)
57. Zhu, X., Gong, S., et al.: Knowledge distillation by on-the-fly native ensemble. Adv. Neural Inf. Process. Syst. **31** (2018)
58. Zou, W., Qi, X., Wu, Z., Wang, Z., Sun, M., Shan, C.: Coco distillnet: a cross-layer correlation distillation network for pathological gastric cancer segmentation. In: 2021 IEEE International Conference on Bioinformatics and Biomedicine (BIBM), pp. 1227–1234. IEEE (2021)

Self-Distillation with the New Paradigm in Multi-Task Learning

Ankit Jha and Biplab Banerjee

Abstract We tackle the problem of multi-task learning (MTL) for solving the correlated visual dense prediction tasks using monocular image source. With further restrictions on the model parameters, the soft-shared MTL Convnets (CNN) feature distinct models for each individual task. Hard-sharing-based models, on the other hand, have shared encoders but individual decoders for every task. MTL models have demonstrated satisfactory performances for pixel-wise dense prediction tasks like semantic segmentation, surface-normal estimation, and depth estimation from the monocular inputs. In general, they impose two inherent drawbacks: (1) such models have constraints with no leverage on the inter-task knowledge, which hinders the performance boost while jointly trained, and (2) explicitly optimization of each task-specific network is required. Incorporating the abovementioned issues into soft and hard-sharing-based MTL models can immensely enhance MTL performance. To that end, we present SD-MTCNN, hard-sharing-based and S^3DMT-Net, soft-sharing-based novel MTL networks. Here, we follow tries to inherit characteristics from deeper CNN layers/feature maps into shallower CNN layers, hence helps in increasing the network's bandwidth. We also utilize the notion of sharing the features of the task-specific encoders, where the task-specific encoder's feature-maps are communicated amongst each other, aka as cross-task interactions. We examine our self-distilled MTLs performance on two different types of visual scenes: Urban (CityScapes and ISPRS) and indoor (NYUv2 and Mini-Taskonomy), and we observe notable gains in all tasks as compared to the referred methodologies.

Keywords CNN · Model compression · Multi-task learning (MTL) · Self-distillation · Soft-sharing · Hard-sharing

A. Jha (✉) · B. Banerjee
Centre of Studies in Resources Engineering, Indian Institute of Technology Bombay, Powai, India
e-mail: ankitjha16@gmail.com

B. Banerjee
Faculty at Center for Machine Intelligence and Data Science (CMInDS), IITB, Powai, India

W. Pedrycz and S.-M. Chen (eds.), *Advancements in Knowledge Distillation: Towards New Horizons of Intelligent Systems*, Studies in Computational Intelligence 1100,
https://doi.org/10.1007/978-3-031-32095-8_6

1 Introduction

With their built-in hierarchy, data-driven feature learning, and data-handling skills, deep learning architectures have transformed the area of computer vision (CV). It's worth noting that deep convolutional neural networks (CNNs) have a character for being data-hungry, requiring huge amounts of labeled training examples to maintain a reasonable degree of generalization. This is a barrier in several different applications, including self-driving vehicles and medical imaging, where accumulating labeled supervision is time-consuming and costly. Multi-task learning (MTL) [1, 16, 19] tries to train several correlated tasks from the same data input and has evolved as one of many techniques to deal with such a situation. Conceptually, MTL models are highly inspired by human learning techniques in which knowledge may be applied across various domains or subjects, resulting in better overall learning. Similarly, in an MTL design, competing tasks offer a regularization impact, allowing the tasks to minimize overfitting. MTL has also been explored in various scenarios in computer vision can be devised [2, 21], such as the dense prediction tasks [16, 20], where cooperative tasks such as pixel-wise classification (segmentation), surface normal estimation and depth prediction are performed concurrently, aids in jointly exploring semantic and geometric visual information (Fig. 1).

Although deep CNN-based MTL models can improve learning for a variety of challenging tasks, they typically encounter design difficulties. Traditionally, an MTL model in the parameter space might be hard-shared and soft-shared [21]. All hard-parameter sharing MTL networks share a feature extractor with distinct task-specific decoders. Whereas, soft-sharing-based MTL models contain an individual network (set of encoders and decoders) for performing respective tasks with a set of disjoint parameters. The correlation between the parameters of the task-specific modules is

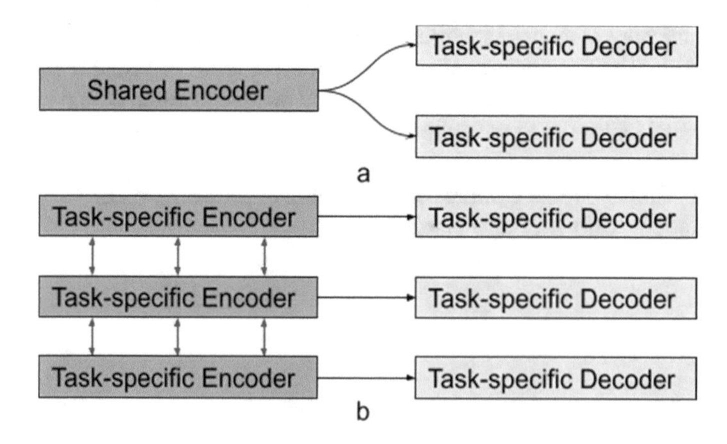

Fig. 1 Overview depicting **a** Hard-sharing-based multi-task learning (MTL) architecture and **b** Sof-sharing-based MTL architecture. Best view in color

deliberately increased to maintain model uniformity. At this stage, we are concentrating on designing improved hard-sharing-based and soft-sharing-based multi-task frameworks for visual dense prediction tasks by increasing the representational capabilities of respective task networks and also improving knowledge sharing within tasks. Similarly, each network employs a typical encoder-decoder design, such as U-Net [21], SegNet [2], and so on. Our work follows the encode-decoder architecture to perform the dense prediction tasks in the MTL setting.

Typically, the l_2-norm is commonly used to regularize model parameters in both hard-sharing-based and soft-sharing-based multi-task architectures. However, this may drastically impair the network's capacity to eavesdrop. Because it has been shown that many tasks can interact complexly with the same set of characteristics, some properties are often easy in learning a task T_1 but challenging to learn for a rival task T_2. Hard-sharing architectures, which use a single feature encoder (shared) for all tasks, circumvent this bottleneck. However, this substantially restricts the encoder's capacity to learn task-specific feature sets. Therefore, if information sharing is done strategically among the tasks, a soft-shared multi-task network provides higher and optimal MTL performance. On the other hand, the network depth and training data volume significantly impact a CNN model's performance. With lots of training samples, a deeper CNN network can significantly attribute feature representations. On the other side, deeper CNN models may only sometimes be deployable. A few more parameters could be added to an existing CNN architecture with modest depth to enhance its performance.

Inspired by the above arguments on MTL performance and its deployment on edge or small devices, we propose novel frameworks (i) Self-Distilled Multi-task CNN (SD-MTCNN), and (ii) Self-Distilled Soft-Sharing Multi-Task CNN (S^3DMT-Net) alleviates the issues faced by the hard-parameter-shared and soft-parameter-shared MTL models, respectively. The hard-parameter-shared MTL architecture, which we refer to as SD-MTCNN [32], takes into account the encoder which is shared and decoders with task-specific modules and contains a distillation path to boost the latent or bottleneck aspects of the MTL.

Additionally, our S^3DMT-Net [33], another MTL model based on a soft-sharing ideology which has an individual set of encoder modules and decoder modules for each task. The layers of the encoder block help in learning the latent features, and the decoder block accounts for task-specific reconstructions. We suggest sharing the feature maps in layer-wise manner between the tasks to ensure the eavesdropping. In this approach, the optimization of all the tasks dictate the task-specific parameters, and each of the encoder layers have access to information from the other correlated tasks. We demonstrate empirically that such an information-sharing strategy outperforms the conventional soft-sharing MTL techniques based on l_2—regularizers.

Next, we propose the concept of *self-distillation*, i.e. distilling information within the network [29] (Fig. 2) in both hard-sharing and soft-sharing MTL setups, where the distilling features from the second-last decoder layer into the encoder's bottleneck of the architecture that improve the performance for the respective the task-specific architecture. We point out that self-distillation (shown in Figs. 3, 4) differs from the

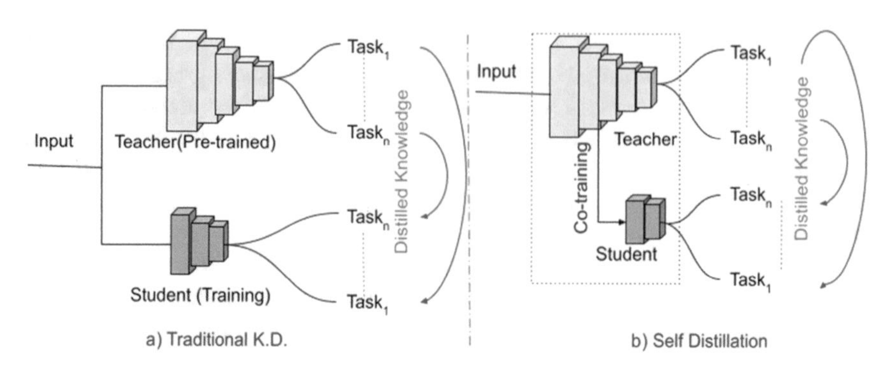

Fig. 2 The general outline of **a** conventional or traditional MTL KD and **b** our proposed MTL SD. Traditional KD specifies the Teacher and Student using two different networks, SD defines the Student and the Teacher networks within the same model. The information shared between Teacher network to Student network during co-training using distillation loss

conventional idea of knowledge distillation (KD) [7], in which a small (shallower) student model extracts the characteristics of an extensive and fully trained (deeper) teacher model. The student and the teacher are specified and designed within the same network architecture in self-distillation (SD), and the Teacher network is incorporated with the Student one. SD enables the latent feature space to be encoded with semantically more significant characteristics, enhancing the task model's overall representational capabilities. Furthermore, because only the encoder module may now generate output for each task, a self-distilled network improves inference time.

We summarized our novel contributions in following points:

- We present the multi-task learning (MTL) models based on hard-sharing and soft-sharing that incorporates the concept of self-distillation for tackling numerous densely predicted tasks from monocular inputs.
- SD-MTCNN aims to transfer information from the decoder's task-specific modules to the task-generic subnetwork, resulting in improved discriminative robust feature embedding for tasks. In SD-MTCNN, the encoder is based on conventional CNN, whereas the decoders use task-specific attention learning.
- By permitting cross-talk among task-specific encoders, the S^3DMT-Net architecture enables eavesdropping in the model. Furthermore, we use the concept of self-distillation to train the individual task models. Consequently, our model can do better and more semantically rich information exchange among tasks.
- Tables 2, 7 provide extensive experimental analysis with thorough ablation studies on the Mini-Tasknomy [25], CityScapes [4], ISPRS [17], and NYUv2 [24] datasets, where increased performance is uniformly found.

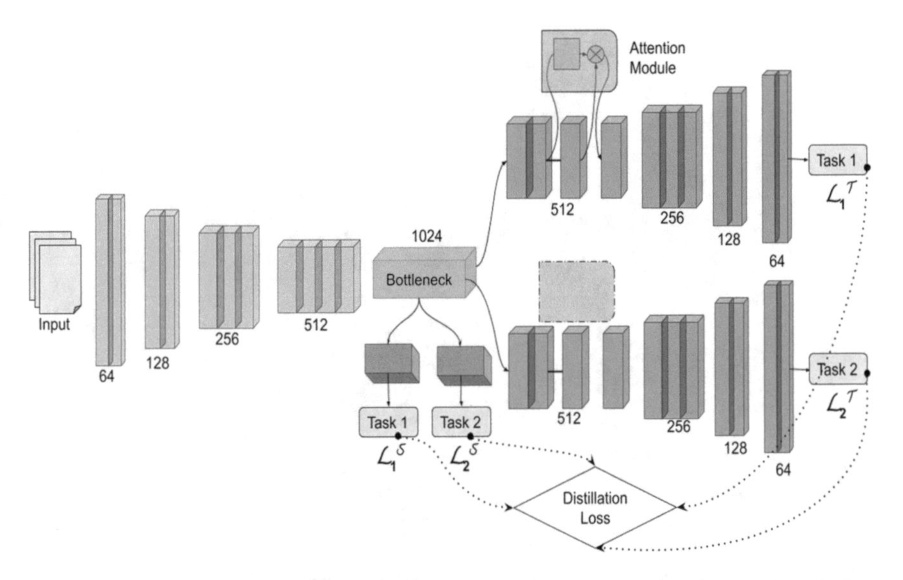

Fig. 3 The block diagram of our SD-MTCNN model is based on encoder-decoder settings. We show the encoder (hard-sharing-based) in Pink, the bottleneck (encoder) in Green, Teacher's decoders in Blue, Student's decoder in Gray, and attention block in Orange. Pink, Gre.en, and Blue sections of the model denote the whole Teacher. In contrast, the Student is depicted with the Pink, Green, and Gray color codes. Best viewed in color

2 Literature Review

Multitask Learning (MTL): Under limited resources or a paucity of adequate supervision in terms of annotated samples, MTL [3, 6, 10] is regarded as one of the cost-effective options. As previously said, complementary information received from several sources frequently aids in the development of a better learning agent. The concept of information sharing aids in the generalization of the machine learning (ML) model and sparks interest in multi-task learning studies. [19, 30] provides a comprehensive overview of MTL. Initially, the MTL problem was approached using feature transformation-based techniques such as matrix decomposition [3], task clustering [9], Bayesian approaches [26], margin-based formulation [31] and so on, which were eventually replaced by CNN architectures [5, 14]. In CNN-based MTL networks, feature extractor's backbone can be shared either soft-parameter-based or hard-parameter-based among the tasks. Cross-Stitch networks [20], attention-based MTL (MTAN) [23], Sluice networks [22], and others are well-studied approaches in this instance. Furthermore, the weight assignment to individual task loss is crucial to handle in multi-task setup. Some solutions related to weight assignment have been employed with the pre-defined loss weights, also few adaptive weighting strategies are taken to leverage the weight assignments to optimize the tasks efficiently, for eg. [12] used the notion of Bayesian uncertainty to leverage the weight assignment issues for the tasks in MTL.

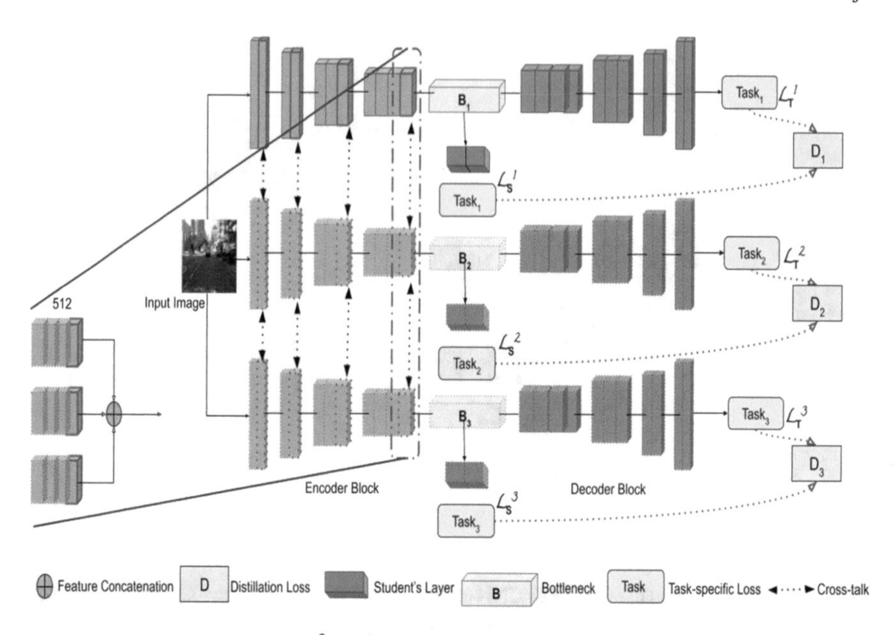

Fig. 4 Architecture diagram of S^{3}DMT-Net with. the encoders connected via a cross-talk (soft-sharing) technique. The red dashed lines represent the final convolutional layer's feature concatenation (512 featu.re map.s) of encoders from each task (similar. to other encoder modules of depth 64, 0.128, and 256). The distillation losses (Eqs. 3 and 4) are incorporated between the decoders of each task of the (Teacher) and the de.convolutional layers (Student)

Knowledge distillation (KD): Using the Teacher-Student information transfer module, the paradigm of knowledge distillation (KD) [7] assists in the compression of deep CNN models. While the Teacher is believed to be a deeper network that has been optimally trained on a huge dataset, the Student is usually constructed as a shallower model, which should fulfill the purpose of propagating the information encoded in the deeper Teacher network to the shallower Student network. Later the testing may be performed directly via Student framework, which provides significant down-sizing in the model's parameters and accelerates without sacrificing performance significantly. There is a plethora of complex KD techniques in the literature, hint learning [18], including attention learning [27], and adversarial (generative) training [15, 16], to mention a few. In order to avoid the need for individual Teacher and Student models, [29] proposed the concept of self-distillation (SD) inside a network, that focuses on distilling information from deeper levels to shallower ones. Such a network generalizes significantly under the SD setting which mainly concentrates to learn the high-level features through shallow layers. In several ways, our models (SD-MTCNN [11] and S^3DMT-Net [32]) varies from the literature. In the soft-shared MTL architecture, we address the hitherto understudied topic of eavesdropping by proposing an intuitive and straightforward mechanism for feature-sharing across task-specific encoders. In contrast, SD-MTCNN is specially built to support MTL with a hard-parameter based feature sharing setup and a more simplified way to combine loss measures and enrich the latent feature space with superior task-specific information. Furthermore, our SD loss assures quality feature learning for task-specific networks. As a result, the task-specific networks can optimize in a satisfactory manner. Through experimental analysis, we demonstrate the efficacy of the suggested approach experimentally.

3 Proposed Methodology

3.1 Problem Formulation

In this section, we define our goal of designing the hard and soft-sharing-based MTL frameworks for solving the correlated tasks from monocular input images. We assume that all the tasks used in the MTL setup resemble some common properties, i.e., they are correlated. Figures 3, 4 depict the broad overview of the proposed self-distilled multitask architecture, i.e., SD-MTCNN and S^3DMT-Net, respectively. In SD-MTCNN, the encoder is shared among the tasks, and for task-specific outputs, it consists of attention-based task-specific decoders to extract prominent task-oriented features. In contrast, S^3DMT-Net consists of task-specific encoder and decoder modules. Here, the encoder module learns abstract features from the given input images, and the task-specific decoders help reconstruct the associated tasks. The vanilla (simple) soft-sharing MTL architectures, one of our baselines, impose no constraints on individual models. The entire proposed MTL architecture is trained under the supervision of combined optimization functions of the correlated tasks and extends the vanilla MTL model with two setups, which are as follows:

a. **Interaction through Cross-task Setup:** Since the information corresponding to each is complimentary, exchanging information is preferred. As proposed in SD-MTCNN, the architecture has the shared encoder space in a hard-parameter-based mechanism that needs to be revised to exchange mutual information in the encoder part. Our proposal for enhancing mutual information sharing among tasks in S^3DMT-Net is to incorporate cross-talk between task encoders. This involves combining all the feature maps received by each task-encoder level, so that all tasks can access the filter maps of their respective task-specific encoder layers. The result is that cross-task knowledge is shared between the encoders of each task, facilitated by cross-task links in the architecture. The impact of such links is reflected in the task-specific loss functions and affects all task-specific encoders during back-propagation.
b. **Self-distillation for Task-specific modules:** In an encoder-decoder architecture, the quality of features extracted by the encoder path significantly affects the decoder's performance. However, as high-level features are typically extracted at deeper levels of the encoder, features extracted at shallower depths may not have the necessary complexity, limiting the model's ability to learn effective representations. To address this issue, we propose a solution that involves including a

distillation loss in each task-specific module. This loss serves as a soft constraint, guiding the encoder feature space towards similarity with the last task-specific decoder or transpose convolutional layer. By improving the latent capability of encoder features in this manner, we anticipate that decoder layers can focus on learning even more semantically relevant characteristics suited to the task at hand. This approach, known as self-distillation (SD), eliminates the need for separate Teacher and Student networks, as knowledge distillation is performed internally within the architecture.

Let us suppose $X = \{x_i^t, y_i^t\}$ for $t = itoT$, dataset consists with multiple tasks T, $x \in X$ is the input image with dimension $\mathbb{R}^{HxWx3}$ and $y^t \in Y$ corresponds to the output labels in the dimension of $\mathbb{R}^{HxWxK}$ where H and W are the height and width of the input image and K denotes number of output channels for the tth task. Major.ly our experiments consider three pixel prediction assignments from monocular input images: Semantic segmentation, surface-normal prediction, and depth estimation. We have an encoder-decoder network for each of the tth tasks, where $f_E^t(\S, \theta_E^t)$ and $f_D^t(\S, \theta_D^t)$ are the encoder and decoder functions with the trainable parameters θ_E^t and θ_D^t in the channel dimension of 1024.

For the tth net.work, we further create an output from the final layers of the task-specific decoders derived directly from.the encoder feature space and similar to the penultimate decoder layer (the layer immediately before the final loss layer) and call the subsequent sub-network as the Student while the whole architecture is termed as the Teacher. Finally, the encoder layers are communicated by both the Teacherand Student, whilst the respective decoder layers of the task-specific modules are particular to the Teacher or the Student architecture. To make it very simple, the Teacher's decoder network can be designated by $f_{TD}^t(\S, \theta_{TD}^t)$ (here, we consider $f_D^t \approx f_{TD}^t$) with θ_{TD}^t be the task-specific parameters, we use $f_{SD}^t(\S, \theta_{SD}^t)$ with θ_{SD}^t as the parameters to denote the task-specific Student's output branches. Finally, we denote the outputs for the tth network at both the Teacher and Student architecture as,

$$\hat{x} = f_E^t(\S, \theta_E^t),\ \widehat{y_T} = f_{TD}^t(\S, \theta_{TD}^t) \text{ and } \widehat{y_S} = f_{SD}^t(\S, \theta_{SD}^t) \tag{1}$$

We consider all the variables as the tensors. In the next sections, we explain the loss functions used in the SD-MTCNN and S^{3}DMT-Net and also discuss the overall objective function.

3.2 *Loss Functions for Task-Specific Modules*

Recall that tasks like semantic segmentation (classification), surface-normal estimation (similarity), and depth estimation (regression) are the three pixel-level prediction tasks taken into account. Hence, we followed MTAN [23] for defining the loss functions for our self-distilled MTL networks, (i) cross entropy (multi-class) loss for semantic segmentation, (ii) l_1 norm i.e. distance based loss for estimating the depth, and (iii) element-wi.se dot-pr.oduct for evaluating surface-normal between the true values and the model predictions, respectively. In order to help with the distillation process, the loss metrics must be assessed at outputs of both the. Student and Teacher models. Specifically, $\mathcal{L}_T^t$ and $\mathcal{L}_S^t$ and denote the loss terms used to train the Teacher and Student for the tth task where $t = \{1, 2, 3\}$ for segmentation, surface-normal estimation and depth-estimation. Mathematically, Eq. 2 defines the loss functions for the Teacher (T) and Student (S) networks ($m \in \{T, S\}$) as follows,

$$\begin{aligned}\mathcal{L}_m^1(\widehat{y}_m^1, y^1) &= E\left[-\frac{1}{HW}\sum_{j=0}^{H-1}\sum_{k=0}^{W-1} y^1(j,k).log\widehat{y}_m^1(j,k)\right]\mathcal{L}_m^2(\widehat{y}_m^2, y^2)\\ &= E\left[-\frac{1}{HW}\sum_{j=0}^{H-1}\sum_{k=0}^{W-1}\left|\left|y^2(j,k) - \widehat{y}_m^2(j,k)\right|\right|_1^1\right]\mathcal{L}_m^3(\widehat{y}_m^3, y^3)\\ &= E\left[-\frac{1}{HW}\sum_{j=0}^{H-1}\sum_{k=0}^{W-1} y^3(j,k).\widehat{y}_m^3(j,k)\right] \end{aligned} \tag{2}$$

for the input image, we denote (W, H) to be the height and the width, respectively. y and $\widehat{y}$ to be the correct and predicted or model's output labels and E denotes the expectation for the loss terms.. As of now, the loss or optimization functions are computed for each. Task at both student (S) and teacher (T), additionally,we propo.se to distillation loss to distill the learned feature space of the bottleneck layers of both S and T networks.

3.3 *Loss for Self-Distillation*

This loss aids in refining a Teacher model's capabilities on top of a Student model. As we jointly learn the classification and regression tasks, we specify the distillation loss for each work independently. The self-distillation losses guarantee that high-level task-specific information is regularized into all tasks' latent space representations. The individual decoders may thus be better focused on reproducing the outputs of each task. In this regard, we employ two different forms of self-distillation loss functions, i.e. Kullback–Leibler divergence (KL) and l_2 distance are used to distill the knowledge between the Teacher and Student networks. In our experimental settings, the classification (segmentation) task is distilled using Eq. 3, whereas regression tasks like dep.th esti.mation and surface-norm.al pre.dicti.on use Eq. 4 for distillation.

$$\mathcal{L}_{KD}^1 = E\left[KL\left(\widehat{y}_S^1, \widehat{y}_T^1\right)\right] \tag{3}$$

$$\mathcal{L}_{KD}^2 = E\left[\left|\left(\widehat{y}_S^2 - \widehat{y}_T^2\right) + \left(\widehat{y}_S^3 - \widehat{y}_T^3\right)\right|_2\right] \tag{4}$$

3.4 Total Loss

While proposing the multitask self-distilled architecture, i.e., S^3DMT-Net and SD-MTCNN, we use two different kinds of training strategies. Firstly, we optimize the model using distillation losses (defined in Eqs. 3 and 4) with the tas.k-speci..fic optimizing function.s for the Teacher network and the Student network.

$$argmin_{(\theta^E,\theta^{SD},\theta^{TD})}\sum\nolimits_{m\in\{T,S\}_{i=1}}^{T}\mathcal{L}_m^i+\lambda\left(\mathcal{L}_{KD}^1+\mathcal{L}_{KD}^2\right) \tag{5}$$

In Eq. (5), T is denoted as the number of assignment or.tasks with $m \in \{T, S\}$ for the Stu.dent (S) or the Teach.er (S) network. λ be the hyper-parameter between.task losses and SD losses. To be precise, Eq. (5) comprises the loss terms corresponding to each task for the Teacher and the Student networks and the SD losses help in attaining the encoder space (both hard-sharing-based and soft-sharing-based mechanisms) to impro.ve the task's performance.

As.in, the fir.st case (Eq. 5) has individual losses for each task for bo.th the Teacher and Student networks; the second follows ensemble learn.ing to train the proposed self-distilled MTL model. Here, the predictions from the net.works (both Student S and Teacher T) are averaged and evaluated with the corresponding ground-truth labels defining the task-specific loss functions. The above described ensembled task-specific loss functions are summed with the task-specific distillation losses to form. the overall optimization, as defined in Eq. 6. Further, we optimize the architecture with Eq. (2) $\mathcal{L}_m^{1-3}$ and $\underline{y}^i = \frac{\hat{y}_T^t+\hat{y}_S^t}{2}$ for $t \in \{1, 2, 3\}$ as the ensembled model predictions.

$$argmin_{(\theta^E,\theta^{SD},\theta^{TD})}\sum\nolimits_{i=1}^{t}\mathcal{L}^i\left(y^i,\underline{y}^i\right)+\lambda\left(\mathcal{L}_{KD}^1+\mathcal{L}_{KD}^2\right) \tag{6}$$

4 Dataset and Experimental Protocols

Here, we define the datasets used in our experimental setups. The training procedures and evaluation metrics used in designing the self-distillation multi-task learning networks are also described in this section. In the end, we explain the inference step followed to examine the performance of our trained self-distilled MTL models.

4.1 Datasets Description

For model experimentation and evaluation of our proposed SD-MTCNN and S^3DMT-Net, we consider four benchmark datasets for MTL: CityScapes [4], NYU.v2 [24], ISPR.S [17], and Mini Taskonomy [25].

NYUv2: There are 1.449 RGB-D scenes (indoor) that exist in the NYUv2 dataset, which are primarily from the video sequences. These scenes present significant complexity, as they exhibit variations in camera viewpoint, occlusion within the scene, and changes in lighting conditions. For our experimental purposes, we focused on three tasks: Segmentation (13-class), depth prediction, and surface normal. To facilitate experimentation, we resized both the images and corresponding labels to dimensions of 128 $\times$ 256.

CitySca.pes: It contains 500.0 street views, resizes all images, and labels 128 $\times$ 256. We perform experiments on seven classes of 30 for semantic segmentation and depth estimation.

Mini Taskonomy: The Taskonomy [25] consists of 26 visual inference problems with more than 4.5. billion high-quality indoor scenes and has pixel-level geometry information. Nevertheless, we only take into account five tasks and refer to them as the Mini Taskonomy: semantic segmentation, edge detection, predicting depth, estimating 2D key-point, and surface normal prediction. The model operates with image.s of 256 $\times$ 2.56 in dimension. The semantic segmentation task has 18 classes for classification.

ISPRS: Digital surface model (DSM) and aerial images of the ground of the city of Germany (Potsdam) are included in the ISPRS [17] dataset. Six classifications and depth estimates are taken into consideration for semantic segmentation (DSM data), respectively. The well-resolved larger scenes' non-overlapping tiles of height and width of 256 $\times$ 256 are considered. Table 1 lists the semantic categories for both outdoor scenarios.

Table 1 The classes for segmentation on scenes of CityScapes (7-class.) and ISP.RS (6-class) datasets

City scapes	ISPRS
Construction	Building
Flat	Car
Human	Cluster/Background
Nature	Impervious surfaces
Object	Low vegetation
Sky	Tree
Vehicle	

Table 2 We show the 3-task experimental results for validating on the 13-class NY.Uv2 datas.et for segmentation, normal estimation and dept.h tasks on SegNet based self-distilled MTL architectures, i.e. SD-MTCNN and S^3DMT-Net. *T* Teacher, *S* Stu.dent, simple MTL as Baseline (BL-I). Equations 5 and 6 are used for our proposed S3DMT-Net architecture, with the best-performing Teacher and Student models selected from the referenced SOTAs. We denote these equations as (5). and (6), respectively. ↓ / ↑ represents the higher./ lower the outperformance. Best results shown in bold

Method	Sem. Seg		Depth Err		Surface				
	↑		↓		Angke distance ↓		Within t ° ↑		
	IoU	mIoU	Abs	Rel	Mean	Median	11.25	22.5	30
Single Task	56.36	17.50	0.6796	0.2930	29.79	23.90	24.39	47.85	59.90
Vanilla MTL	51.88	15.59	0.6177	0.2663	32.08	26.93	21.16	43.01	55.00
Dense [8]	52.73	16.06	0.6488	0.2871	33.58	28.01	20.07	41.50	53.35
Cross-Stitch [20]	50.23	14.71	0.6481	0.2871	33.56	28.58	20.08	40.54	51.97
Split (Wide) [23]	51.19	15.89	0.6494	0.2804	33.69	28.91	18.54	39.91	52.02
Split -Deep	41.17	13.03	0.7836	0.3326	38.28	36.55	9.50	27.11	39.63
MTAN [23]	55.32	17.72	0.5906	0.2577	31.44	25.37	23.17	45.65	57.48
SD-MTCNN^T (5)	56.90	22.44	0.5857	0.2483	29.66	25.02	22.66	45.84	58.39
SD-MTCNN^S	56.43	21.90	0.5986	0.2532	30.76	26.98	19.52	44.20	57.32
SD-MTCNN^T (6)	56.61	22.31	**0.5864**	**0.2458**	30.04	25.24	21.36	45.40	58.20
SD-MTCNN^S	56.11	22.59	0.5899	0.2506	30.42	25.63	21.22	44.79	57.21
S^3DMT-Net^T (5)	**59.96**	**24.42**	0.5922	0.2511	**28.31**	**22.62**	**25.69**	**50.10**	**62.34**
S^3DMT-Net^T	59.10	24.18	0.6059	0.2673	29.28	23.78	24.00	47.97	60.07
S^3DMT-Net^T (6)	59.47	24.10	0.5928	0.2553	28.76	22.94	25.20	49.82	61.55
S^3DMT-Net^S	58.83	23.96	0.6106	0.2645	29.40	23.85	24.14	48.11	60.28

4.2 Model Architecture

1. **Self-distil.led Multi-.task CNN (SD-MTCN.N)**: It primarily consists of the SegNet model (encoder-decoder setup). The shared encod.er has four C.NN blocks. As shown in Fig. 3, two CNN layers are used in the first and second blocks, whereas we use three and four layers of CNN blocks in the third and fourth blocks. All of the CNN blocks in our architecture have a kernel size of 3 × 3. Following each CNN block, we include a max-pooling layer with a kernel striding of 2. To ensure training stability, we apply non-linearity through ReLU and BatchNorm layers after each CNN block. The encoder space produces feature maps of depths 64, 128, 256, and 512.

   The bottleneck layer, which serves as the shared features space, contains feature maps of dimension 1024. To construct the decoder module, we follow a symmetric architecture to that of the encoder module. Additionally, we include attention modules in each CNN block of the decoder module. These

attention modules consist of two convolutional layers with a kernel size of 3 x 3, each followed by a batch normalization layer. We also apply a sigmoid activation function to ensure that the mask values fall within the range of [0, 1].
To enable the distillation process in MTL, we implement a separate deconvolutional layer for the Student network. We also configure the entire SD-MTCNN model, also known as the Teacher network, to derive task-specific outputs directly from the bottleneck layer, as depicted in Fig. 3.

2. **S^3DMT-Net**: Our S^3DMT-Net architecture, similar to the hard-parameter-based self-distilled MTL network (SD-MTCNN), utilizes an encoder-decoder-based SegNet model. For each task, the architecture comprises individual encoder and decoder blocks. To accommodate cross-task interactions, we use a soft-sharing strategy among all encoders and self-distillation at the decoder. The encoder blocks consist of four CNN blocks: the first two blocks each have two layers, while the third and fourth blocks have three and four CNN layers, respectively, with a kernel size of 3×3. A max-pool layer with a kernel striding of 2 follows the convolutional blocks. To ensure uniformity and avoid overfitting during training, we include batch-norm (BN) layers and rectified linear unit (ReLU) layers. In the encoder portion, cross-task connections exist among the last layers of each CNN block, as illustrated in Fig. 4.

   The encoder block and decoder block are coupled by the bottleneck or latent layer, which evaluates the depth of 1024 features. The design of the decoder module is similar to that of the encoder modules. The deeper network in the suggested S3DMT-Net is referred to as the Teacher. To perform task-specific learning, the Student or shallow network comprises deconvolutional (transpose convolution) layers fro.m the bottlene.ck and we refer t.o it as soft-sharing-based multi-task learning for dense prediction tasks.

4.3 Training Protocol and Evaluation Metrics

Training Protocol: The training for SD-MTCNN is iterated with an initial learning rate of 10^{-4} for 200 epochs and we optimize the models using Adam optimizer in our setup. For the NYU.v2, Mini-.Taskonomy, CityScapes, and ISPRS, we select a batch of size 2, 4, 8, and 8, respectively. λ is the hyper-parameter used in Eq. (5). and (6) and is set t.o 1 in all the training experiments. Also, considering that our main goal is to demonstrate the efficacy of the self-distillation technique, we assume equivalent weights for the task-specific loss functions. Similar strategies have been followed to train the soft-sharing based SD MTL network, i.e., S^3DMT-Net. We select the hyper-parameter λ by ablating the setup for multiple times and observe that λ with values [0.8–1] provide optimal performance. We train our methods on a 12 GB NVIDIA GTX 2080 Ti GPU card and 64 GB RAM mounted with Intel Xeon Processor.

Evaluation Protocols/Metrics P: We use the standard protocols to evaluate the proposed SD-MTCNN and S^3DMT-Net, defined in [23]. The performance of the pixel-wise classification (segmentation task) is evalua.ted using I.oU and m.IoU. In contrast, we calculate the absolute and relative error between the true and predicted pixels for regression tasks, i.e., key-point estimation, depth estimation, and edge detection. Finally, we incorporate cosine similarity to analyze the network performance for the surface-normal task.

5 Results and Discussion

5.1 Competitors

Our proposed SD-MTCNN and S^{3}DMT-Net architectures have been evaluated against several state-of-the-art (SOTA) competitors, including DenseNet, MTAN, and Cross-Stitch network. We compare the performance of SD-MTCNN and S^3DMT-Net with the defined baselines in Eq. 5 and 6, as described in Sect. 3. In our methodologies, we use unity weights for task-specific loss terms.

Our experimental results in Tables 2, 3, 4] demonstrate significant progress in all of the given tasks. We train self-distilled MTL models, i.e., SD-MTCNN and S^3DMT-Net, with Eq. 5 for the NYUv2 3-task. SD-MTCNN outperforms referred models for segmentation (IoU and mIoU) by at least (+2.8% and + 21.1%), depth estimation (absolute and relative errors) by at least (-0.8% and -3.8%), and norm.al estimation (me.an, m.edian, and t angles) by minimum of (–6%, –1.4%, and 1.6%).

On the other hand, our model with soft-shared encoders (S3DMT-Net) significantly gains for all the tasks. We note the performance difference for the segmentation task (mIoU and IoU) by (+27.44% and + 20.73%), depth esti.mation (relative er.ror) by (–2.63%), and sur.face norm.al predicti.on (m.ean, m.edian and $t\circ$ angles) at least b.y (–11.06%, –12.1.6% and + 7.8%) on the Tea.cher network/branch. We also note the performance of the Student network, where it o.utperforms by (+26.72% and + 4.64%), (–0.1%), and (–1.74.%, –0.5% and + 0.3%) for the respec.tive three tasks. Com.paring the perfo.rmance between SD-MTC.NN and S^3DMT-Net, we can certainly observe a high margin in the pixel wise classification accuracy (IoU) of around 5% for S^{3}D.MT-Net, also shown in Fig. 5.

Based on our experiments, we found that the self-distilled MTL architectures we proposed (SD-MTCNN and S^3DMT-Net achieved almost all of the best results across various metrics on the NYUv2 2-task, CityScapes 2-task, ISPRS 2-task, and Mini Taskonomy 5-task datasets. The visualization outputs in Figs. 7, 8 show that our models' predictions are very similar to the true maps for the CityScapes and ISPRS datasets. Furthermore, we observed that both the Teacher and Student modules (as defined in Eqs. 5–6) of our models outperformed relevant methods. We highlight the scores of the networks that performed better than others.

Table 3 We show the 5-task experimental results for validating on the 18-class Mini-Taskono.my da.taset for seman.tic segm.ent.ation, edge detection, depth estimation, key-point, and surface normal prediction tasks on SegNet based self-distilled MTL architectures, i.e. SD-MTCNN and S^3DMT-Net. T Teacher, S Stu.dent, simple MTL as Baseline (BL-I). Equations 5 and 6 are used for our proposed S3DMT-Net architecture, with the best-performing Teacher and Student models selected from the referenced SOTAs. We denote these equations as (5). and (6), respectively. ↓ / ↑ represents the higher./ lower the outperformance. Best results shown in bold

Method	Sem. Seg.↑		Depth Err.↓		Surface↑	Key ↓	Edge ↓
	IoU	mIoU	Abs	Rel	CS	Abs	Abs
Single Task	88.62	47.40	0.0284	0.4469	0.9217	0.0219	0.0122
Vanilla MTL	88.85	49.21	0.0307	0.4216	0.9285	0.0173	0.0094
Dens.e [8]	87.63	44.76	0.0379	0.6329	0.8677	0.0459	0.0329
Cross-Stitch [20]	86.87	35.66	0.0375	0.5481	0.8601	0.0488	0.0396
.Split -Wi.de [23]	88.72	49.26	0.0447	0.5125	0.8631	0.0455	0.0411
Split -D.eep	88.89	47.52	0.0453	0.4909	0.8674	0.0408	0.0390
MTAN [23]	88.38	48.91	0.0361	0.6440	0.8707	0.0481	0.0211
SD-MTCNN^{T} (5)	89.36	55.36	0.0290	**0.3572**	0.9058	0.0138	0.0097
SD-MTCNN^{S}	89.02	54.78	0.0370	0.4010	0.8608	0.0327	0.0118
SD-MTCNN^{T} (6)	89.21	55.26	0.0327	0.4318	0.8907	0.0147	0.0101
SD-MTCNN^{S}	88.93	53.68	0.0419	0.5221	0.8616	0.0340	0.0129
S^{3}DMT-Net^{T} (5)	**89.69**	**58.35**	**0.0269**	0.3906	**0.9288**	**0.0121**	**0.0087**
S^{3}DMT-Net^{T}	89.43	57.47	0.0299	0.4652	0.9123	0.0203	0.0113
S^{3}DMT-Net^{T} (6)	89.50	58.14	0.0282	0.4108	0.9244	0.0126	0.0093
S^{3}DMT-Net^{S}	89.03	57.66	0.0334	0.4515	0.9111	0.0255	0.0167

5.2 *Ablation Studies*

Here, we define the ablation setups for the proposed architectures and also analyze the effect of the number of tasks on both the self-distilled MTL models.

A. **Ablation on Self-distilled MTL Networks:** We study the impact of cross-talk (soft-sharing) and SD, starting from a vanilla multi-task setup to our proposed two self-distilled MTL networks, i.e., SD-MTCNN and S^3DMT-Net. We opt three baseline approaches for the comparison of the proposed self-distilled MTL architectures.

   (i) **Baseline BL-I** refers to the vanilla SegNet MTL model, which has separate encoders and decoders for each task, similar to our Teacher architecture.
   (ii) **Baseline BL-II** is a modified version of BL-I, where a soft-sharing mechanism is introduced between all the encoders of each task during training. Soft-sharing connections are provided in the encoder feature space by concatenating task-specific encoded features, as shown in Fig. 3.

Table 4 We show the 2-task experimental results for validating on the 13-class NYUv2 dataset, 7-class CityScapes and 6-class ISPRS for segmentation, normal, and depth tasks on SegNet based self-distilled MTL architectures, i.e. SD-MTCNN and S^3DMT-Net. T Teacher, S Stu.dent, simple MTL as Baseline (BL-I). Equations 5 and 6 are used for our proposed S^3DMT-Net architecture, with the best-performing Teacher and Student models selected from the referenced SOTAs. We denote these equations as (5) and (6), respectively. ↓ / ↑ represents the higher/lower the outperformance. Best results shown in bold

Method	NYUv2				City scapes				ISPRS			
	Sem. Seg ↑		Depth Err. ↓		Sem. Seg ↑		Depth Err. ↓		Sem. Seg ↑		Depth Err. ↓	
	IoU	mIoU	Abs	Rel	IoU	mIoU	Abs	Rel	IoU	mIoU	Abs	Rel
Single Task	56.30	17.84	0.6796	0.2930	91.06	51.70	0.0134	24.95	80.19	38.62	0.1762	1.3871
Vanilla MTL	56.79	18.50	0.6889	0.2909	91.22	52.05	0.0139	26.37	80.31	39.15	0.1896	1.4949
Dense [8]	55.59	17.22	0.6002	0.2654	90.89	51.91	0.0138	27.21	78.92	35.92	0.1746	1.3936
Cross-Stitch [20]	53.99	17.01	0.6095	0.2671	90.33	50.08	0.0154	34.49	78.69	34.83	0.1794	1.5041
Split -Wide [23]	55.83	18.13	0.6126	0.2584	90.63	50.17	0.0167	44.73	79.34	35.98	0.1896	1.633
Split -Deep	46.39	13.40	0.7321	0.3057	88.69	49.85	0.0180	43.86	66.60	24.46	0.3712	3.4296
MTAN [23]	56.24	18.32	0.5931	0.2562	91.11	53.04	0.0144	33.63	79.02	35.98	0.1867	1.3072
SD-MTCNN^T (5)	57.18	23.01	0.5847	0.2466	92.54	56.70	0.0131	27.68	82.44	45.91	0.1675	1.2907
SD-MTCNN^S	55.80	23.19	0.6033	0.2588	91.60	54.09	0.0150	33.23	81.91	45.64	0.1733	1.4112
SD-MTCNN^T (6)	56.29	22.63	0.6042	0.2544	91.57	54.63	0.0134	31.11	82.23	45.21	0.1728	1.2990
SD-MTCNN^S	5609	22.50	0.6103	0.2674	90.99	53.27	0.0151	34.82	81.88	44.72	0.1760	1.4108
S^{3}DMT-Net^T (5)	**59.57**	**23.13**	**0.6142**	**0.2182**	**92.61**	**57.56**	**0.0124**	**23.51**	**82.62**	**46.48**	**0.1641**	**1.2853**
S^{3}DMT-Net^T	58.51	22.86	0.6234	0.2542	92.08	55.63	0.0138	28.31	82.41	45.13	0.1685	1.3700
S^{3}DMT-Net^T (6)	58.81	23.11	0.6148	0.2566	92.36	56.79	0.0129	24.85	82.34	46.20	0.1702	1.2997
S^{3}DMT-Net^S	57.97	22.49	0.6276	0.2601	91.91	55.42	0.0125	27.63	82.16	45.24	0.1777	1.4018

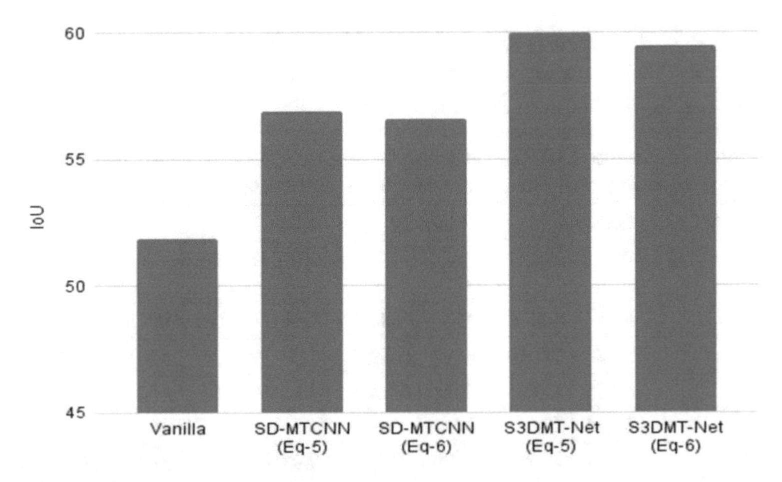

Fig. 5 Bar graph for comparing the pixel accuracy (IoU) for Vanilla MTL, SD-MTCNN [11] and S^3DMT-Net [32] on 3-task NYUv2 dataset

(iii) **Baseline BL-III** incorporates Student networks at each task-specific bottleneck, but without soft-sharing among the encoders. The Student and Teacher architectures share the latent encoded feature space, and the predictions of the two are distilled using distillation losses.

Tables 5, 6, 7 show that there was a significant improvement in performance by changing the baseline designs (Baseline I-III), although they still fell short of specific state-of-the-art (SOTA) evaluation measures. The S3DMT-Net architecture was designed using Baseline (BL-II) and Baseline (BL-III) at the end of the encoder and decoder, and the same training process as described in Sect. 4.2 was followed. Both Eqs. (5 and 6) described in Sect. 3.4 were used to train this setup. The edge detection, depth, and 2D key-point prediction tasks outperformed all other tasks described for the created Mini-Taskonomy (Sect. 4.1), and the proposed architecture surpassed all of the prior baselines and referenced SOTAs in terms of evaluations. The ablations on the architecture design demonstrated the importance of cross-talk (soft-sharing) and self-distillation. Figure 5 compares the pixel accuracy (IoU) performance of vanilla MTL, SD-MTCNN, and S^3DMT-Net architectures.

B. **Effect of Number of Tasks**: Our proposed MTL models' effectiveness is evaluated by altering the numb.er of tasks. We decide to opt the classification task i.e. segmentation as the primary task and rest as the auxiliary ones. The Mini Taskonomy has the most tasks of any of the datasets listed in Sect. 4.1. On the basis of Eq. 5, we train the S^3DMT-Net while disregardi.ng the performance of all auxiliary tasks. For the time being, we are only concerned with the incorporation of auxiliary/supplementary tasks rather than permuting with the primary task. For task-specific loss terms, we tra.in the following stated setting with weight equals to 1. The X and Y axis in Fig. 6 reflect the amount of tasks and accuracy (IoU), respectively. It is possible to demonstrate that increasing

Table 5 We show ablation analysis on 3-task 13-class N.YUv2 dataset for segmentation, normal estimation, and depth tasks on SegNet based self-distilled MTL architectures, i.e. SD-MTCNN and S^3DMT-Net. T Teacher, S Stu.dent, simple MTL as Baseline (BL-I). Equations 5 and 6 are used for our proposed S3DMT-Net architecture, with the best-performing Teacher and Student models selected from the referenced SOTAs. We denote these equations as (5) and (6), respectively. ↓ / ↑ represents the higher./ lower the outperformance. Best results shown in bold

Method	Sem. Seg		Depth Est		Surface				
	↑		↓		Angke Distance ↓		Within t °↑		
	IoU	mIoU	Abs	Rel	Mean	Median	11.25	22.5	30
Vanilla MTL	51.88	15.59	0.6177	0.2663	32.08	26.93	21.16	43.01	55.00
BL - II	57.65	19.02	0.6221	0.2745	30.16	23.16	24.57	47.99	59.80
BL - III^T	59.34	23.57	0.6451	0.2677	29.51	23.87	25.16	49.67	61.28
BL - III^S	58.97	23.88	0.6589	0.2734	29.97	23.51	24.29	48.48	60.94
Traditional KD	52.55	15.54	0.6136	0.2542	31.94	26.47	21.96	43.72	55.48
SD-MTCNN^T (5)	56.90	22.44	**0.5857**	**0.2483**	29.66	25.02	22.66	45.84	58.39
SD-MTCNN^S	56.43	21.90	0.5986	0.2532	30.76	26.98	19.52	44.20	57.32
SD-MTCNN^T (6)	56.61	22.31	0.5864	0.2458	30.04	25.24	21.36	45.40	58.20
SD-MTCNN^S	56.11	22.59	0.5899	0.2506	30.42	25.63	21.22	44.79	57.21
S^3DMT-Net^T (5)	**59.96**	**24.42**	0.5922	0.2511	**28.31**	**22.62**	**25.69**	**50.10**	**62.34**
S^3DMT-Net^T	59.10	24.18	0.6059	0.2673	29.28	23.78	24.00	47.97	60.07
S^3DMT-Net^T (6)	59.47	24.10	0.5928	0.2553	28.76	22.94	25.20	49.82	61.55
S^3DMT-Net^S	58.83	23.96	0.6106	0.2645	29.40	23.85	24.14	48.11	60.28

the number of auxiliary tasks enhanc.es the main task considerably. In a few cases, the segmentation task performs almost identically, as shown in Fig. 6 (like, S^3DMT-Net (2-task) and S^3DMT-Net (3-task)).

C. **Computational Protocols**: The proposed self-distilled MTL models are computationally less expensive than standard MTL architectures because we use transposed convolutional layers for decoders in the Student network. In the inference stage, we can separately evaluate the Student and Teacher networks. Such self-distilled MTL setups provide better optimization in fewer training iterations.

5.3 *Visualization*

For Fig. 9, outputs from filter maps of size 0.64 from the first CNN block of each task for respective encoders and.penultimate CNN block with filter size 0.64 from each of the task-.specific decoder. We utilize only 3 feature maps (channels) in order to view the learned feature maps conveniently. At the decoder end, the decoded visualization for the depiction of depth and semantic tasks justify a strong correlation in task-specific aspects. As seen in Fig. 4, the task-specific encoded characteristics are

Table 6 We present an ablation analysis of SegNet-based self-distilled multi-task learning (MTL) architectures for 5 different tasks including semantic segmentation, normal estimation, depth estimation, key-point prediction, and edge detection on the Mini-taskonomy dataset, i.e. SD-MTCNN and S^3DMT-Net. T Teacher, S Stu.dent, simple MTL as Baseline (BL—I). Equations 5 and 6 are used for our proposed S3DMT-Net architecture, with the best-performing Teacher and Student models selected from the referenced SOTAs. We denote these equations as (5) and (6), respectively. ↓ / ↑ represents the higher./ lower the outperformance. Best results shown in bold

Method	Sem. Seg.↑		Depth Est.↓		Surface↑	Key ↓	Edge ↓
	IoU	mIoU	Abs	Rel	CS	Abs	Abs
Vanilla MTL	88.85	49.21	0.0307	0.4216	0.9285	0.0173	0.0094
BL - II	88.18	46.63	0.0322	0.5303	0.9020	0.0150	0.0102
BL - III^T	89.39	55.13	0.0306	0.6350	0.9243	0.0141	0.0099
BL - III^S	88.90	53.88	0.0343	0.9016	0.9206	0.0174	0.0134
SD-MTCNN^T (5)	89.36	55.36	0.0290	**0.3572**	0.9058	0.0138	0.0097
SD-MTCNN^S	89.02	54.78	0.0370	0.4010	0.8608	0.0327	0.0118
SD-MTCNN^T (6)	89.21	55.26	0.0327	0.4318	0.8907	0.0147	0.0101
SD-MTCNN^S	88.93	53.68	0.0419	0.5221	0.8616	0.0340	0.0129
S^{3}DMT-Net^T (5)	**89.69**	**58.35**	**0.0269**	0.3906	**0.9288**	**0.0121**	**0.0087**
S^{3}DMT-Net^T	89.43	57.47	0.0299	0.4652	0.9123	0.0203	0.0113
S^{3}DMT-Net^T (6)	89.50	58.14	0.0282	0.4108	0.9244	0.0126	0.0093
S^{3}DMT-Net^S	89.03	57.66	0.0334	0.4515	0.9111	0.0255	0.0167

exchanged via cross-talk (concatenation of the respective encoder block's la.st.layer) and finally fused to the relevant representation space. The CityScapes dataset is used to visualize these feature maps.

6 Summary and Future Directions

This chapter concludes our contribution to dense prediction multi-task learning with the self-distillation scenario. Our novel self-distillation methodology infused the knowledge within the network for MTL. We especially intend to distill information from decoders's task-specific modules to the layers of the encoder using the hard-sharing-based and soft-sharing-based MTL CNN architectures. To train the network endlong, two training strategies are used: One consists of combining the Teacher-Student task-specific losses with our proposed SD losses between the tasks, and the second compiles the strategy based on ensemble learning for each of the task loss as well as the distillation losses. Our experimental evaluations claim that the proposed self-distillation MTL beats the conventional way of distilling the information and referred MTL methods. Also, self-distillation in multi-task can be leveraged to deploy the well trained over edge devices to do multiple tasks simultaneously. Our present

Table 7 We show the ablation analysis on 2-task for 13-class NYUv2, 7-class CityScapes and 6-class ISPRS datasets on SegNet based self-distilled MTL architectures, i.e. SD-MTCNN and S^3DMT-Net. T Teacher, S Stu.dent, simple MTL as Baseline (BL-I). (5). and (6) used for Eqs. 5 and 6 respectively. We also highlight the Teacher and Student which.ever performs best from the referred SOTA methods. ↓ / ↑ represents the higher./ lower the outperformance. Best results shown in bold

Method	NYUv2				CityScapes				ISPRS			
	Sem. Seg ↑		Depth Est. ↓		Sem. Seg ↑		Depth Est. ↓		Sem. Seg ↑		Depth Est. ↓	
	IoU	mIoU	Abs	Rel	IoU	mIoU	Abs	Rel	IoU	mIoU	Abs	Rel
Vanil.la MTL	56.79	18.50	0.6889	0.2909	91.22	52.05	0.0139	26.37	80.31	3.9.15	0.1.896	1.4949
BL - I.I	56.30	17.84	0.6239	0.2683	91.0.39	52.75	0.0145	31.53	81.17	3.9.93	0.1.761	1.5841
BL - II.I^T	58.62	22.59	0.6843	0.2885	91.54	55.68	0.0132	29.45	81.55	4.5.76	0.0.1753	1.5449
BL - II.I^S	58.08	21.80	0.6990	0.2927	91.72	54.51	0.0139	33.87	81.28	4.5.15	0.1.784	1.5685
SD-MTCNN^T (5)	57.18	23.01	**0.5847**	0.2466	92.54	56.70	0.0131	27.68	82.44	45.91	0.1675	1.2907
SD-MTC.NN^S	55.80	**23.19**	0.6033	0.2588	91.60	54.09	0.0150	33.23	81.91	45.64	0.1733	1.4112
SD-MTCNN^T (6)	56.29	22.63	0.6042	0.2544	91.57	54.63	0.0134	31.11	82.23	45.21	0.1728	1.2990
SD-MTC.NN^S	5609	22.50	0.6103	0.2674	90.99	53.27	0.0151	34.82	81.88	44.72	0.1760	1.4108
S^{3}DMT-Net^T (5)	**59.57**	23.13	0.6142	**0.2182**	**92.61**	**57.56**	**0.0124**	**23.51**	**82.62**	**46.48**	**0.1641**	**1.2853**
S^{3}DMT-.Net^T	58.51	22.86	0.6234	0.2542	92.08	55.63	0.0138	28.31	82.41	45.13	0.1685	1.3700
S^{3}DMT-Net^T (6)	58.81	23.11	0.6148	0.2566	92.36	56.79	0.0129	24.85	82.34	46.20	0.1702	1.2997
S^{3}DMT-.Net^S	57.97	22.49	0.6276	0.2601	91.91	55.42	0.0125	27.63	82.16	45.24	0.1777	1.4018

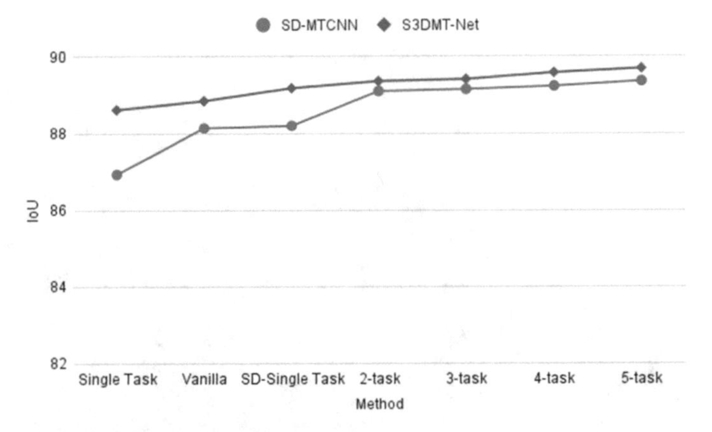

Fig. 6 Analysis on the effect of the amount of tasks in multitask learning and compare the perf.ormance between SD-MTCNN and S^3DMT-Net

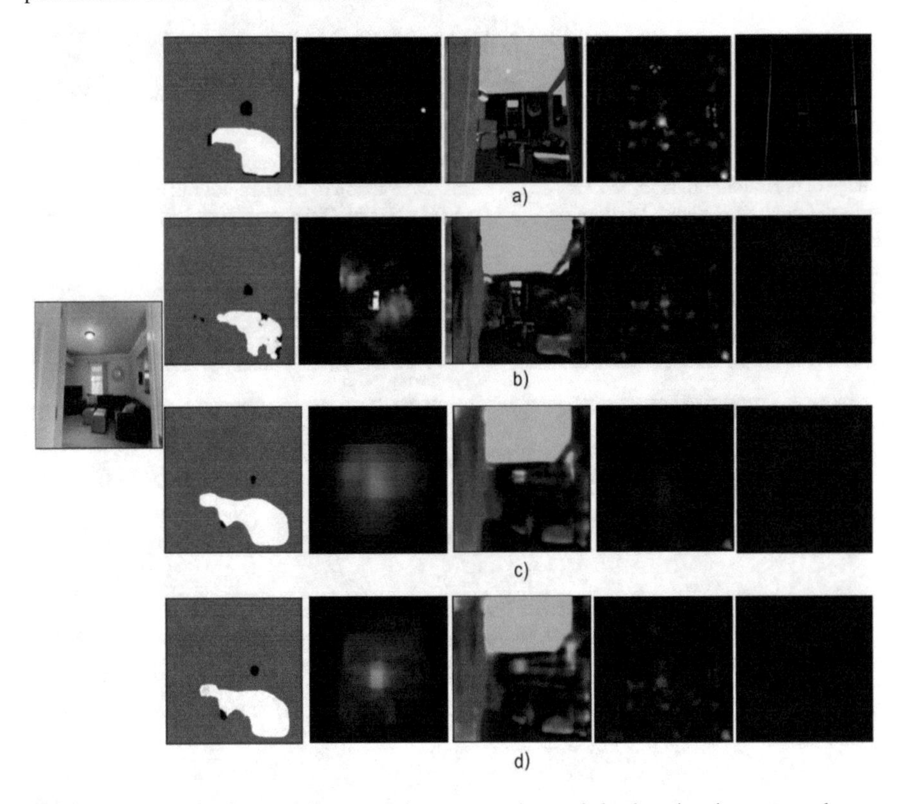

Fig. 7 Visualizing and comparing semantic segmentation and depth estimation outputs by our proposed self-distilled MTL models for the given input monocular image of the Mini-Taskonomy dataset. Top to botto.m depicts: **a** Ground Truth, **b** Conventional Knowledge Distillation (KD), **c** SD-MTCNN (Teacher T), and **d** S^3DMT-Net (Teacher T). Best view in color

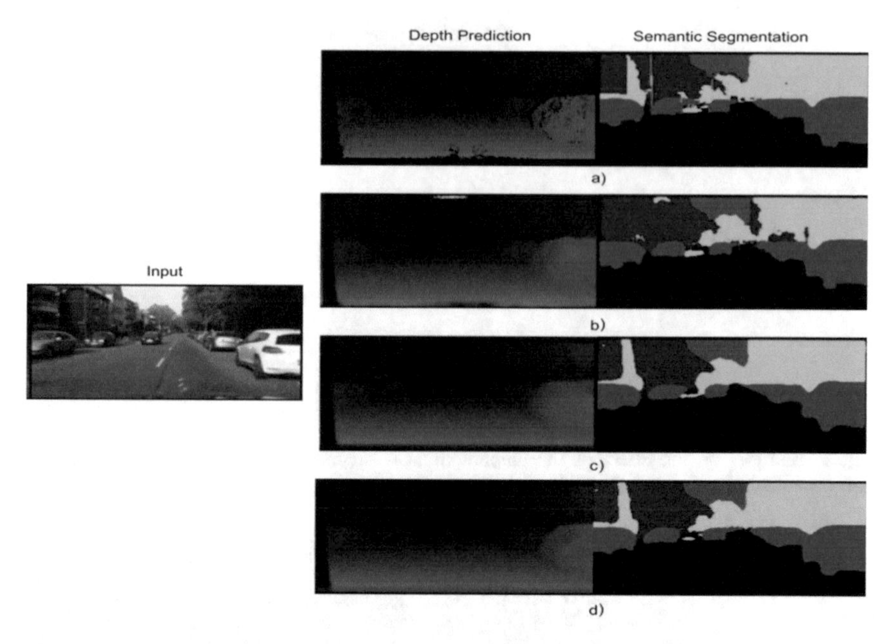

Fig. 8 The visualization of estimation the depth and segmentation of each pixel for the given input image of the outdoor dataset, i.e., CitySc.apes. Images from top to down represent: **a** Ground.Truth, **b** Traditional Knowledge Distillation (KD), **c** SD-MTCNN (Teacher T), and (**d**) S^3DMT-Net (Teacher T). Best view in color

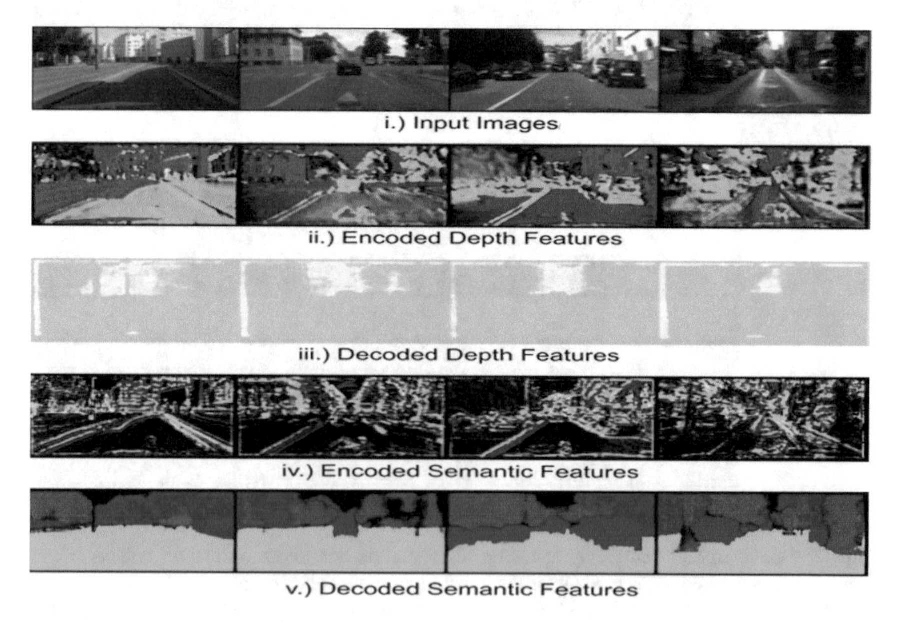

Fig. 9 The visualized feature maps for the encoded and decoded convolutional blocks of S^3DMT-Net for the respective input images

research involves adding a variety of students from decoders between the Student and the Teacher in order to accomplish a principal continuous self-distillation.

References

1. Caruana, R.: Multitask learning. Machine learning **28**(1), 41–75 (1997)
2. Vijay Badrinarayanan, Alex Kendall, Roberto Cipolla.: SegNet: A deep convolutional encoder-decoder architecture for image segmentation. CoRR abs/1511.00561 (2015). arXiv:1511.00561 http://arxiv.org/abs/1511.00561
3. Jianhui Chen, Jiayu Zhou, Jieping Ye.: Integrating low-rank and group- sparse structures for robust multi-task learning. In Proceedings of the 17th ACM SIGKDD international conference on Knowledge discovery and data mining. ACM, 42–50 (2011)
4. Marius Cordts, Mohamed Omran, Sebastian Ramos, Timo Rehfeld, Markus Enzweiler, Rodrigo Benenson, Uwe Franke, Stefan Roth, Bernt Schiele.: The cityscapes dataset for semantic urban scene understanding. In: Proceedings of the IEEE conference on computer vision and pattern recognition. 3213–3223 (2016)
5. David Eigen and Rob Fergus.: Predicting depth, surface normals and semantic labels with a common multi-scale convolutional architecture. In Proceedings of the IEEE international conference on computer vision. 2650–2658 (2015)
6. Theodoros Evgeniou, Massimiliano Pontil.: Regularized multi–task learning. In: Proceedings of the tenth ACM SIGKDD International Conference on Knowledge Discovery and Data Mining. ACM, 109–117 (2004)
7. Geoffrey Hinton, Oriol Vinyals, Jeff Dean.: Distilling the knowledge in a neural network. arXiv:1503.02531 [stat.ML] (2015)
8. Gao Huang, Zhuang Liu, Laurens Van Der Maaten, Kilian Q Weinberger.: Densely connected convolutional networks. In: Proceedings of the IEEE Conference on Computer Vision and Pattern Recognition. 4700–4708 (2017)
9. Laurent Jacob, Jean-philippe Vert, Francis R Bach.: Clustered multi-task learning: A convex formulation. In: Advances in Neural Information Processing Systems. 745–752 (2009)
10. Ankit Jha, Awanish Kumar, Biplab Banerjee, Subhasis Chaudhuri.: AdaMT-Net: An Adaptive Weight Learning Based Multi-Task Learning Model For Scene Understanding. In: 2020 IEEE/CVF Conference on Computer Vision and Pattern Recognition Workshops (CVPRW). 3027–3035. https://doi.org/10.1109/CVPRW50498.2020.00361 (2020)
11. Ankit Jha, Awanish Kumar, Biplab Banerjee, Vinay Namboodiri.: SD-MTCNN: Self-Distilled Multi-Task CNN. In BMVC (2020)
12. Alex Kendall, Yarin Gal, and Roberto Cipolla.: Multi-task learning using uncertainty to weigh losses for scene geometry and semantics. In: Proceedings of the IEEE Conference on Computer Vision and Pattern Recognition. 7482–7491 (2018)
13. Diederik P Kingma and Jimmy Ba.: Adam: A method for stochastic optimization. arXiv preprint arXiv:1412.6980 (2014)
14. Abhishek Kumar, Hal Daume III.: Learning task grouping and overlap in multi-task learning. arXiv preprint arXiv:1206.6417 (2012)
15. Peiye Liu, Wu Liu, Huadong Ma, Tao Mei, Mingoo Seok.: KTAN: Knowledge transfer adversarial network. CoRR abs/1810.08126 (2018)
16. Zhiqiang Shen, Zhankui He, Xiangyang Xue.: MEAL: Multi-Model Ensemble via Adversarial Learning. CoRR abs/1812.02425 (2018). arXiv:1812.02425 http://arxiv.org/abs/1812.02425 (2018)
17. Rottensteiner, F., Sohn, G., Gerke, M., Wegner, J.D., Breitkopf, U., Jung, J.: Results of the ISPRS benchmark on urban object detection and 3D building reconstruction. ISPRS J. Photogramm. Remote. Sens. **93**(2014), 256–271 (2014)

18. Adriana Romero, Nicolas Ballas, Samira Ebrahimi Kahou, Antoine Chassang, Carlo Gatta, Yoshua Bengio.: FitNets: Hints for Thin Deep Nets. arXiv:1412.6550 [cs.LG] (2014)
19. Sebastian Ruder.: An overview of multi-task learning in deep neural networks. arXiv preprint arXiv:1706.05098 (2017)
20. Ishan Misra, Abhinav Shrivastava, Abhinav Gupta, Martial Hebert.: Cross-Stitch Networks for Multi-Task Learning. In: The IEEE Conference on Computer Vision and Pattern Recognition (CVPR) (2016)
21. Olaf Ronneberger, Philipp Fischer, and Thomas Brox.:. U-net: Convolutional networks for biomedical image segmentation. In International Conference on Medical Image Computing and Computer-assisted Intervention. Springer, 234–241 (2015)
22. Sebastian Ruder, Joachim Bingel, Isabelle Augenstein, Anders Søgaard.: Sluice networks: Learning what to share between loosely related tasks. ArXiv abs/1705.08142 (2017)
23. Shikun Liu, Edward Johns, Andrew J. Davison.: End-To-End Multi-Task Learning With Attention. In The IEEE Conference on Computer Vision and Pattern Recognition (CVPR) (2019)
24. Nathan Silberman, Derek Hoiem, Pushmeet Kohli, Rob Fergus.: Indoor segmentation and support inference from rgbd images. In European Conference on Computer Vision. Springer, 746–760 (2012)
25. Trevor Standley, Amir R Zamir, Dawn Chen, Leonidas Guibas, Jitendra Malik, Silvio Savarese.: Which tasks should be learned together in multi-task learning? arXiv preprint arXiv:1905.07553 (2019)
26. Chao Yuan.: Multi-task learning for bayesian matrix factorization. In: 2011 IEEE 11th International Conference on Data Mining. IEEE, 924–931 (2011)
27. Sergey Zagoruyko, Nikos Komodakis.: Paying More Attention to Attention: Improving the Performance of Convolutional Neural Networks via Attention Transfer. CoRR abs/1612.03928. arXiv:1612.03928 http://arxiv.org/abs/1612.03928 (2016)
28. Amir R Zamir, Alexander Sax, William Shen, Leonidas J Guibas, Jitendra Malik, Silvio Savarese.: Taskonomy: Disentangling task transfer learning. In Proceedings of the IEEE Conference on Computer Vision and Pattern Recognition. 3712–3722 (2018)
29. Linfeng Zhang, Jiebo Song, Anni Gao, Jingwei Chen, Chenglong Bao, Kaisheng Ma.: Be your own teacher: improve the performance of convolutional neural networks via self distillation. CoRR abs/1905.08094. arXiv:1905.08094 http://arxiv.org/abs/1905.08094 (2019)
30. Yu Zhang and Qiang Yang.: A survey on multi-task learning. arXiv preprint arXiv:1707.08114 (2017)
31. Jun Zhu, Ning Chen, Eric P Xing. Infinite latent SVM for classification and multi-task learning. In Advances In Neural Information Processing Systems. 1620–1628 (2011)
32. Ankit Jha, Biplab Banerjee, Subhasis Chaudhuri.: S3 DMT-Net: improving soft sharing based multi-task CNN using task-specific distillation and cross-task interactions. In Proceedings of the Twelfth Indian Conference on Computer Vision, Graphics and Image Processing (ICVGIP '21). Association for Computing Machinery, New York, NY, USA, Article 16, 1–9. https://doi.org/10.1145/3490035.3490274. (2021)

Knowledge Distillation for Autonomous Intelligent Unmanned System

Anatolii Kargin and Tetyana Petrenko

Abstract The need for more advanced Unmanned Systems (US) is supported by the growing demand not only of the military industry but also of the civil community. The efforts of AI-enabled US developers are focused on increasing the level of autonomy of future systems. Known AI models can't comprehensively solve the problem of US autonomy due to gap between the two paradigms "processing data from sensor" and "decision-making based on domain expert knowledge". Through introduced universal models of external and internal meaning of Knowledge Granule (KG), approach proposed allows to move from the tasks of distilling data from sensors and distilling semantic knowledge to the single task of sense distilling. The distillate as a meaning of the spatio-temporal segment of data from the sensors is presented with the word sense of a high level of abstraction. The task of knowledge distillation divided into two subtasks: distillation the knowledge about the US environment by abstracting and about temporal events stream by convolution. Ways of using distillate in decision-making and control of the US are considered. An example of decision-making and control based on a traditional Fuzzy Logic System (FLS) is given, the input variables of which are the meaning of the US data from sensors obtained by the sense distillation model. A new goal-driving control model based on distillate is discussed, too. The results of the experiments demonstrated an increase in US autonomy due to the ability to abstract and flexible to switch of action plans are given.

A. Kargin (✉) · T. Petrenko
IT Department, Faculty of Information Systems and Technology, Ukrainian State University of Railway Transport, Kharkiv, Ukraine
e-mail: kargin@kart.edu.ua

T. Petrenko
e-mail: petrenko_tg@kart.edu.ua

A. Kargin
Department of Electronic Computers, Faculty of Computer Engineering and Control, Kharkiv National University of Radio Electronics, Kharkiv, Ukraine

W. Pedrycz and S.-M. Chen (eds.), *Advancements in Knowledge Distillation: Towards New Horizons of Intelligent Systems*, Studies in Computational Intelligence 1100,
https://doi.org/10.1007/978-3-031-32095-8_7

Keywords Autonomous intelligent unmanned system · Knowledge distillation · Knowledge granule · Computing with words · Multilevel abstracting · Fuzzy logic systems · Data from sensors · Goal-driving control

1 Introduction

US passed through three stages in their development [1–3]. At the first stage, programming-based automatic US for autonomous operation required a completely ordered environment, precise adherence to time intervals and sequence of operations. At the second stage, the requirements of ordering the environment were partially removed and intelligent US could perform some functions under conditions of interference and partial uncertainty. Robots providing various kinds of services are examples. Today, the Autonomous Intelligent US (AIUS) are relevant, which are hoped to replace a human in various areas of his activity [4–9]. The widespread deployment of US has actualized the problem of US autonomy. We will consider the autonomy as the ability to carry out a US mission without human intervention. Autonomy is supported by many US components. This chapter focuses on Decision Making and Control (DM&C) functions of the AIUS.

DM&C must take into account the limitations of the US domain. Consider these features of AIUS:

(1) AIUS belongs to the class of embedded systems that operate in real time and directly communicate with sensors and actuators. Therefore, the DM&C computational model should be compact and efficient [2, 3].
(2) AIUS uses data from a large number of heterogeneous sensors, which characterizes the US perception system as multimodal and heterogeneous [10]. This requires that the perception models be able to represent the results of processing heterogeneous data in a uniform way.
(3) AIUS operates in an environment that characterized by uncertainty and variability, noisy data, and decisions are made in conditions of incomplete knowledge. This feature is taken into account in many approaches to primary data processing [11–13].
(4) AIUS feature of decision-making is a multi-stage process in the real time. Decision is make considering a plan of actions for achieving the goal. At each stage of plan, a current state of the plan implementation and a current situation are taken into account.
(5) AIUS should be transparent [14]. This means that the AI must be able to explain and demonstrate the decision made earlier when any events occur that require investigation, including for obtaining a legal opinion.

Let's dwell on the main issue for AIUS. What features of the US are the causes of violation of DM&C autonomous functioning and what requirements do they place on AI models? We have identified four main causes which violate the AIUS autonomy, due to DM&C limited capacity, and for each of them we recommend the remedies.

This information is summarized in Table 1. The remedies in Table 1 are presented below as a list of AI capabilities that support DM&C autonomy:

(1) Models of all AI-enabled DM&C components should be based on a universal model of the meaning of data from sensors. The sense model should be able to represent the data meaning by concepts of different levels of generalization.
(2) The formal foundation of the sense model should be Fuzzy Set (FS).
(3) Dimensionality reduction of the decision-making space in real time should be done as a sense distillation which decides two tasks. The extraction of the meaning of a large dataset from sensors and representing it with a few concepts, and the extraction the meaning of the temporal sequence of events and representing it with a concept of a high level of abstraction.
(4) The FLS-based DM&C can overcome the problem of a large space dimension if it has the ability to use the sense values obtained in the distillation process as FLS inputs variables.

Table 1 Factors that violate the AIUS autonomy

N	Cause	Violation	Remedy
1	Large dimension of the decision-making space represented by a set of granule data from sensors	The solution cannot be obtained for some areas of this space	Reducing the decision-making space while preserving the meaning of the original sensors data set
2	Threshold boundaries of granulation of data from sensors	Incorrect decision because that the real data from sensors not assigned to the corresponding granules due to distorted by noises	Fuzzy representation and processing data granules
3	Knowledge incompleteness due to the complexity of the US environment, generated by a large number of objects and connections between them as also by dynamics of their changes	A stage of the action plan cannot be completed due to low confidence in its feasibility in the current situation	Using the experience of implementing the plan in various similar situations
			Using an environmental exploration strategy to obtain missing information
		Multi-stage plan can't fully implement due to a predetermined planning strategy with strong restrictions on plan stages	Using the strategy of continuous conditional goal-setting planning with fuzzy restrictions on the stages of the plan
			Using the mechanism of the data meaning correction, taking into account the cognitive assessment of the situation
4	Different US components interpret and understand the meaning of the same data in different ways	Solutions compete for the activation of the US actuator, but their comparison is impossible	Using the universal model of data meaning for all types of sensors modalities and knowledge

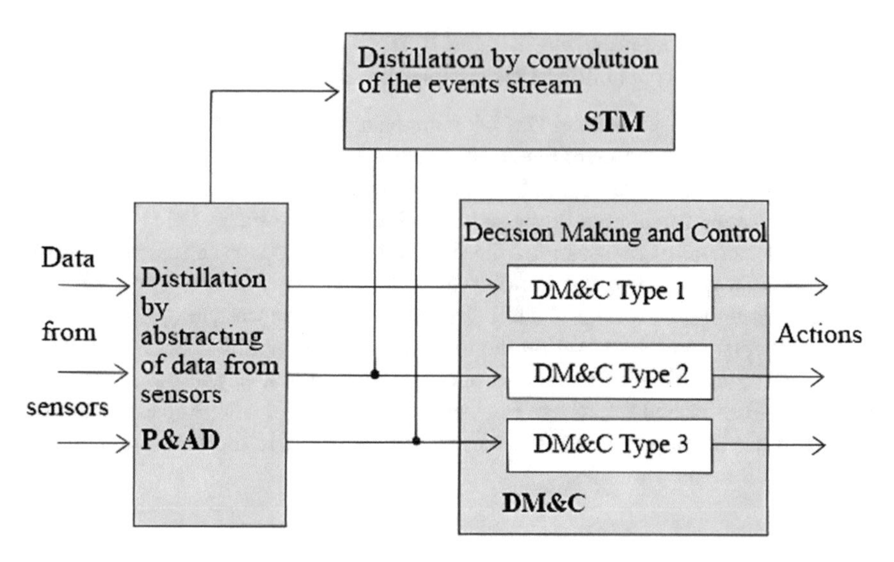

Fig. 1 Composition of DM&C supported by AI

(5) The DM&C can overcome the challenges of plan execution if it has the ability to correct the meaning of data from sensors based on a cognitive assessment of the situation.

Figure 1 shows the structure of DM&C supported by AI with the above abilities. Three components have been identified. The Perception and Abstraction Data (P&AD) component in real-time processes data from sensors and builds US environment state model. The environment state is presented by concepts, more specifically, by the sense of words of different levels of abstraction. The P&AD component solves the sense distillation problem by extracting the meaning of a large dataset from sensors. The second component in Fig. 1 is Short-Term Memory (STM) tasking the construction of US history as an events stream. STM solves the sense distillation problem too by extracting the meaning of the temporal sequence of events and representing it with a concept of a high level of abstraction. The DM&C component itself (Fig. 1) includes three different types of decision-making tasks. DM&C Type 1 is the task of making the current control decision, considering the states of the environment and US only at the time of the decision. Many AI-enabled IoT tasks are of this type [15]. DM&C Type 2 suggests that to find the current control decision, information is used not only about the current state, but about the process's history [16]. The area of autonomous vehicles is rich in examples of this type of problems [8, 9]. DM&C Type 3 is the control of the implementation of a multi-stage plan, taking into account the changes of environment state and the US target. Examples of such AI-based DM&C Type 3 tasks presented in [17–20]. DM&C components as inputs use quantitative estimates of the meaning of data from sensors obtained from the sense distillation components P&AD and STM (Fig. 1).

This chapter discusses models of both types of sense distillation and its use in decision-making problems in US.

2 Sense Distillation Problem

In general, the DM&C initial model can be represented as (1)

$$\begin{aligned}\mathbf{u}(t_0) = \mathbf{f}^0(x_{ijk}(t),\ k = 1, 2, \ldots, n_j,\ j = 1, 2, \ldots, m_i,\ i = 1, 2, \ldots, h,\\ t = t_{-l}, t_{-(l+1)}, \ldots, t_{-1}, t_0, t_{+1}, \ldots, t_{+r}),\end{aligned} \tag{1}$$

where $\mathbf{u}$ is US control decisions vector; $\mathbf{x} = (x_{ijk}(t))$ is input vector representing at time t the value of the kth data granule from the jth sensor of ith modality; t_0 is current moment of time, t_{-l} is previous and t_{+l} subsequent lth points of time; h is number of sensory modalities; m_i is number of sensors of ith modality and n_j is the number of data granules that cover the range of possible values of the jth sensor.

As noted above, the large-scale complicated model (1) does not allow it to be implemented either analytically or numerically, so it is transformed into several lightweight models (2).

$$\begin{aligned}&\mathbf{u}(t_0) = \mathbf{f}^3(\mathbf{q}(t_0), \mathbf{p}(t_0), \mathbf{w}(t_0));\\ &\mathbf{q}(t_0) = \mathbf{f}^{21}(\mathbf{w}(t),\ t = t_{-l}, t_{-(l+1)}, \ldots, t_{-1}, t_0);\\ &\mathbf{p}(t_0) = \mathbf{f}^{22}(\mathbf{w}(t),\ t = t_0, t_{+1}, t_{+2}, \ldots, t_{+r});\\ &\mathbf{w}(t_0) = \mathbf{f}^1(x_{ijk}(t_0),\ k = 1, 2, \ldots, n_j,\ j = 1, 2, \ldots, m_i,\ i = \overline{1, h}).\end{aligned} \tag{2}$$

In (2) $\mathbf{f}^3$ there is a fuzzy model of DM&C component Type I (Fig. 1) for computing the control decisions vector $\mathbf{u}$ based on the numerical values of input vectors $\mathbf{q}$, $\mathbf{p}$, $\mathbf{w}$ of small dimension. A $\mathbf{f}^{21}$and $\mathbf{f}^{22}$ are models of the sense distillation (component STM, Fig. 1) of time events sequences, respectively, preceding the moment of decision making t_0 and the upcoming plan of actions. The $\mathbf{f}^1$ is model of sense distillation (component P&AD, Fig. 1) with small dimension output vector $\mathbf{w}$ which represents the numerical values of the meaning of the input vector $\mathbf{x}$.

Nova day the rich experiences of solving distillation problem on the basis of Artificial Neural Networks (ANN) are accumulated [21, 22]. However, they cannot be used as sense distillation models $\mathbf{f}^{21}$, $\mathbf{f}^{22}$ and $\mathbf{f}^1$ due to the requirement which were previously formulated, especially lack transparency decision-making process. Another knowledge-oriented approach has excellent transparency in explaining how decisions are made, but the works devoted to the knowledge distillation cannot be used as sense distillation model, too. In this works, the meaning of knowledge is considered within the semantic networks theory and is given only by semantic relations [23, 24]. This shows that there is a problem is known as the gap between the two paradigms "Data from sensors" and "Computing With Words" (CWW). The bridge between these paradigms is established by semiotics. Semiotics conceptual model gives word representation by the triangle. The word belongs to the sign system as its element. The word reflects sensation and perception of the essences of the real-world. And, thirdly, the word gives the concept to the sign. Three models of word were introduced, accordingly [25, 26]: sign, internal meaning and external meaning of the word.

Figure 2a illustrate these three models of word. The External Meaning of Word (EMW) is semantic model defining the sense of a word N using the meaning of other words. Formally, EMW is a semantic relationship graph: the vertex N is linked to the vertices M_i, which depict the words revealed the sense of the word N [26]. The graph arcs have the parameters that indicate the semantic relationship type. The cloud around N depicts the sign model of word presented in natural language. In Fig. 2a, arrows shaded inside depict the Internal Meaning of Word (IMW). This emphasizes the IMW is computational sense of the word, that expresses the perception of data from sensors. A numerical assessment of the IMW expresses the degree of conformity of the EMW with the situation represented by the spatio-temporal segment of data from sensors [25, 26]. For unambiguous understanding, we will use the term Knowledge Granule (KG) to denote a portion of both data from sensors and verbal knowledge received from an expert in the form of words. In this regard, instead of the abbreviations EMW and IMW, we will use EMKG and IMKG, respectively.

Thus, the representation of sense distillation model is the EMKG, and the distillation process is IMKG computing. Figure 2b shows an example of representation of sense distillation model type $\mathbf{f}^1$ in (2). The Knowledge Base (KB) fragment presents the sense distillation model as a hierarchically ordered set of KGs. Such structure describes the expert's knowledge about the domain. At the lower (zero) level are KGs, whose EMKG is determined through granular data from sensors of an arbitrary type (numbers, codes or text, subsets of elements of different nature, etc.) [27]. Above

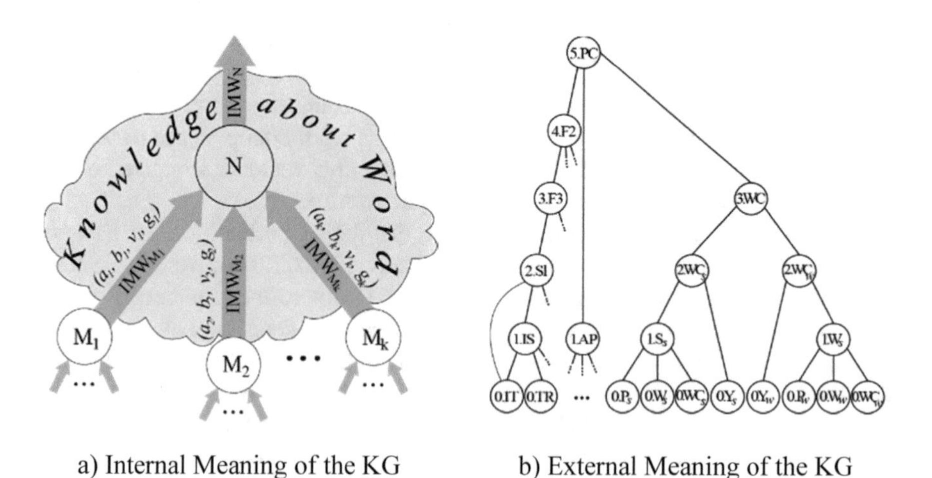

Fig. 2 Graphic illustration of the concept of the KG meaning

are KGs, the understanding meaning of which requires reference to the concepts of lower levels and possibly directly to granular data from sensors. At the upper levels of the knowledge structure are KGs, which represent the meaning of not individual data, but the meaning of the objects, processes or whole situation or scene. Such a granular structure of knowledge is consistent with the general concept of information granules [28]. In Fig. 2b, the five-level KB fragment of the Smart Traffic-Light (STL) domain is presented [26]. The fragment shows how the sense of *5.PC* KG (comfort conditions of a pedestrian) of the fifth level of abstraction is defined based on the a dozen *0*th level granules of data from sensors. The distillation process is the computation of the IMKG in accordance with the EMKG knowledge structure and represents the bottom-up process of abstracting and generalization. Obtained according to model (2), distillate $\mathbf{w}(t_0)$ is a numerical assessment of conformity the definition of pedestrian comfort with current data from the sensors, represented by zero-level KGs in Fig. 2b. In this case, the distillate as a numerical assessment of the date sense is consistent with the DM&C knowledge representation model (Fig. 1), and the IMKG can be directly used as an input variable of the DM&C.

Objective of this study is sense distillation model that maps the spatio-temporal segment of data from sensors into a sense of KGs presented by a high-level abstraction word, with the aim of subsequent use this distillate as input of DM&C. This chapter discusses an original approach to transforming a large-scale complicated decision-making model $\mathbf{f}^0$ based on a set of data from sensors into lightweight decision-making models $\mathbf{f}^3$ based on the meaning of this data set. The meaning of the dataset is the distillate produced by the new sense distillation models $\mathbf{f}^{21}$, $\mathbf{f}^{22}$ and $\mathbf{f}^1$ which will be discussed in this chapter.

3 Related Work

The AI for AIUS domain can be viewed from different perspectives.

The conceptual issues of organizing AI-enabled tasks in AIUS are discussed in many works devoted to the creation of AI for AIUS [1–3, 5–9, 29]. The classification of AI proposed for AIUS have done with respect to US specificity. Mechanical, thinking, and feeling AI are distinguished [3, 30, 31]. Restrictions of "pure" ANNs for using in AIUS are discussed in [32] and a hybrid approach is proposed for using ANNs in conjunction with knowledge-based models [33]. A Thrill-K, as architecture of universal cognitive AI is proposed [33]. Its knowledge is distributed across three levels of abstraction. At the lower level, instantaneous knowledge, represented by ANN parameters, tuned to solve specific problems. At the second and third levels are the standby and external knowledge sources. In works dedicated to AIUS, AI-enabled tasks are considered [34]. Such cognitive functions as learning [29, 35], emotions [36], perception and emotions [37], and adaptation are discussed. The MicroPsi architecture based on the theory of General AI is proposed [38]. General AI comprehensively combines such cognitive functions as emotions, motives, and needs.

Data distillation is mainly used not at the Edge Computing level for decision making in AIUS [21], but at the Cloud Computing level, where the limitations on computing resources are not critical and do not require coordination with other AIUS components. Special methods for overcoming the incompleteness of knowledge at sparse fuzzy-rule-based systems are developed [39]. Dimension reduction of datasets from sensors is discussed in instance selection (dataset reduction, or dataset condensation), data pre-processing, machine learning and data mining [40, 41]. However, these methods assume that features that are either redundant or irrelevant are pruned out and thus can be removed without significant loss of information. At the same time, the key issue of data distillation is not solved, namely, the creation of a new feature that generalizes the features of the original data necessary for decision making.

Approaches, models, and AI frameworks for creating AI-enabled components at the Edge Computing level are considered when creating robotics, the Internet of things, smart machines. Decision making carries out Rules Engine (RE) in this domain [42–44]. RE is based on different inference models ranging from symbolic inference to Bayesian inference. The use of FLS as a fuzzy RE is not considered since "pure" FLS has a problem. There is a limited capability. The number of input variables must not exceed 5–7 [45]. Scalability is challenge, when new input numerical variables is adding or the linguistic variables definitions is changing [44]. Additionally, the problem of empty space of sparse fuzzy-rule-based systems, where the FSs do not cover the whole input universe of discourse is known [39]. The possibility of using RE based on fuzzy rules is considered in [25, 46–48]. To do this, the Abstraction Engine is included in AIUS, which preliminarily performs the function of data distillation, and the distillate is already used as the input data of the RE. Here, the distillation of data from sensors, in contrast to the traditionally used ANN approaches [22], is based on the abstraction and categorization paradigm, the idea and foundations of which are established in [49–51]. The idea of abstraction and categorization to reduce the dimensionality of a data set by creating new knowledge that aggregates the properties stored in the data was the impetus for the creation of a granular computing approach [52, 53]. It is convenient to implement the abstraction mechanism on the base of information granules. Granular computing provides fast and flexible rule processing [28, 52, 54–56]. A set of design methods has been developed that converts data into information granules, for example, FSs, interval, rough, and intuitionistic FSs [13, 27, 28, 52, 53].

The categorization served as the basis for the development of another AI paradigm, namely CWW, founded by Zadeh [57] and developed in [53, 58–60]. The CWW idea is that computing systems built on CWW principles can reason and make decisions based on concepts. CWW uses variables that have qualitative values and takes into account the vagueness, imprecision and semantic vagueness of the information. Unlike classical theories, which use probabilistic numerical interpretations of qualitative concepts and semantic uncertainty, CWW operates with improbable characteristics based on the degree of truth. The full range of CWW models can be found in review [61].

The mathematical basis on granular computing and CWW are FSs [62, 63], which have convincing generalization properties: the sense of data expressed by a FS, inference using words, the meaning of which is also determined by FSs, and the result of inference either in the form of words [64] or numerical data [65]. These CWW capabilities satisfy the requirements previously listed, however, the computational resources required for this constrain the direct application of CWW models in AIUS. In tasks that are critical from the point of view of computational resources, parametrized fuzzy numbers are used [66–68]. Fuzzy arithmetic operations on parameterized fuzzy numbers are much faster than fuzzy inference, including CWW.

4 Distillation of Sense of Data from Sensors

In this section, we give a mathematical definition of the fuzzy Certainty Factor (CF), as the basis on which all subsequent models are built. Then we will give a formal definition of the knowledge granule by introducing the EMKG and IMKG and their examples. The structure of the KB, the sense distillation by abstracting in the space of data from sensors algorithm and examples are also given.

4.1 *Fuzzy Certainty Factor*

The fuzzy CF model based on Stanford's theory of the CF [51]. In rule-based systems, CF represents the change in an expert's belief in a hypothesis when new evidence is obtained. The CF universe of discourse is $[-1.0 \leq cf \leq +1.0]$. A CF between 0 and +1 means that given a piece of evidence increases the expert's belief in hypothesis, whereas a CF between *–1* and *0* means that the expert's belief decreases. CF values close to zero means that the evidence obtained by the expert does not carry any useful information to change his confidence in the hypothesis. CF has been introduced into expert systems for modeling human reasoning under uncertainty as an alternative to the probabilistic Bayesian method. There are known attempts to combine the fuzzy CF with original numerical CF, but these approaches have not found wide distribution. In rough set, intuitionistic FS, q-rung orthopair FS, special Membership Function (MF) are introduced for a more complete and accurate account of uncertainty [69]. The works [25, 26] introduced CF as a fuzzy characteristic, considering the fuzziness, partial observability, relevance of evidence, which decreases due to aging of the information. Information aging means that information about evidence is either obtained a long time ago or has not changed for a long time. To do this, the MF must depend on time.

Thus, the CF introduced, is the fuzzy L-R number [25] with three parameters: ($-1.0 \leq \alpha \leq +1.0$) certainty; t_L is the time interval elapsed since the change of evidence; t_R is the time interval elapsed since the receipt of the evidence

$$\mathbf{X} : \{x | m_{\mathbf{X}}(x),\ \forall x \in [-q, +q],\ q \geq +1\} \tag{3}$$

with Gaussian L-R MF

$$m_{\mathbf{X}}^{L} = \exp(-(x-\alpha)^2/2t_L^2), \forall x \in [-q, \alpha],$$
$$m_{\mathbf{X}}^{R} = \exp(-(x-\alpha)^2/2t_R^2), \forall x \in [\alpha, +q]. \quad (4)$$

The introduction of t_L and t_R time parameters in the CF is important for real-time AIUS applications because it allows automatically "bind" time to all the characteristics of the model.

Based on the (3), (4), the fuzzy CF numerical characteristic, which expresses the result of data aging, is determined. It is the Presumed Certainty (PC). The PC is calculated

$$cf = \alpha \cdot k_t, \quad (5)$$

$$k_t = 1 - \left(v_L \cdot \frac{\sum\limits_{\forall x \in [-q,\alpha]} m_{\mathbf{X}}^{L}(x)}{Card([-q,\alpha])} + v_R \cdot \frac{\sum\limits_{\forall x \in (\alpha,+q]} m_{\mathbf{X}}^{R}(x)}{Card([\alpha,+q])} \right). \quad (6)$$

In (5), v_L and v_R are the normalized aging rates, *Card(***X***)* is scalar cardinality of a FS **X**. Coefficient k_t consider the influence of both time parameters $t = t_L + t_R$ on the PC. Figure 3 shows examples of CF.

Figure 3a reflects the case when information about the evidence has not been received for a long time, naturally, the evidence has not changed. Figure 3b depicts another situation when the information about the evidence has just arrived, but the evidence has not changed the certainty. According to (5) and (6), $cf \to 0$ for outdated information, and $cf \approx \alpha$ for fresh evidence.

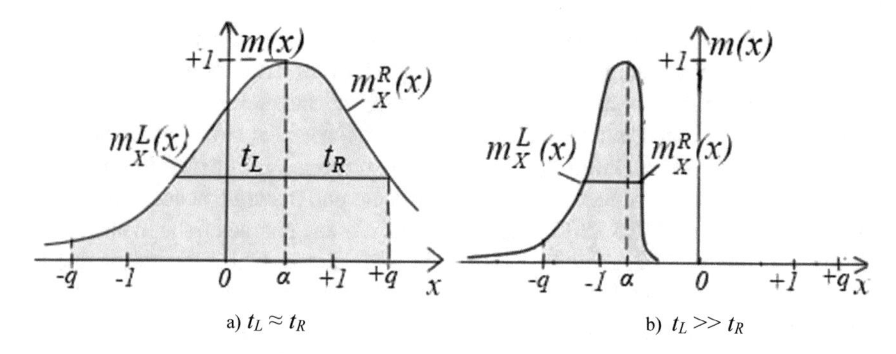

Fig. 3 Two CF cases with different values of time intervals from the moment of data change (t_L) and acquisition (t_R)

4.2 External Meaning of the KG

As mentioned earlier, the EMKG model represents the KG sense through the meaning of the words of lower levels of abstraction. The EMKG model is based on semantic relationships. Four semantic relations "object-property" (*consist_of*), "whole-part" (*part_of*), "genus-species" (*is_a*), and temporal relationship (*before*) are suggested for Smart Rules Engine [46, 47]. In traditional semantic networks are used first three relationships for knowledge representation. To present knowledge about dynamical situations, for example, temporal events or sequence of actions, the last relationship is introduced. Relationships are parameterized by following parameters. Certainty (a), temporary delay (b), information completeness (g), and information aging rate (v) are these parameters. Below, the EMKG model is given.

$$<N, know, \{<M_i, (a_i, b_i, v_i, g_i)>, M_i \in \Omega_N\}>, \tag{7}$$

where N is a KG identifier; *know* is a KG sign model; $\Omega_N = \{M_i\}_{i=1,2,\ldots,I}$ is a set of KGs of the lower levels of abstraction, which are used for the definition of the meaning of the N KG; M_i is a low-level KG identifier; a_i, b_i, v_i, g_i are the parameters.

According to Mendel's theory, the KG presented by word N, has a different sense (means different things) for different people [64]. Therefore, each expert (group of experts) has own belief that the word M_i suites to explain the sense of the word N. In the future, we will use only KG instead of "Word that presents KG" or "KG expresses the sense of word". The CF is base of quantitative assessment of the certainty. In (7), the certainty is pointed by parameter $-1.0 \leq a \leq +1.0$. For cases when the disclosing of sense of the KG M_i is necessary to point either the presence, either the absence of the component presented by KG M_i, the value of the parameter sets $a_i \approx +1.0$ or vice versa $a_i \approx -1.0$, respectively. In (7), the parameter b is a dynamic situation characteristic. Dynamic characteristics of temporal events and processes are presented by time interval parameter $0 \leq b < \infty$. For example, the situation presented KG N is following. The event presented by the KG M_j, should appear later by τ time units than the event presented the KG M_i. For these case in (7) $b_i = 0$, $b_j = \tau$. Knowledge relevance presented by the IMKG of M_i, is lost during the time. The information aging rate $0.0 \leq v_i \leq 1.0$ is set according to the how fast is it. Aging rate is $v_i = 0.0$ for static situations and $v_i = 1.0$ for fast processes. Information completeness parameter $0.0 \leq g_i \leq 1.0$ numerically characterizes to what degree knowledge about only one M_i KG is enough to explain the meaning of the N KG. When the KG N definition has done by examples enumeration $\{M_i, i = 1, 2, \ldots, I\}$, then for each M_i g_i is 1.0. When the KG N definition has done by set of KGs $\{M_i, i = 1, 2, \ldots, I\}$, then information completeness limited by condition $g_1 + g_2 + \cdots + g_I \geq 1.0$. It is possible to represent any of the above relationships based on the model (7) [47]. Following it is an example of presentation of the EMKGs distributed on different levels of generalization of data from sensors.

As an AIUS example we will consider the STL. The STL control task is set taking into account the following criterions: (1) Safety, (2) Local situation, (3) Pedestrian comfort, (4) Global transport situation, (5) Emergency cases. In this section, the EMKG presentation in the form (7) is considered on the example of one of the criteria, namely, pedestrian comfortable conditions. A KB is represented by fragment included 28 KGs distributed across 5 levels. The fragment of this hierarchical structure is shown Fig. 2b. The definition "Pedestrian comfort" KG by sequential disclosure of the concepts meaning, starting from the 5th level of abstraction and ending with the zero level at which data granules from sensors are presented, is given below (8).

$$\begin{aligned}
&1. < 5.PC, \mathit{pedestrian_comfort}, \{< 3.WC, (0.9, t,\ 0.01,\ 0.45) >, < 1.AP, (0.8, t,\ 0.1,\ 0.3) >,\\
&\qquad\qquad < 4.F2, (0.9, t,\ 0.01,\ 0.25) >;\\
&2. < 4.F2, \mathit{factor\ 2nd\ degr\ import}, \{< 0.Il, (0.9, t, 0.1, 0.4) >, < 0.Ni, (-0.75, t, 0.1, 0.35) >,\\
&\qquad\qquad < 3.F3, (0.75, t, 0.1, 0.25) >\} >;\\
&3. < 3.WC, \mathit{weather\ comfort}, \{< 2.WCs,\ (0.95, t,\ 0.01,\ 1.0) >, < 2.WCw,\ (0.95, t,\ 0.01,\ 1.0) >\} >;\\
&4. < 3.F3, \mathit{factor\ 3rd\ degr\ import}, \{ < 2.Sl,\ (-0.75, t, 0.1, 0.8) >, < 0.RR, (-0.75, t,\ 0.1,\ 0.3) >\} >;\\
&5. < 2.WCs, \mathit{weather\ comfort\ spr-sum}, \{< 1.Ss, (0.9, t, 0.1, 0.5) >, < 0.Ys, (0.95, t,\ 0.1,\ 0.5) >\} >;\\
&6. < 2.WCw, \mathit{weather\ comfort\ fall-wint}, \{< 1.Ws, (0.9, t, 0.01, 0.5) >, < 0.Yw, (0.95, t, 0.01, 0.5) >\} >;\\
&7. < 2.Sl, \mathit{sleet}, \{< 0.Pd, (0.75, t,\ 0.01,\ 0.4) >, 0.IT, (0.9, t, 0.5, 0.5) >, < 1.IS, (0.75, t, 0.5, 0.35) >\} >;\\
&8. < 1.AP, \mathit{air\ exhaust\ pollution}, \{< 0.CH, (0.75, t, 0.01, 0.3) >, < 0.CO, (0.75, t, 0.01, 0.3) >,\\
&\qquad\qquad < 0.NO, (0.75, t, 0.01, 0.4) >\} >;\\
&9. < 1.Ss, \mathit{spr-sum}, \{ < 0.Ps, (0.75, t, 0.01, 0.3) >, < 0.Ws, (0.75, t, 0.01, 0.3) >, < 0.WCs, (0.75, t, 0.01, 0.4) >\} >;\\
&10. < 1.Ws, \mathit{fall-wint}, \{< 0.Pw, (0.75, t, 0.1, 0.4) >, < 0.Ww, (0.75, t, 0.1, 0.3) >, < 0.WCw, (0.75, t, 0.1, 0.3) >\} >;\\
&11. < 1.IS, \mathit{snow\ sticking}, \{< 0.Pd, (0.75, t, 0.01, 0.45) >, < 0.TR, (0.9, 10, 0.5, 0.2) >, < 0.IT, (0.9, 0, 0.001, 0.5) >\} >;\\
&12. < 0.Ps, \mathit{precipit\ spr-sum},\ \{< \mathit{dry}, (0.75, t, 0.1, 1.0) >, < \mathit{drizzling\ rain}, (0.75, t, 0.1, 1.0) > \} >;\\
&13. < 0.Pw, \mathit{precipit\ fall-wint},\ \{< \mathit{dry}, (1.0, t, 0.05, 1.0)\ \} >;\\
&14. < 0.Ws, \mathit{wind\ spr-sum}, \{< \mathit{calm}, (0.75, t, 0.1, 1.0) >, < \mathit{gentle}, (0.75, t, 0.1, 1.0) >, < \mathit{refresh}, (0.75, t, 0.1, 1.0) >\} >;\\
&15.\ < 0.Ww, \mathit{wind\ fall-wint}, \{< \mathit{calm}, (0.75, t,\ 0.1,\ 1.0 >, < \mathit{gentle}, (0.75, t,\ 0.1,\ 1.0) >\} >;\\
&16.\ < 0.WCs(\mathit{wind\ chill\ spr-sum}, \{< \mathit{chilly}, (0.75,\ t,\ 0.1,\ 1.0) >, < \mathit{warm}, (0.75,\ t,\ 0.1,\ 1.0) >\} >.
\end{aligned} \tag{8}$$

In (8), belonging of KG to certain abstraction level indicates by first digit of a KG identifier (Fig. 2b). At the bottom level, the sense of "Pedestrian comfort" KG is explained by 17 of *0*th level KGs defined on a set of precipitation, wind, temperature, air pollution, illumination, noises granules and season data. The EMKG definitions of *0*th level KGs not included in (8) are done below. These are such KGs: "Air pollution CH" (***0.CH***), "Air pollution CO" (***0.CO***), "Air pollution N_{oy}" (***0.NO***), "Illumination" (***0.Il***), "Noises" (***0.Ni***), "Road repair work" (***0.RR***), "Precipitation" ***0.Pd***, "Icing temperature" (***0.IT***), "Temperature reduction" (***0.TR***), "Season of the year, spring–summer" (***0.Ys***), "Season of the year, fall-winter" (***0.Yw***).

Definition (8) represents the domain expert knowledge. The dynamic characteristics are not important from the point of view of the domain therefore, the (8) do not specify values of the *t* parameter. The complete model includes a dynamical property, such as "temperature reduction". The choice of the value of *a* parameter, in addition to the semantic knowledge of the expert about how to explain the meaning of the concept, is influenced by the data source characteristics. For noisy data, the certainty in their localization may not be high. Therefore, in order not to underestimate their contribution to the certainty of the situation at higher abstraction levels, the value of a parameter is not indicated as high $a = 1.0$. For example, in the definitions given in lines 7–16 in (8), $a = 0.75$ is specified.

4.3 Knowledge Base Structure

Knowledge, on the basis of which the sense distillation by abstracting solved, is organized into a structure in accordance with L. Zadeh Restriction-Centered Theory [60] and distributed over the following layers [48, 51]:

(1) Quantitative Abstraction (QA) layer. This is the sensors data granulating based on restrictions on the accuracy of the solution
(2) Definitive Abstraction (DA) layer. This is a mapping of a data quantitative constraint presented by Information Granules (IGs) into a word semantic constraint presented by KGs
(3) Abstraction by Generalization (GA) layer. Step by step words-based abstracting according to IMKG computing model.

On Fig. 4 layers are divided into levels of abstraction.

On the Data from Sensors layer AIUS sensors are grouped by modalities. The sensory modality, for example, “Temperature” includes one temperature sensor, and the sensory modality “Air pollution” includes three CH, CO, and N_{oy} sensors. For each AIUS sensory modality, at the QA layer (Fig. 4) there is a set of IGs, the external meaning of which is given by the restrictions imposed by the CF-function $\alpha(x)$ on the domain universe. Figure 5 graphically shows the definition of three IGs: *IGd* (*dry*), *IGdr* (*drizzling rain*) and *IGrf* (*heavy rainfall*) for the “Precipitation” modality. They are determined by the piecewise-linear CF-function $\alpha(p)$ at the universe of the precipitation amount per hour (Fig. 5). The constraint specified as a CF-function is fuzzy, like the definition of terms (fuzzy granules) for linguistic variables in FLS. The difference lies in the range of possible values (for the MF it is $[0, 1]$, and for the CF-function it is $[-1, 1]$). It should be noted that IGs may be defined using any of the

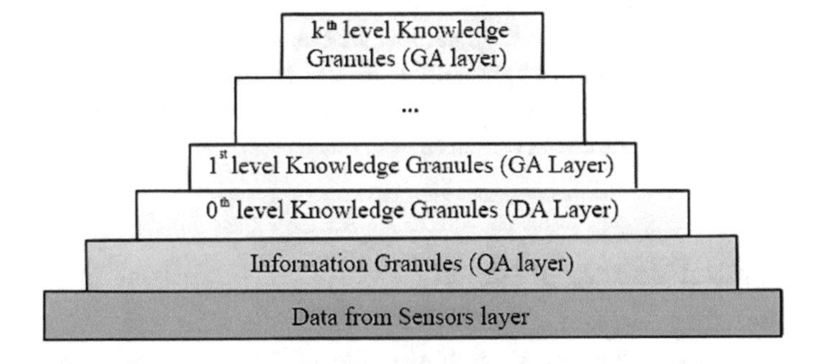

Fig. 4 Abstraction levels of the sense distillation task

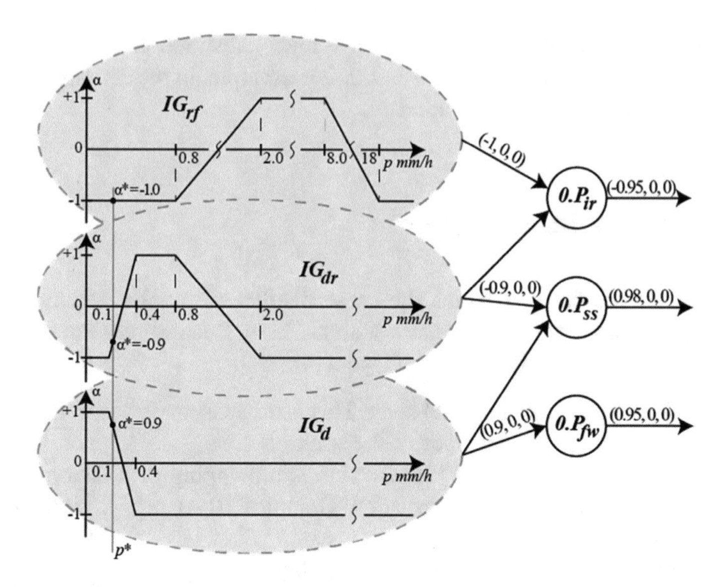

Fig. 5 An example of defining and representing the sensory modality "Precipitation" at two QA and DA abstraction layers

granulation methods and algorithms [27]. The DA knowledge layer is represented by KGs of the zero level, whose EMKG is defined on a limited by one modality of IG set of QA layer. For example, *0.Pss* KG in (8) is defined on the following two IGs: *dry* and *drizzling rain* (Fig. 5). This knowledge layer is the bridge between the two paradigms "data" and "word", as mentioned in the introduction to this chapter. The meaning of the zero level KG is the verbal expression of perceptions, which are represented by data from sensors (IGs of QA layer). This layer in (8) includes KGs whose identifier begins with zero. The GA layer is represented by a multi-levels structure formed by KGs from the *1*st to the *k*th level (Fig. 4). At the *l*th level there are KGs to determine the EMKG of which requires at least one KG of the *l-1*st level and an arbitrary number of KGs of lower levels. The P&AD KB includes the set of KGs of the DA and GA layers: $\Omega_{KG} = \Omega^0{}_{KG} \cup \Omega^1{}_{KG} \ldots \cup \Omega^k{}_{KG}$, where $\Omega^i{}_{KG}$ is a subset of KGs of the *i*th level (Fig. 4).

In conclusion, note that for the example discussed above, a data set from 11 different data sources (7 sensory modalities and 11 sensors) represented at domain universes using 27 IGs is mapped into one abstract concept "Weather comfort" of 3*rd* abstraction level (granule ***3.WC*** in (8)).

4.4 Internal Meaning of the KG

The IMKG expresses the degree of conformity of EMKG (7) with the situation represented by the spatio-temporal data segment. A numerical estimate of the IMKG

(Fig. 2a) depends, firstly, on the EMKG parameters in (7), and, secondly, on the IMKG of KGs $\{M_i \in \Omega_N\}$ calculated for the same date segment. The IMKG is an assessment of the matching a verbal representation (7) of EMKG to data from sensors on which the KG sense is determined. Below is the formal model of IMKG.

The IMKG computing model inputs and output are shown in Fig. 2a. The IMKG is calculated based on its EMKG definition (7) and IMKG_{M1}, IMKG_{M2}, … , IMKG_{Mk}.

$$\mathbf{X}_N = f(X_{M1}, X_{M2}, \ldots, X_{Mk}), \tag{9}$$

where X_{M1}, X_{M2}, …, X_{Mk} are the IMKGs of the M_1, M_2, …, M_k KGs in the form (3).

The IMKG calculation based on operations with fuzzy L-R numbers (3), (4). In (9), to calculate $\mathbf{X}_N$, the EMKG knowledge (7) is used. The essence of computing is comparison the EMKG definition (7) with IMKGs inputs. The procedure of computation is divided into three steps: *matching* is pair-wise comparison of the EMKG parameters (a_i, b_i), presented in the EMKG definition (7), with the IMKG_{Mi}, of the corresponding input variable; *aggregation* is summary estimates of similarities obtained in the first step for all IMKG_{M1}, …, IMKG_{Mk} input variables; *actualization* is a procedure for calculating the CF output parameters IMKG_N (Fig. 2a).

Matching is comparing two fuzzy L-R numbers operation. The fuzzy L-R number $X_i = (a_i,\ b_i, 0)$ obtained from (7) is compared with L-R number $\mathbf{X}_i^{in} = (\alpha_i, t_{Li}, t_{Ri})$ represents IMKG_{Mi} corresponding input. The result is a new L-R number $\mathbf{X'}_i = (\alpha_i' = \alpha_{\rho i}, t_{Li}' = t_{Li}, t_{Ri}' = t_{Ri})$. It is a fuzzy certainty that the compared fuzzy numbers are close $\alpha_{\rho i} = \rho\big(X_i, X_i^{in}\big) = X'_i$. Computational model of the closeness of two fuzzy L-R number is given in [25].

Aggregation is operation of summation of k weighted fuzzy L-R numbers $\{X_i'\}_{i=1,2,\ldots,k}$ to obtain the fuzzy closeness estimate. The weighting coefficient g_i is the information completeness parameter from the definition (7). The operation result is again a fuzzy L-R number.

$$\mathbf{X}'' = (\alpha'', t_L'' = \sum_{i=1,2\ldots k} t_{Li}', t_R'' = \sum_{i=1,2\ldots k} t_{Ri}'), \tag{10}$$

where

$$\alpha'' = \begin{cases} \alpha''', if\ -1 \le \alpha''' \le +1, \\ +1,\ if\ \alpha''' > +1, \ \text{and}\ \alpha''' = g_1 \cdot \alpha_{\rho 1} + \cdots + g_k \cdot \alpha_{\rho k} \\ -1,\ if\ \alpha''' < -1, \end{cases}$$

Actualization of IMKG_N value is the final phase of computing. The *cf* is computed by the (5) on the basis of the X'' (10) and then value *cf* is used to find the parameters of the L-R number $\mathbf{X}_N$ in (9).

$$\alpha_N = cf, \tag{11}$$

$$t_{L_N} = \begin{cases} 0, \; if|cf - {}^{-}cf| \geq \varepsilon, \\ {}^{-}t_{L_N} + 1, \; other\;wise, \end{cases} \quad (12)$$

$$t_{R_N} = MAX(t_{R_1}, t_{R_2}, \ldots, t_{R_k}), \quad (13)$$

where ${}^{-}t_{R_N}, {}^{-}cf$ are the parameter values in the previous calculation step; $t_{R_1}, t_{R_2}, \ldots, t_{R_k}$ are the IMKG parameters of input variables.

Sense Distillation based on Abstracting algorithm (SDbA) is shown in Fig. 6.

Input data from sensors modality, domain KB in the form of a structure Fig. 4, in which the KG is the definition of "What is it" in the form of prototypes (7).

Output set of $cf(KG_{il})$, $\forall\, KG_{il} \in \Omega^{l}{}_{KG}$, $\forall\, \Omega^{l}{}_{KG} \in \Omega^{0}{}_{KG} \cup \Omega^{1}{}_{KG} \ldots \cup \Omega^{k}{}_{KG}$, where $cf(KG_{il})$ is a numerical assessment of the confidence that the data from sensors of given modality are consistent with the verbal definition of the situation presented by the KG_{il}.

The SDbA algorithm is based, firstly, on the multi-level organization of KB (Fig. 5), secondly, on the EMKG representation model, discussed in Sect. 4.4, and, thirdly, on the IMKG calculation model, considered in this subsection. The algorithm calculates the IMKG of the granules by levels, starting from the bottom. The KB structure has "good" connections from a computational point of view: to calculate

Step 1. Quantitative Abstraction (QA)

while $r \leq R$, where R is quantity of modality sensors **do**

get x^* data from sensor

while $q \leq Q$, where Q is quantity of sensor IGs **do**

Calculate CF: $\alpha^* = \alpha(x^*)$;

$$t_R = \begin{cases} 0, & if\; x^* \neq error; \\ {}^{-}t_R + 1, & if\; x^* = error. \end{cases}$$

$$t_L = \begin{cases} 0, & if\;\; |\alpha - {}^{-}\alpha| \geq \varepsilon; \\ {}^{-}t_L + 1, & other\;\; wise; \end{cases}$$

end while

end while

Step 2. Definitive Abstraction (DA)

while $q \leq Q^0$, where Q^0 is quantity of zero level KGs which belongs to modality **do**

while $i \leq I_j$, where I_j is quantity of IGs which explain the meaning of KG_j **do**

Calculate CF of KG_j sequentially using formulas (11)-(13)

end while

end while

Step 3. Abstraction by Generalization (GA)

while $l \leq L$, where $l = 1,2,\ldots,L$, L is quantity of levels of KB **do**

Calculate CF of KG_j sequentially using formulas (11)-(13)

end while

end DbA

Fig. 6 The SDbA algorithm

the IMKG of the *l*-level KG, only IMKGs of the lower levels are used. Therefore, the calculations of IMKGs are deployed strictly by layers from the bottom up.

In conclusion, we give an example of IMKG calculations based on SDbA for one time sample for the knowledge fragment (8) shown in Fig. 2b. Calculations are given for four sensory modalities: "Precipitation" ($p^* = 0.15$ mm/h), "Day_of_year" ($day^* = 35$), "Wind_speed" ($v^* = 5.5$ m/sec), and "Air_temperature" ($T^* = 12$ °C). The calculation results are shown in Table 2. The rows of the table correspond to the layers shown in Fig. 4. The calculation results are shown by levels from bottom to top, in accordance with the layers shown in Fig. 4. First, according to the SDbA algorithm, the QA step was performed for the data from the sensors (sensors layer in Table 2).

Granulation results are shown in the QA layer line. As an example, the Table 2 shows the results for two sensory modalities "Precipitation" and "Wind", the EMKG of which is shown in Fig. 5. For these data, the situation of correctly obtaining values that have changed compared to the previous control moment was simulated. Therefore, for them $t_L = 0$, $t_R = 0$. Similarly, you can continue to analyze the results for the next levels.

The SDbA algorithm has computed the IMKG for all granules of the DA and GA layers of the KB fragment in the Table 2. Thus, for the example under consideration, the result of sense distillation is the sense of the situation presented by data from 11 sensors. Sense of the situation is represented at different levels of abstraction by the set of IMKGs with positive values, for example, $cf(KG) > 0.5$. So, the sense of the situation at a pedestrian crossroad is described at the first level of abstraction as comfortable conditions in terms of air exhaust pollution ($cf(1.AP) = 0.98$), at the third level as comfortable weather conditions ($cf(3.WC) = 0.85$) and at the fifth level as comfortable conditions for a pedestrian ($cf(5.PC) = 0.8$).

Table 2 Results of SDbA calculations on the KB fragment (8)

Layers	(ID: α, t_L, t_R)
GA level 5	(5.PC: +0.8, 0, 0)
GA level 4	(4.F2: +0.85, 0, 0)
GA level 3	(3.WC: +0.85, 0, 0), (3.F2: +0.9, 0, 0)
GA level 2	(2.WCs: −0.75, 0, 0), (2.WCw: +0.85, 0, 0),
GA level 1	(1.Ss: +0.95, 0, 0), (1.Ws: +0.85, 0, 0), (1.AP: +0.98, 20, 25)
DA level 0	(0.P_s: +0.98, 0, 0), (0.P_w: +0.95, 0, 0), (0.P_{ir}: −0.95, 0, 0), (0.Y_w: +0.95, 0, 0), (0.Y_s: −0.98, 0, 0), (0.WC_w: +0.75, 0, 0), (0.W_s: +0.95, 0, 0), (0.WC_s: +0.6, 0, 0)
QA layer	(IG_d: +0.9, 0, 0), (IG_{dr}: −0.9, 0, 0), (IG_{rf}: −1.0, 0, 0), (IG_{Yw}: +0.95, 0, 0); (IG_{cl}: −1.0, 0, 0), (IG_{gl}: +0.95, 0, 0), (IG_{rh}: +0.45, 0, 0)
Sensors layer	p* = 0.15 mm/h, day* = 35, v* = 5.5 m/sec, T* = 12 °C

5 Distillation of Sense of Events Footprint

In this section we describe a model of events stream representation by the Events stream Footprint (EF) and the Footprint Blur algorithm which upgrade the EF. A $\mathbf{f}^{21}$ and $\mathbf{f}^{22}$ sense distillation models (2) based EF convolution algorithm is discussed.

5.1 *Events Sequence Footprint Model*

Let $\Omega^{ev}{}_{KG} \subset \Omega_{KG}$ be the KGs subset of different P&AD KB levels whose CF changes form the stream of interest events. The events set $\boldsymbol{W} = \{w_1, \ldots, w_i, \ldots, w_M\}$ is fuzzy defined on the intersection of the KGs of $\Omega^{ev}{}_{KG}$. It was shown [70], sense of the events streams built on lower-level abstraction KGs of the P&AD KB (Fig. 4) is retained in sparse event stream built on the upper-level KG. The present chapter adopted an interval event model, in which the appearance of an event is identified by $cf \approx +1.0$, and the disappearance of $cf \approx -1.0$. In [25], the model for calculating the CF of an event based on the CF parameters of KGs from the P&AD KB is given. In this subsection, we will assume that IMKG change of $KG_i \subset \Omega^{ev}{}_{KG}$ with parameters $(\alpha_{KGi}, t_{Ri}, t_{Li})$ generates an event w_i, which will be denoted by $(\alpha_{wi}, t_{Ri}, t_{Li})$.

Events Sequence Footprint consists of the two parts. Let, $t_{-q}, t_{-(q+1)}, \ldots, t_{-1}, t_0, t_1, \ldots, t_{b-1}, t_b$ are time instants, where t_0 is current, $t_{-q}, t_{-(q+1)}, \ldots, t_{-1}$ are previous and $t_1, \ldots, t_{b-1}, t_b$ are the future US decision-making moments of time. An event w_i is associated with certain moment of time. Therefore, in the corresponding sequence of events

$$Fpr = ((w_s, t_{-q}), \ldots, (w_j, t_{-1}), (w_i, t_0), (w_j, t_{+1}), \ldots, (w_p, t_{+b})) \tag{14}$$

two subsequences can be distinguished. To making decision at time t_0, the finite sequence of $q + 1$ events occurred at this and earlier times $t_{-q}, t_{-(q+1)}, \ldots, t_{-1}$ is needed to know as US history. The subsequence $(w_j, t_{+1}), \ldots, (w_p, t_{+b})$ is US action plan what is needed to know too when making a decision at time t_0. The sequence of events (14) representing different temporal situations can have different lengths. The representation of the events sequence footprint model in the form (14) is problematic due to the explicit use of the values of time points. The events footprint model will first be considered using the example of representing and processing events about US history, and then extended to the action plan. The footprint model problem is posed as follows.

At current time t_0, STM store the sequence of characteristics of $q + 1$ recent events $\{w^0{}_i, w^{-1}{}_j, w^{-2}{}_p, \ldots, w^{-q}{}_s\}$ (14). The $r = 0, -1, -2, \ldots, -q$ indicate the ordinal number of events in the sequence (14) with respect to the current event $w^0{}_i$. The event $w^{-1}{}_j$ happened before the event $w^0{}_i$, and so on. The event fuzzy characteristic is CF with parameters $(\alpha_{wi}, t_{Ri}, t_{Li})$.

A dynamic discrete flat vector field is numerical model of the footprint (Fig. 7). The x-axis indicates the elements of the events set $\mathbf{W} = \{w_1, w_2, \ldots, w_k\}$. The y-axis is event ordinal numbers r, and the z-axis is event CF. A special discrete vector field with vector $\boldsymbol{cf}(w_i, r)$ associated with each point of the two-dimensional space (w_i, r) is considered. Shown in Fig. 7, a vector can have positive or negative directions along the z-axis, only. A vector field considered is dynamic. The CFs values of all previously occurred events are changed at each event occurrence moment. Events are shifted "deep into memory" (the event $w^r{}_i$ parameter r is replaced by $r - 1$, $w^{r-1}{}_i$) and, secondly, the vector modulus is reduced by "forgetting" coefficient $exp(-v)$, where $0 \leq v \leq 1$ is the data aging rate in EMKG definition (7). Such simulation based on the results of STM studies in cognitive psychology and show the gradual "blur" of event footprint in the memory [70].

Thus, the STM content at arbitrary time represents an EF. Foundation of the STM characteristics updating algorithm is formal representation of the vector field. We represent the vector field by a matrix and denote (15).

$$\mathbf{cf} = \begin{pmatrix} cf(w_1, 0) & cf(w_1, -1) & \cdots & cf(w_1, -q) \\ cf(w_2, 0) & cf(w_2, -1) & \cdots & cf(w_2, -q) \\ \cdots & \cdots & \cdots & \\ cf(w_k, 0) & cf(w_k, -1) & \cdots & cf(w_k, -q) \end{pmatrix}. \tag{15}$$

The vector field model above introduced has only a limited set of vectors directed only along the z-axis in the positive or negative directions. It significantly lighted model. In (15), instead of vectors, scalars are indicated in the form of signed numbers $cf(wi, r)$. This presentation of footprint fuzzy characteristics is used by updating algorithm. Following is footprint characteristics updating algorithm.

Footprint Blur (FB) algorithm:

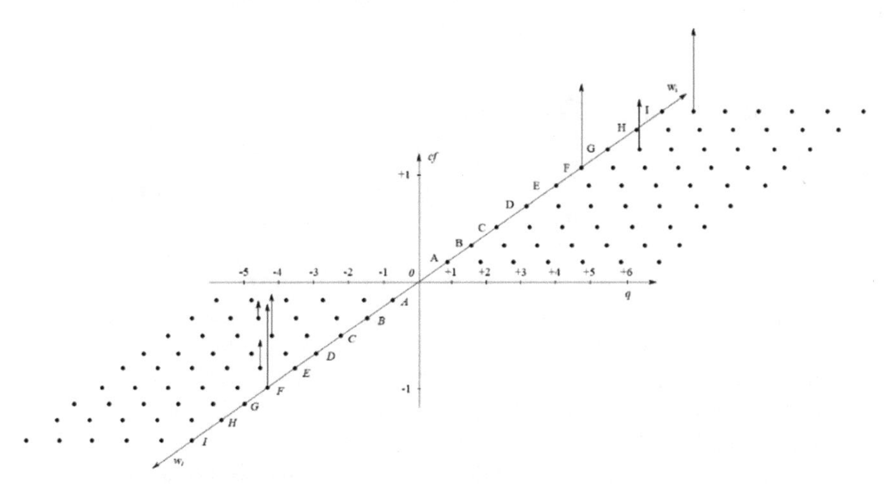

Fig. 7 Dynamic discrete vector field

(1) Begin when the event w_p occurs
(2) Remove the $(q + 1)$th column of matrix (15) and multiply the resulting matrix $k \times q$ by the forgetting coefficient

$$\mathbf{cf}' = \exp(-\nu) \cdot \mathbf{cf}(:, m + 1). \tag{16}$$

(3) Perform horizontal concatenation of a column vector $\mathbf{cf}''$ (17) in which, apart from w_p, all other elements are *0* with the resulting matrix $\mathbf{cf}'$ in (16).

$$\mathbf{cf} = [\mathbf{cf}'', \mathbf{cf}'], \ \mathbf{cf}'' = \begin{pmatrix} 0 \\ \cdots \\ cf(w_p, 0) \\ 0 \end{pmatrix}. \tag{17}$$

According to algorithm, a new matrix has calculated. The *1*th column is the characteristics of the last event that triggered its memorization, the *2*nd column represents the event that was in the *1*th column in the original matrix (15), and so on, The last $(m + 1)$st column of matrix (17) represents the event that was in the mth column of matrix (15). The last column $(m + 1)$ of matrix (15) is removed.

Thus, the EF at an arbitrary event localization moment contains an event history footprint, the memory depth of which is limited by the coefficient of "forgetting" v. For $v = 0$, there is no forgetting, and EF theoretically remembers the entire history as a sequence of events. As v increases, the memory depth r decreases due to the blurring of the event, since $cf(w_i, r) \rightarrow 0$, that is, the confidence that this was an event w_i decreases. The case $cf(w_i, r) \approx 0$, as mentioned in Sect. 4.1, describes the situation when there is no information about the event at all.

Similarly, the FB algorithm performs blurring of the plan footprint. The blur propagates from the current event in the direction of the target event.

5.2 Events Footprint Convolution

AIUS DM&C makes decisions based on behavior's history precedents, which are represented by events stream prototypes. The flow chart of events footprint convolution is shown in Fig. 8.

At each time clock of receiving data from sensors, P&AD maps data into sense of corresponding KGs. Analysis of the CF values changes of KGs ($KG_i \subset \Omega^{ev}{}_{KG}$) gives set of events currently. The events localized are stored into EF and footprint is modified. The FB algorithm described above do this. In the next phase, the certainty that the EF is belonged to each prototype of events streams are calculated. In Fig. 8, the CF_{Pr} calculation block shows this phase. The IMKGs calculated at this phase are transferred at the last phase, to the DM&C as an input's numerical variables.

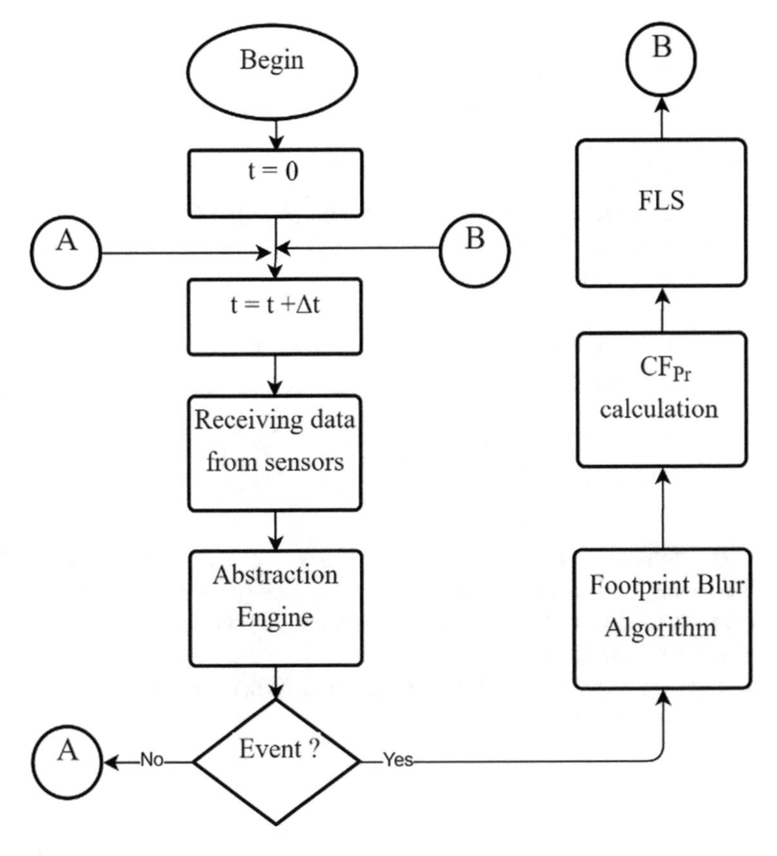

Fig. 8 The flow chart of events footprint convolution for decision making

Let for domain under consideration, N events streams prototypes are allocated. To make decisions, the DM&C have to know what confidence the EF suited with each of the N prototype $\mathbf{Pr}_l$, $l = 1, 2, \ldots, N$. Setting the task of assigning a control decision is reduced to ranking N prototypes according to the proximity criterion of a prototype with EF of different length. This is the task of calculating the IMKG of N granules of a new type. To do this, the STM KG^{ev} was introduced, whose EMKG is the prototype of events stream $\mathbf{Pr}_l$, and the IMKG calculation is reduced to EF convolution. In the events footprint convolution model, the prototype is represented by matrix $\mathbf{Pr}_l^{\alpha} = |\alpha_{Prl}(w, r)|$ and the row vector $\mathbf{Pr}_l^{\tau} = |\tau_{Prl}(r), r = 1, 2, \ldots, q|$, like a vector field. Depend on what case of interval event w required, two values $\alpha_{Prl}(w, r) = +1$ or $\alpha_{Prl}(w, r) = -1$ are pointed in $\mathbf{Pr}_l^{\alpha}$. When event isn't considered in the prototype the value $\alpha_{Prl}(w, r) = 0$ is indicated at the rth spot of the sequence. Taking into account this feature, the $\mathbf{Pr}_l^{\alpha}$ matrix is convoluted into a row-vector (18). In (18), w_i is event identifier, and r is event order number in EF. The last occurred event has $r = 0$.

$$\mathbf{Pr} = (\alpha_{\mathrm{Pr}\,l}(w_i, r) = q^{-r},\ i = 1, \ldots, k,\ q \in \{0, -1, +1\}). \tag{18}$$

The following algorithm compute the IMKG (certainty that EF matches the prototype $\mathbf{Pr}_l$).

Footprint-to-prototype Matching (FM) algorithm:

(1) Compute the first (vertical) and second (horizontal) projections of the pre-processed prototype matrix $\mathbf{Pr}_l^{\alpha} = |\alpha_{Prl}(w, r)|$. A modified matrix convolution operation is used.

$$\begin{aligned}
&\boldsymbol{\alpha}^1 = (\alpha^1(r) = \alpha'(w_m, r) \cdot \alpha_{Prl}(w_m, r),\ r = 0, -1, \ldots, -q_l),\\
&where\, \alpha'(w_m, r) = \underset{i=1,k}{MAX}(|\alpha(w_i, r)| \cdot |\alpha_{Prl}(w_i, r)|).\\
&\boldsymbol{\alpha}^2 = (\alpha^2(w_i) = \alpha(w_i, r'),\ i = 1, \ldots, k),\\
&where\, \alpha(w_i, r') = \underset{r=-1,-2,\ldots,-q}{MAX}(|\alpha(w_i, r)|).
\end{aligned} \tag{19}$$

(2) Compute the CFs $\mathbf{cf}^1 = (cf^1(r),\ r = 0, -1, \ldots, -q_l)$ by (5), (6) based on the values $\boldsymbol{\alpha}^1$ from (19), and $t_R = 0$, and $t_L = |\tau(r) - \tau_{Prl}(r)|$, where $\tau(r)$ is corresponded value from the renew footprint (15) and $\tau_{Prl}(r)$ is corresponded element of matrix $\mathbf{Pr}_l^{\tau}$.

(3) Compute $cf_{\mathbf{Pr}l}$ by weighted sum operation. In (20), the r^* first projections (19) of vector field (15) are summing.

$$cf_{\mathrm{Pr}\,l} = \frac{1}{r*} \sum_{r=-1,-2,\ldots,-r*} cf^1(r). \tag{20}$$

The value $-1 \le cf_{\mathrm{Pr}\,l} \le +1$ is a IMKG$_l$ as a numerical measure of the certainty that EF (15) matches the prototype $\mathbf{Pr}_l$.

5.3 Examples and Simulation Results

We will consider, as AIUS example, the mobile transport robot or co-bot. Co-bot shifts the goods between spots of automated warehouse [7]. We will consider the situation when at position *1* AIUS decides where to continue moving either to position *A* either to position *4*. Figure 9 shows AIUS Artificial Environment (AE). The decision making require to know the history: from what position either *B* or *2* and by via positions *5, 4* or *3, 2* AIUS arrived at position *1*. To decide, the AIUS uses the storing sequence of events reflected steps of movement, goods loading and unloading, their storage, and others. Let, for the example under consideration, the events set is denoted similar the positions symbols of the routes $\boldsymbol{W} = \{A, B, C, D, 1, 2, 3, 4, 5\}$ (Fig. 9). For example, the event *B* sense is "AIUS stands at position *B*, the container has been loaded, and shipment preparation is completed now". AIUS route through four

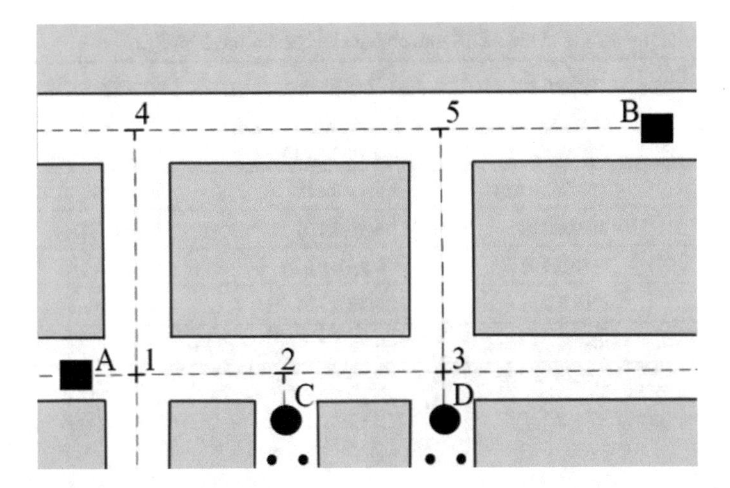

Fig. 9 AIUS artificial environment

positions *B*, *5*, *3*, *1* and its service at two positions *B*, *3* have been simulated (Fig. 9). For three cases when AIUS make decisions at position 5, 3 and 1, simulation results are discussed. For initial input data $cf(w_i, 0)$, the events stream is shown in Table 2 in the format <$\boldsymbol{cf(w_i,0)}$; $(cf(w_i, -1), cf(w_i, -2), cf(w_i, -3), cf(w_i, -4))$> for each input event w_i. The constant $cf(w_i,0) = 0.9$ is used for all events w_i for the first example. For the second example, the influence of the initial data fuzziness is considered. The simulation STM parameters $q = 4$, $v = 0.1$ were used.

Two KG^{ev}'s are considered, represented by prototypes $\boldsymbol{Pr_1}^{\alpha} = |\alpha(1, 0) = +1, \alpha(3, -1) = +1, \alpha(4, -2) = -1, \alpha(B, -3) = +1|$, $\boldsymbol{Pr_1}^{\tau} = |\tau(0) = 5, \tau(-1) = 10, \tau(-2) = 8|$, $\boldsymbol{Pr_2}^{\alpha} = |\alpha(1, 0) = +1, \alpha(2, -1) = +1, \alpha(5, -2) = -1, \alpha(4, -3) = +1|$, $\boldsymbol{Pr_2}^{\tau} = |\tau(0) = 10, \tau(-1) = 15, \tau(-2) = 8|$. The first prototype presents AIUS history: loading at positions *B*, then not loading at position *4*, then arriving at position *3*, and arriving at position *1* just now. The second prototype describes dynamic situation which different of above: loading at positions *4* and *2* and not loading at position *5*. In Table 3, the column "*Position 5*" shown the data generated at the time when AIUS arrived at position *5* from position *B*. In each line, first digit in bold is IMKG value $cf(w_i, 0)$ calculated by SDbA based on data from the P&AD. In Table 3, four numbers in parentheses is history associated with corresponded event. For example, for event *B* the row "*Position 5*" is displayed **0**; (0.8, 0, 0, 0) that means $cf(B, 0) = 0.0$, $cf(B, -1) = 0.8$, $cf(B, -2) = 0.0$, $cf(B, -3) = 0.0$. At the moment of occurrence of event *3*, the EF is represented by data from column "*Position 3*" in Table 3.

It reflects that the previous was event *5*, because $cf(5, -1) = 0.8$, and before this the event *B* happened, because $cf(B, -2) = 0.7$. When new event appear, the CFs of all earlier events is blurred (compare $cf(B, -1) = 0.8$ in the column "*Position 5*" with $cf(B, -2) = 0.7$ in the column "*Position 3*"). In Table 3, the lower darkened part shows the first projections CF calculated by (19) for two prototypes $\boldsymbol{Pr_1}$ and $\boldsymbol{Pr_2}$ introduced above. The CFs are calculated at moments of inputting new events in

Table 3 Modeling results of the AIUS movement along the route *B, 5, 3, 1*

w_i	*Position 5*	*Position 3*	*Position 1*
4	**0**; (0,0,0,0)	**0**; (0,0,0,0)	**0**; (0,0,0,0)
B	**0**;(0.8,0,0,0)	**0**; (0,0.7,0,0)	**0**; (0,0,0.6,0)
5	**−0.9**;(0,0,0,0)	**0**; (−0.8,0,0,0)	**0**; (0, −0.7,0,0)
2	**0**; (0,0,0,0)	**0**; (0,0,0,0)	**0**; (0,0,0,0)
3	**0**; (0,0,0,0)	**0.9**; (0,0,0,0)	**0**; (0.8,0,0,0)
1	**0**; (0,0,0,0)	**0**; (0,0,0,0)	**0.9**; (0,0,0,0)
Pr_i	Step 1	Step 2	Step 3
1	$\mathbf{cf^1} = (0, 0, 0, 0)$ $cf_{Pr1} = 0$	$\mathbf{cf^1} = (0, 0, 0, 0)$ $cf_{Pr1} = 0$	$\mathbf{cf^1} = (0.8, 0.7, 0.6, 0)$ $cf_{Pr1} = 0.7$
2	$\mathbf{cf^1} = (0, 0, 0, 0)$ $cf_{Pr2} = 0$	$\mathbf{cf^1} = (0, 0, 0, 0)$ $cf_{Pr2} = 0$	$\mathbf{cf^1} = (0, 0.7, 0, 0)$ $cf_{Pr2} = 0.23$

STM. Table 3, for both prototypes each row contains first projection vectors $\mathbf{cf^1}$ and certainty $cf_{\mathbf{Pr2}}$ that EF matches the prototype $\mathbf{Pr}_l$. In corresponding rows "*Position 3*" and "*Position 5*" can be seen $cf_{Pr1} = cf_{Pr2} = 0$ because no enough history data for classification at the moments of completion of the AIUS service. When the event *1* appeared, the EF has been classified with high certainty $cf_{Pr1} = 0.7$, $cf_{Pr2} = 0.23$. These results allow us to conclude that EF is more consistent with KG_1 than KG_2 since $IMKG_1$ has $cf_{Pr1} = 0.7$ compared to $IMKG_2$ which has $cf_{Pr2} = 0.23$. Thus, the distillate of $EF = (cf(1, 0) = 0.9, cf(3, -1) = 0.8, cf(5, -2) = -0.7, cf(B, -3) = 0.6)$ has a sense of prototype $\boldsymbol{Pr_1}$ "The AIUS has now arrived for service at position *1*, moved through positions *B, 5, 3,* was serviced at positions *B, 3* and was not serviced at position *5*" and is presented by the convolution of EF into $cf_{Pr1} = 0.7$.

6 AIUS Decision Making and Control Engines

In this section, three types of DM&C engines are described. Shown, that the IMKGs as distillates from P&AD and STM can be inputs variables of DM&C engines (Fig. 1). For the first and second types, the role of DM&C engine is performed by the traditional FLS. For the first type, the P&AD distillate represents the state of the AIUS environment and is used as input FLS variable. For the second type, the FLS input variables are IMKGs, coming from both the P&AD and STM engines (Fig. 1). And for the third type, IMKGs are used as input variables of goal-driving granules in multi-purpose continuous planning and control task. This is a new model of multi-purpose continuous planning and goal-driving control, which, in addition to P&AD and STM engines, is supported by four cognitive engines: EMotion Engine (EME),

ATtention Engine (ATE), Goal-Driving Engine (GDE) and PLanning Engine (PLE). In the following subsections, examples of solving the above three tasks are described.

6.1 Fuzzy Logic System as DM&C Engine of AIUS

The features of the AIUS tasks indicated in Sect. 1, due to the requirements of autonomy, did not allow the implementation of the FLS-based DM&C, despite its capabilities such as generalization, overcoming data noise, and others. The reason, as mentioned earlier, is the difficulty of tuning the knowledge base, which system developers face when the number of input variables reaches 5–7. Distilling sense of data by abstracting overcomes this problem. Instead of using many variables from sensors as FLS inputs, it is possible to use, for example, one FLS input variable, which is the IMKG in a generalized form representing the meaning of this data set. The construction of the FLS, which uses the distillate obtained by abstracting, is explained below using the example of a STL discussed in Sect. 4.

The example uses the FLS of the T1 FS Mamdani model. The IMKGs coming directly from the upper abstraction levels of P&AD KB (8), namely *pedestrian comfort*, *local situation*, *global transport situation*, *emergency case* and *safe*, are FLS engine inputs. Two numerical variables are used coming directly from the sensors, additionally. There are *pedestrian waiting time* and *traffic light*. The FLS output (*switch light*) is numerical certainty $-1 \leq \alpha \leq +1$. Fuzzy rules are built on the Linguistic Variables (LVs) whose names are the same as the names of KGs of the P&AD KB. Besides *pedestrian waiting time* and *traffic light*, other outputs and inputs LVs are defined in the PC universe $-1 \leq cf \leq +1$ and are not related to the domain. All LVs have three terms, each represented by trapezoidal MF. The MF parameters are *HIGH* (*+0.1*, *+0.5*, *+1.0*, *+1.0*)}, *UNKNOWN* (*−0.5*, *−0.1*, *+0.1*, *+0.5*) and {*LOW* (*−1.0*, *−1.0*, *−0.5*, *0.0*). An example of a fuzzy rule is given below (21).

$$\begin{aligned}
&\textbf{IF} safe\, is\, HIGH\, \textbf{and}\\
&\quad local\, situation\, is\, UNKNOWN\, \textbf{and}\\
&\quad pedestrian\, comfort\, is\, LOW\, \textbf{and}\\
&\quad global\, transport\, situation\, is\, HIGH\, \textbf{and}\\
&\quad emergency\, case\, is\, UNKNOWN\, \textbf{and}\\
&\quad pedestrian\, waiting\, time\, is\, MIDDLE\, \textbf{and}\\
&\quad traffic\, light\, is\, GREEN-CAR,\ RED-PEDESTRIAN\\
&\textbf{THEN}\, turn\, on\, yellow-car\, and\, red-pedestrian\, is\, HIGH.
\end{aligned} \tag{21}$$

The numerical values of IMKG, in the PC form, calculated by the SDbA are the input values of the FLS engine. These IMKGs numerical values are then used in the FLS inference. For example, when processing rule (21), the $cf_{7.SF} = 0.9$, $cf_{6.LS} = -0.35$, $cf_{3.GTS} = 0.6$, $cf_{5.EC} = 0.25$ values were used, some of them $cf_{5.PC} = 0.8$, from Table 2 and *pedestrian waiting time* ($t = 42$ s), *traffic light* ($TL = 12$) values from sensors. After the accumulation of the results of 23 rules of type (21) and defuzzification, the rank of all control decisions is obtained. For the given inputs, for example, PC of turn on the light combination "yellow for car and red for pedestrian" is $cf_{\text{turn on yellow-car and red-pedestrian}} = 0.65$.

Making decisions when managing a STL, taking into account the criteria considered in the example, requires the use of more than three dozen input variables [26]. This does not allow using the FLS without preliminary reducing the number of input variables. Distillation by abstracting made it possible to reduce this number to four, while retaining the meaning of the primary situation, represented on a full set of 30 variables.

6.2 Using AIUS History in Fuzzy Logic System Engine

This section will show how the distillate obtained by AIUS history convolution creates an opportunity for the implementation of DM&C Type 2 based on the FLS engine. The presentation of an approach is illustrated by the example of transport mobile co-bot as AIUS considered in Sect. 5.

Below, as an example, there is a fuzzy rules fragment of the FLS KB. The rules define the decision-making knowledge needed to continue moving AIUS arrived to position 1 (Fig. 9). Rules describe situation when co-bot loaded cargo on positions *B* and *3*. Then co-bot should move to position *A*, if cargo was loaded at positions *3* and 2, or co-bot should move to position *D*, or no move at all. The knowledge about the first case is presented by the $\mathbf{Pr}_1 = \{1^0, 3^{-1}, 5^{-2}, B^{-3}\}$ prototype, and about the second case by $\mathbf{Pr}_2 = \{1^0, 2^{-1}, 5^{-2}, 4^{-3}\}$ prototype (Fig. 9). Below are the rules in which the IMKGs reflect the compliance of the STM EF with the indicated prototypes.

$$R_1 : \textbf{IF}\ event\ (1)\ \textbf{and}$$
$$CF\_1\ is\ HIGH\ \textbf{and}$$
$$CF\_Pr_1\ is\ HIGH\ \textbf{and}$$
$$Energy\_Reserve\ is\ HIGH$$
$$\textbf{THEN}\ CF\_Move\ is\ A$$
$$R_2 : \textbf{IF}\ event\ (1)\ \textbf{and}$$
$$CF\_1\ is\ HIGH\ \textbf{and}$$
$$CF\_Pr_2\ is\ HIGH\ \textbf{and}$$
$$Energy\_Reserve\ is\ NOT\ LOW$$
$$\textbf{THEN}\ CF\_Move\ is\ D$$
$$R_3 : \textbf{IF}\ event\ (1)\ \textbf{and}$$
$$CF\_1\ is\ HIGH\ \textbf{and}$$

$$CF\_Pr_1\ is\ LOW\ \textbf{and}$$
$$CF\_Pr_2\ is\ LOW$$
$$\textbf{THEN}\ CF\_Move\ is\ NO \tag{22}$$

In rules (22), IF condition has the term *event (1)*. It points that the rule may processed by the FLS engine only when the event *1* occurred. At the current step, rule is disabled from the processing when event does not occur (in AI classical production systems this rule is struck out). This significantly reduces the number of rules to be processed and, as a result, significantly reduces the time calculations, what is critical for the AIUS. In rules (22), inputs LVs have names *CF_1*, CF_Pr_1 and CF_Pr_2. Each of them has three terms *LOW*, *NO* and *HIGH* with trapezoidal and triangular MFs defined on the universe $-1 \leq cf \leq +1$. The LV *Energy_Reserve* is also defined by the three terms LOW, NO and *HIGH*, but already on the domain scale *[0, 100]* of the level of charge of the AIUS accumulator battery. Output LV *CF_Move* is set by three singletons *4*, *A*, *NO*. The rules use CF_Pr_1 and CF_Pr_2 LVs which are the IMKGs of Pr_1 and Pr_2 KGs described the fuzzy belief that EF matches the Pr_1 and Pr_2 prototypes, respectively. The input numerical variables of these LVs are cf_{Pr}, which are calculated by FM algorithm.

This example shows that the capabilities of the FLS can be extended to solve problems that require data about events stream of arbitrary length for logical inference. Distillation by convolution of the events footprint while preserving the meaning of the events stream opens up such possibilities.

6.3 Model of Multi-purpose Continuous Planning and Goal-Driving Control

The models discussed in Sects. 4 and 5 distillate knowledge that, at a high level of generalization, answer the question "What Is This". Sections 6.1 and 6.2 above show that such distillates open up the possibility of using FLS for AIUS decision making. However, the FLS supported by P&AD does not solve the main task of AIUS: continuous situational planning and management of implementing a multi-stage plan [26]. Solving this problem requires combining two types of knowledge "What Is This" and "How Do It". This subsection introduces a new KG type model that defines "How Do It" and proposes an approach to reconcile its inputs with the distillation results provided by the "What Is This" KG type.

The "How Do It" knowledge is structured and presented by KG, too. The knowledge portion describes elementary action that leads to the achievement of a local goal under conditions satisfied. This type of KG has called Goal-Driven KG (GDKG). The GDKG model has difference from KG because GDKG defines the causal relationship. Causal relationship describes what the AIUS does with environmental objects in order to achieve a local goal. The GDKG model rely on representation of binary

Table 4 EMKG of GDKG: prototypes definition

ID	Goal	Object	US State	Action
Move_to	DtOIt, FSCr	LPA	CsMF, TbS, Angl0	DvCsFQ
Load	LPA	A, ConLU	ChS, TbS	Wait
Unload	LPA	B, ConLU	ChS, TbS	Wait
Search	LPA, DtOF	FSCr	ChS, TbL	TrnTbL
Align	LPA, RcFs	HsIdI, DtOF	ChR, TbS	DvCsL

relationship between the object of environment and AIUS. Three components represent relationship in EMKG of GDKG. These are the object of environment, the AIUS state as a second object, and the relation itself. In the EMKG model, directly relationship is represented by the prototype $\boldsymbol{Pr}_{Goal}$ of the desired (target) state of the relationship. The object, as a second component, is represented by a prototype $\boldsymbol{Pr}_{Obj}$, too. It identifies the object interacting, and, as an option, a fragment of the history of object state changes specified by $\boldsymbol{Pr}_{STM\_obj}$ prototype. The AIUS state required to achieve the local goal, as the third component, is a prototype $\boldsymbol{Pr}_{UState}$, and, as an option, prototype of control actions history $\boldsymbol{Pr}_{STM\_Act}$ may be pointed. Beside components of relationship, the EMKG specifies the AIUS action $\boldsymbol{Pr}_{Act}$ that results in the local target. Thus, an EMKG model of GDKG with N sign model is the tuple (23).

$$<N, Pr_{Goal}, (Pr_{Obj}, [Pr_{STM\_obj}]), (Pr_{UStat}, [Pr_{STM\_Act}]), Pr_{Act}>. \quad (23)$$

A prototypes in (23) are vectors $\boldsymbol{Pr} = (cf^*(g_i), i = 1, 2, \ldots, k, g_i \in \Omega_{WITKB})$, in which, in contrast to (7), only the CF $cf^*(g_i)$ values are indicated. What is important, the all prototypes include references to KG from only "What Is This" KB.

The GDKG model is illustrated in the example of mobile robot, discussed in Sect. 5, additionally equipped with sensors unit placed on turntable. Sensors unit includes video camera and ultrasonic distance sensor. On the robot chassis bottom is fixed a reflection sensors matrix. In Fig. 9, the AE of AIUS shown. The route intersections are identified by numbers from 1 to 5. The AE ground is marked. The center lines of road and intersections are the marks along which the robot moved. The mission of AIUS is carrying of cargo from position *A* to position *B*. The explanation of GDKG model will be done on the following subset of KGs from WITKB. Data is given as a pairs (sign model, verbal explanation): *(**FSx**, Crossroad/Charger Floor Sign), (**RcFs**, Road center Floor sign), (**HsIdI**, Hanging sign of the Intersection), (**X**, Identifier of Intersection), (**DtOx**, Linguistic estimation of Distance to an Object is In touch/Close/Near/Middle/Far), (**Anglx**, Angle between the rails of the robot chassis and the turntable)*. Below is an example of KGs which EMKG are AIUS action: *(**DvCsxyz**, Drive the chassis forward/backward, slowly/quickly, turn right/left), (**TrnTbx**, Turn the turntable right/left), (**Csxyz**, Chassis is stopping/moving, forward/backward, slowly/quickly), (**Tbxy**, Turntable is* *stopping/rotating, right/left)*. In Table 4, the EMKG of GDKG representation examples are shown according to (23). In Table 4 in prototypes, the GDKG names are indicated only, and the values of $cf^*(g_i)$ are omitted for the compactness.

The plan of experiments provided that the AIUS mission performance require the following needs. In order to obtain missing information by studying environment is required the "scrutiny" need. "Self-preservation" need helps AIUS to avoid damage of its equipment, and the "consumption" need will allow to maintain the level of recharging the batteries. Table 4 shows GDKGs of mission (*Move_to*, *Loading*, *Unloading*) and two granules *Search* and *Align* of scrutiny need. The GDKG *Move_to* controls the implementation of robot movement between two adjacent intersections. In Table 4 in line *Move_to*, the EMKG describes the moving along the marking line under the PID control algorithm (*DvCsFQ* action granule). To implement this action, it is required that AIUS state is: (1) chassis is moving forward (*CsMF*), and (2) angle between the chassis rails and the turntable is *0* (*Angl0*), and (3) turntable is stop (*TbS*).

Until the local goal in the "*Goal*" column in Table 4 specified is reached, action will be active. Distance from intersection, spotted both ground (*FACe*) and overhead (*DrOIt*) signs, to robot is local goal specified by linguistic estimation.

The computation model of the IMKG of GDKG is presented by $\boldsymbol{IMKG}_{Goal \cup Obj \cup LPA}$, $\boldsymbol{FP}_{STM\_obj \cup STM\_Act}$ inputs and $\boldsymbol{IMKG}_{Action}$ output vectors, and $\boldsymbol{State}_{GDKG} = (cf_{State}, cf_{Goal}, cf_{Attention}, cf_{Cond}, cf_{Action})$ state vector. All these fuzzy characteristics are CFs and defined on the $(-1 \le cf(g_i) \le +1)$ universe. The GDE, PLE, ATE and EME, when computing IMKG of GDKG, uses the system state vector, which includes T1 FSs characterizing emotions (***Emotion***), context (***Context***) and needs (***Need***).

The IMKG of GDKG (IMGD) algorithm is given below.

(1) **Turn on attention**. In living nature, attention mechanism plays important role in parallel processing data from sensors. It localizes data from certain fragment of environment required to achieve the local goal. This attention mechanism is borrowed and used for changing the GDKG state. In depend of two factors, the GDKG attention is excited. Firstly, GDKG must be relevant to meet the current need. As the AIUS needs are given by FSs ***Need***, the attention value depends on the its MF. Secondly, GDKG must be relevant to meet the context. Based on the state of the current implementation of the AIUS action plan, the context presented by MF of the FS ***Context*** characterizes the next plan stage suited to situation. The CF of attention computes by the expression (24).

$$\alpha_{Attention} = -1 + (\mu_{Need} + \mu_{Emotin^+} - \mu_{Need} \cdot \mu_{Emotin^+}) + \mu_{Context}. \quad (24)$$

In brackets in (24), the algebraic sum operation carries out over a pair of FSs ($\boldsymbol{Need} \oplus \boldsymbol{Emotion}^+$).

(2) **Compute the GDKG state**. Preliminarily, the distances $\alpha_\rho(g)$ are computed by (10) for all granules g listed in (25). Distance $\alpha_\rho(g)$ characterizes matching of the current situation represented by the IMKG of the gth KG of WITKB

to the prototype specified in (23). The CF of the GDKG state is computed by (25).

$$\alpha_{State} = MIN(-1 \cdot \alpha_{Goal}, \alpha_{Cond}), \tag{25}$$

where

$$\alpha_{Goal} = MIN(\alpha_{Attention}, ((\sum_{\forall g \in \Omega_{Goal}} \kappa_g \cdot \alpha_\rho(g)) \oplus \mu_{Emotion\_3})),$$

$$\alpha_{Cond} = MIN(\alpha_{Attention}, ((\sum_{\forall g \in \Omega'} \kappa_g \cdot \alpha_\rho(g)) \oplus \mu_{Emotion\_2})),$$

$$\sum_{\forall g \in \Omega'} \kappa_g = 1, \sum_{\forall g \in \Omega_{Goal}} \kappa_g = 1, \Omega' = \Omega_{Obj} \cup \Omega_{LPA} \cup \Omega_{UStat} \cup \Omega_{STM}.$$

According to (25), the relevant GDKG ($\alpha_{Attention} \approx +1$) get in an excited state when the IMKG of all KGs from WITKB match to the prototypes (23) ($\alpha_{Cond} \approx +1$), and the goal is not reached ($\alpha_{Goal} \approx -1$). When the goal is reached, GDKG get out from an excited state and get in an inhibited one. Goal achievement is identified by high IMKG value of all KGs from WITKB describing the goal (23). Figure 10 shows these cases.

(3) **Compute the action state**. The IMKG of KGs belonging to the set of control actions Ω_{Act} of the EMKG defined in (23) are computed by (26).

$$\alpha_{Action}(g) = \alpha_{IMKG}, \alpha_{IMKG} = \alpha_{State} \cdot (1 - \mu_{Emotin^-}). \tag{26}$$

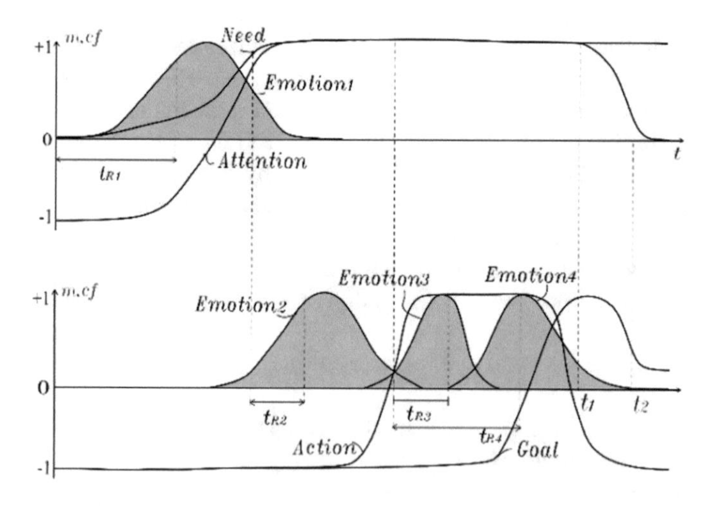

Fig. 10 Graphic illustration of the conditions for the emergence of emotions

The states KGs not belonged to Ω_{Act} have not changed. Since other GDKGs independently "offers its" control actions, the resulting control is chosen according to the principle "winner-take-all". So, if only no actions with a large value of $cf_{Action}(g)$, the action excited by GDKG under consideration will be implemented. In (26) shown, that negative emotion corrects the IMKG by the value of MF of $Emotion^{-}$. Emotional linear correction of IMKG for both positive and negative values of α_{State} causes a reducing confidence in the decision $\alpha_{IMKG} \rightarrow 0$ when emotion increase $\mu_{Emotion} \rightarrow 1$.

As can be seen from (24)–(26), the IMKG implementation is supported by several engines: GDE, PLE, ATE and EME. They perform independently and separately under certain conditions. Each engine checks the needs, context and GDKG states vectors and under satisfied conditions implements its function. Below we will briefly consider the role and functioning of the two PLE and EME engines, since the functions of GDE and ATE can be seen directly from (24)–(26). PLE produces plan of actions in accordance with the actual need and situation. Its functions include to manage implementation of plan and to check the vector of context. PLE manages plans independently for each need. The precedent of AIUS behavior leading to achievement of global goal is organized as an action plan consisting of individual stages [26]. One GDKG providing elementary action to achieve the local goal, implements one stage of plan. A plan is represented by a tuple $<ID, (GDKG_j, KG_i, n)>, n = 0, 1, 2, \ldots, l>$, where ID is the plan identifier, n is a serial number of plan stage, KG_i is the name of KG in WITKB, which is pointed as LPA in EMKG of $GDKG_j$ and specified explicitly in the plan. Example under consideration, the plan of mission implementation (Fig. 9) as an ordered sequence $(Load\ CrgL)^0 \rightarrow Move\_to\ A^1 \rightarrow Move\_to\ 4^2 \rightarrow Move\_to\ 5^3 \rightarrow Move\_to\ B^4 \rightarrow (Unload\ CrgU)^5$ are shown in Table 5 in the form of a set of tuples. In mentioned above plan, the GDKG names are from Table 4, and the KG IDs are from WITKB, and the superscript shows the stage number in the plan. At each moment of time, fuzzy number $\boldsymbol{Context}_{PLi}$ characterizes the current active stage of the plan. The MF of this fuzzy number is defined on the universe of ordered numbers $0, 1, \ldots, l$ of stages of plan. The PLE manages the $\boldsymbol{Context}_{PLi}$ values according to the changing of stages of the plan. Before computing $\alpha_{Attention}$ in (24), PLE calculates $\mu_{Context} = \mu_{ContextPLi}(n)$ as a fuzzification operation for number n which is the ordinal number of the GDKG stage in the plan.

EME plays main role in supporting of autonomy of DM&C in AIUS. In situations when there is a danger of violating any phase of achieving a local goal, the EME helps to switch AIUS behavior. We have introduced six kinds of emotions for different

Table 5 AIUS action plans: mission (1st line) and scrutiny need (2nd and 3rd lines)

ID	Plan stages
1	(Load, CrgL, 0), (Move_to, A, 1), (Move_to, 4, 2), (Move_to, 5, 3), (Move_to, B, 4), (Unload, CrgU, 5)
2	(Search, HsIdI, 0), (Align, Angl0, 1)
3	(Move_to, HsIdI, 0)

situation violating the implementation of plan. In Fig. 10, shown the emergence conditions of four of the six emotions. The model of emotion is FS with MF defined in the universe of the time delay. It means, the numerical assessment of emotion depends on time delay until the new decision may be found overcame violation. Thus, the emotion assessment is value of MF $\mu_{Emotion}(t_R)$ of FS *Emotion*, where t_R is time delay parameter of CF of KG (3), (4). The FS of emotion is defined by expert.

The *Emotion_1* (impatience, restlessness) emergences when a long time the need slowly grows in a range of small values and there are no other exciting needs. In this situation according (24), there is no possibility to turn on attention, and, consequently, there is no GDKG excited. The *Emotion_1* "helps" turn on attention that opening the opportunity satisfy the need in advance. In (24), the value of the emotion MF is added to the need CF $\mu_{Emotion+} = \mu_{Emotion\_1}(t_{R1})$, where t_{R1} is the CF parameter of ***Attention*** fuzzy number (time interval since the last change of the value of $\alpha_{Attention}$, Fig. 10).

The *Emotion_2* (anxiety) emergences when attention is turned on but the data from the sensors is not enough to identify the condition Ω' in (25) required to excite the GDKG. Two reasons cause this situation. Firstly, the lack or incompleteness of data about object Ω_{Obj} or relation component Ω_{LPA}, Ω_{UStat} in (25) produce this situation. In this case, to obtain additional data from environment the emotion initiates the search behavior. To do this, the scrutiny need is grown by $\mu_{Scrutiny} = \mu_{Emotion\_2}(t_{R2})$). Secondly, the prototype of situation isn't identified due to the violation of data from sensors by noises. The lack of certainty of data obtained is compensated by the emotion $\mu_{Emotion\_2} = \mu_{Emotion\_2}(t_{R2})$ in (25).

Emotion_3 (goal anticipation) occurs when GDKG is waiting for data describing the goal, and data about environment, object and AIUS state have matched with situation prototype (23), and actions have activated. In other words, AIUS is in process of achieving the goal. To compensate for the low certainty in target localization due to insufficient or noisy sensor data, *Emotion_3* increases the CF value in (25) by $\mu_{Emotion\_3} = \mu_{Emotion\ 3}(t_{R2})$.

The other three emotions *Emotion_4* (worry in goal anticipation), *Emotion_5* (frustration) and *Emotion_6* (satisfaction) are defined similarly.

In Table 6, shown some experiment results with the multi-purpose continuous planning and goal-driving control of AIUS. The situation under describing, the AIUS has arrived at intersection *1* from loading position *A*. The *1*st stage of plan *1* in Table 5 is implementing. In Table 6, two last columns are data of mission, need *Scrutiny*, and the first three emotions discussed above. For the sake of compactness, in Table 6 doesn't include MFs of full history of mission, but only those that demonstrate the switching behavior of AIUS. The MFs values are rounded to one digit after zero and zero is omitted. For example, the notation *–0.9* corresponds to the *–0.92*.

Table 6 in 1st row, data corresponding to implementation of the 1st stage of the mission plan have shown. The mission and need MFs have the values $\mu_{Mission} = 0.8$, $\mu_{Scrutiny} = 0$. The GDKG *Move_to* has excited and the values of parameters computed by IMGD algorithm are $\alpha_{Attention} = 0.8$, $\alpha_{Goal} = -0.9$, $\alpha_{Cond} = 0.8$, $\alpha_{State} = 0.8$, $\alpha_{Act} = 0.8$. Movement along the marking line is realizing under control of excited action, and GDKG is in process of anticipation of the goal. When the local

Table 6 Fragment of the experiment with the multi-purpose continuous planning and goal-driving control of AIUS

Step	*Move_to*[a]	*Search*[a]	*Align*[a]	Need[b]	Emotion[c]
1	0.8, –0.9, 0.8, 0.8, 0.8	–1, –1, –1, –1, –1	–1, –1, –1, –1, –1	0.8, 0	0, 0, 0
2	0.8, 0.8, 0.8, –0.8, –0.8	–1, –1, –1, –1, –1	–1, –1, –1, –1, –1	0.8, 0	0, 0, 0
3	0.8, –0.9, 0.02, 0.02, 0.02	–1, –1, –1, –1, –1	–1, –1, –1, –1, –1	0.8, 0	0, 0, 0
4	0.8, –0.9, 0.02, 0.02, 0.02	0.7, –0.7, 0.7, 0.7, 0.7	–1, –1, –1, –1, –1	0.8, 0.4	0, 0.5, 0
5	0.8, –0.9, 0.02, 0.02, 0.02	–0.6, 0.9, 0.7, –0.5, –0.5	0.6, –0.9, 0.9, 0.6, 0.6	0.8, 0.6	0, 0.5, 0
6	0.8, –0.9, 0.8, 0.9, 0.9	7, –0.7, 0.7, 0.7, 0.7	–1, 0, 0, 0, 0	0.8, 0	0, 0, 0

[a] $\alpha_{Attention}$, α_{Goal}, α_{Cond}, α_{Stat}, α_{Act}
[b] $\mu_{Mission}$, $\mu_{Scrutiny}$
[c] $\mu_{Emotion\ 1}$, $\mu_{Emotion\ 2}$, $\mu_{Emotion\ 3}$

goal has been reached $\alpha_{Goal} = 0.84$ (local goal presents by data from hanging sign $ID_{Intersection} = 1$, $\alpha_{DtOIt} = 0.9$ and ground intersection sign $\alpha_{FSCr} = 0.8$), the GDKG state is set to $\alpha_{State} = -0.8$. That deactivates the control $\alpha_{Act} = -0.8$ (2nd row of the Table 6). In 3rd row shown the computing results of 2nd stage of the mission plan. At the beginning stage the situation doesn't satisfier prototype because according to (25) $\alpha_{Cond} = 0.02$ (video camera is directed to intersection *2* but no *4*, then $\alpha_{LPA=4} = -1$, and the rest of the data from the perception system are $\alpha_{CsMF} = -0.9$, $\alpha_{TbS} = 1$, $\alpha_{Angl0} = 1$). No actions can be initiated due to GDKG is in neutral state $\alpha_{Stat} = 0.02$. This GDKG state kept for a certain time interval (Fig. 9) until the $\mu_{Scrutiny}$ and $\mu_{Emotion\_2}$ reach the level shown on row 4. The *0*th stage of the scrutiny need plan *1* has activated. This excites the action *TrnTbL* as $\alpha_{Act} = 0.7$. The left rotation of the turntable is continuing until the goal $\alpha_{HsIdI} = 0.9$, $\alpha_{DtOF} = 0.9$ will be met. In Table 6 the *5*th row, the goal is reached and PLE change the context so that next *1*st stage of plan *1* is relevant now $\mu_{ContextPL2}(0) = -1$, $\mu_{ContextPL2}(1) = 1$. In 5th row, to terns on the action of chassis rotation without changing the turntable direction, the GDKG *Align* is excited. Execution of this action continues until their directions coincide, that is until goal will be achieved (*Angl0*). In Table 6, 6th row show data that simultaneously satisfies the conditions of both plans: mission (GDKG *Move_to* of the 2nd stage) and the scrutiny need (GDKG *Search* of the 1st stage). Action excite level of mission higher than scrutiny need (6th row), the GDE choose action of GDKG *Move_to* of mission plan. On this step the MF value of emotion $\mu_{Emotion\_2} = 0$ (Fig. 10), and hence the need has value $\mu_{Scrutiny} = 0$. It is the PLE condition for deactualization of the scrutiny need plan.

7 Conclusion and Future Works

In this chapter, for AIUS decision-making in unordered environment, have been proposed a novel knowledge distillation model based on integrating two paradigms "data from sensor" and "computing with words". Through introduced universal models of external and internal meaning of knowledge granule, our approach allows to move from the tasks of distilling data from sensors and distilling semantic knowledge to the single task of sense distilling. This approach can significantly reduce the AIUS decision-making model, while maintaining the quality of decisions in complex dynamic situations. The experiments on the using distillate in the form of the internal meaning of knowledge in the tasks of FLS decision-making, continuous planning, and goal-driving control are performed. The results of the experiments demonstrate an increase in AIUS autonomy due to the ability to abstract and generalize, and flexible switching of action plans.

Further development of the sense distillation approach is planned in the following directions. At first, it is expansion autonomy AIUS due to possibility of learning in the operational (online real-time) mode in order to automatically form new KGs in the KB. Secondly, to create a conceptual and computational model of the Feeling AI to support the autonomy of AIUS decision making not only through the sense distillation and learning, but also through the implementation of the cognitive functions of attention, context and emotions. And, thirdly, to conduct a study of the possibilities and advantages of using the Feeling AI for decision-making applications of Edge Computing level that require maintaining a high level of autonomy.

References

1. Zhang, T., et al.: Current trends in the development of intelligent unmanned autonomous systems. Front. Inf. Technol. Electron. Eng. **18**, 68–85 (2017). https://doi.org/10.1631/FITEE.1601650
2. Chen, J., Sun, J., Wang, G.: From unmanned systems to autonomous intelligent systems. Engineering **12**, 16–19 (2022). https://doi.org/10.1016/j.eng.2021.10.007
3. Reis, J., Cohen, Y., Melao, N., Costa, J., Jorge, D.: High-tech defense industries: developing autonomous intelligent systems. Appl. Sci. **11**(11) (2021). https://doi.org/10.3390/app11114920
4. Shakhatreh, H., et al.: Unmanned aerial vehicles (UAVs): a survey on civil applications and key research challenges. J. IEEE Access **7**, 48572–48634 (2019). https://doi.org/10.1109/ACCESS.2019.2909530
5. Unmanned Systems: NovAtel (2022). https://novatel.com/industries/unmanned-systems. Accessed 20 Sept 2022
6. Lockheed, M.: The future of autonomy. Isn't human-less. It's human more (2022). https://www.lockheedmartin.com/en-us/capabilities/autonomous-unmanned-systems.html. Accessed 25 Sept 2022
7. Rasmussen, S., Kingston, D., Humphrey, L.: Brief Introduction to Unmanned Systems Autonomy Services (UxAS). International Conference on Unmanned Aircraft Systems (ICUAS) (2018). https://doi.org/10.1109/ICUAS.2018.8453287

8. Litman, T.A.: Autonomous Vehicle Implementation Predictions: Implications for Transport Planning. Victoria Transport Policy Inst. Rep. (2022). https://www.vtpi.org/avip.pdf. Accessed 25 Sept 2022
9. Joseph, L., Mondal, A.K. (eds.): Autonomous Driving and Advanced Driver-Assistance Systems (ADAS). Applications, Development, Legal Issues, and Testing, 1st edn. CRC Press, Boca Raton (2021). https://doi.org/10.1201/9781003048381
10. Yasumoto, K., Yamaguchi, H., Shigeno, H.: Survey of real-time processing technologies of IoT data streams. J. Inf. Process. **24**(2), 195–202 (2016). https://doi.org/10.2197/ipsjjip.24.195
11. Klein, L.A.: Sensor and Data Fusion: A Tool for Information Assessment and Decision Making, 2nd edn (2012). SPIE Press, Bellingham. https://doi.org/10.1117/3.928035
12. Gou, J., Yu, B., Maybank, S.J., Tao, D.: Knowledge distillation: a survey. Int. J. Comput. Vision **129**, 1789–1819 (2021). https://doi.org/10.1007/s11263-021-01453-z
13. Hu, X., Shen, Y., Pedrycz, W., Li, Y., Wu, G.: Granular fuzzy rule-based modeling with incomplete data representation. IEEE Trans. Cybernet. **52**(7), 6420–6433 (2022). https://doi.org/10.1109/TCYB.2021.3071145
14. Winfield, A.T., et al.: IEEE P7001: a proposed standard on transparency. Front. Robot. AI **8** (2021). https://doi.org/10.3389/frobt.2021.665729
15. Siemens: Next-Gen AI Manufacturing/Industrial Sector (2021). https://siemens.com/innovation. Accessed 10 Sept 2022
16. Michels, K., Klawonn, F., Kruse, R., Nürnberger, A.: Fundamentals of control theory. In: Fuzzy Control. Studies in Fuzziness and Soft Computing, vol. 200. Springer, Heidelberg, pp. 57–234 (2006). https://doi.org/10.1007/3-540-31766-x_2
17. Russell, S.J., Norvig, P.: Artificial Intelligence. A Modern Approach, 3rd edn. Pearson Education, Upper Saddle River (2010)
18. Peralta, F., Arzamendia, M., Gregor, D., Reina, D.G., Toral, S.: A comparison of local path planning techniques of autonomous surface vehicles for monitoring applications: the Ypacarai Lake case-study. Sensors **20**(5) (2020). https://doi.org/10.3390/s20051488
19. Leonetti, M., Iocchi, L., Stone, P.: A synthesis of automated planning and reinforcement learning for efficient, robust decision-making. Artif. Intell. **241**, 103–130 (2016). https://doi.org/10.1016/j.artint.2016.07.004
20. Liu, H., Li, X., Fan, M., Wu, G., Pedrycz, W., Suganthan, P.N.: An autonomous path planning method for unmanned aerial vehicle based on a tangent intersection and target guidance strategy. IEEE Trans. Intell. Transport. Syst. **23**(40), 3061–3073 (2022). https://doi.org/10.1109/TITS.2020.3030444
21. Cheng, X., Rao, Z., Chen, Y., Zhang, Q.: Explaining knowledge distillation by quantifying the knowledge. In: 2020 IEEE/CVF Conference on Computer Vision and Pattern Recognition (CVPR) (2020). https://doi.org/10.1109/CVPR42600.2020.01294
22. Pedrycz, W., Chen, S.M.: Deep Learning: Concepts and Architectures. Studies in Computing Intelligence, vol. 866. Springer, Cham (2020). https://doi.org/10.1007/978-3-030-31756-0
23. West, P., et al.: Symbolic Knowledge Distillation: From General Language Models to Commonsense Models (2022). https://doi.org/10.48550/arXiv.2110.07178
24. Ramamurthy, P., Aakur, S.N.: ISD-QA: iterative distillation of commonsense knowledge from general language models for unsupervised question answering. In: 26th International Conference on Pattern Recognition (ICPR), pp. 1229–1235 (2022). https://doi.org/10.1109/ICPR56361.2022.9956441
25. Kargin, A., Petrenko, T.: Spatio-temporal data interpretation based on perceptional model. In: Mashtalir, V., Ruban, I., Levashenko, V. (eds.) Advances in Spatio-Temporal Segmentation of Visual Data. Studies in Computational Intelligence, vol. 876, pp. 101–159. Springer, Cham (2020). https://doi.org/10.1007/978-3-030-35480-0
26. Kargin, A., Petrenko, T.: Multi-level computing with words model to autonomous systems control. In: Pakstas, A., Hovorushchenko, T., Vychuzhanin, V., Yin, H., Rudnichenko, N. (eds.) CEUR Workshop Proceedings, vol. 2711, pp. 16–30 (2020). http://ceur-ws.org/Vol-2711/paper2.pdf

27. Zhu, Z., Pedrycz, W., Li, Z.: Construction and evaluation of information granules: from the perspective of clustering. IEEE Trans. Syst. Man Cybernet. Syst. **52**(30), 2024–2037 (2022). https://doi.org/10.1109/tsmc.2020.3035605
28. Pedrycz, W.: From data to information granules: an environment of granular computing. In: 2021 IEEE 20th International Conference on Cognitive Informatics & Cognitive Computing (ICCI*CC) (2021). https://doi.org/10.1109/iccicc53683.2021.9811327
29. Huang, M., Rust, R.: Artificial intelligence in service. J. Service Res. **21**(2), 155–172 (2018). https://doi.org/10.1177/1094670517752459
30. Czerwinski, M., Hernandez, J., Mcduff, D.: Building an AI that feels: AI systems with emotional intelligence could learn faster and be more helpful. IEEE Spectr. **58**(5), 33–38 (2021). https://doi.org/10.1109/MSPEC.2021.9423818
31. Chen, H., et al.: From automation system to autonomous system: an architecture perspective. J. Marine Sci. Eng. **9**(6) (2021). https://doi.org/10.3390/jmse9060645
32. Thompson, N.C., Greenewald, K., Lee. K., Manso, G.F.: The Computational Limits of Deep Learning. Cornell University (2022). https://doi.org/10.48550/arXiv.2007.05558
33. Singer, G.: Thrill-K: a blueprint for the next generation of machine intelligence. Towardsdatascience.com (2022). https://towardsdatascience.com/thrill-k-a-blueprint-for-the-next-generation-of-machine-intelligence-7ddacddfa0fe. Accessed 10 Sept 2022
34. Nordhoff, S., Kyriakidis, M., Arem, B., Happee, R.: A multi-level model on automated vehicle acceptance (MAVA): a review-based study. Theor. Issues in Ergonom. Sci. **20**(6), 682–710 (2019). https://doi.org/10.1080/1463922X.2019.1621406
35. Bachute, M.R., Subhedar, J.M.: Autonomous driving architectures: insights of machine learning and deep learning algorithms. Mach. Learn. Appl. **6** (2021). https://doi.org/10.1016/j.mlwa.2021.100164
36. Macoir, J., Hudon, C., Tremblay, M., Laforce, R.J., Wilson, M.A.: The contribution of semantic memory to the recognition of basic emotions and emotional valence: evidence from the semantic variant of primary progressive aphasia. Soc. Neurosci. **14**(6), 705–716 (2019). https://doi.org/10.1080/17470919.2019.1577295
37. Al-Shihabi, T., Mourant, R.: Toward more realistic driving behavior models for autonomous vehicles in driving simulators. Transp. Res. Rec. J. of the Transp. Res. Board **1843**(1) (2003). https://doi.org/10.3141/1843-06
38. Bach, J.: Principles of Synthetic Intelligence: Psi: An Architecture of Motivated Cognition, 1st edn. Oxford University Press, New York (2009)
39. Chen, S.M., Ko, Y.K.: Fuzzy interpolative reasoning for sparse fuzzy rule-based systems based on α-cuts and transformations techniques. IEEE Trans. Fuzzy Syst. **16**(6), 1626–1648 (2008). https://doi.org/10.1109/TFUZZ.2008.2008412
40. Garcia, G., Luengo, J., Herrera, F.: Data Preprocessing in Data Mining. Intelligent Systems Reference Library (2015). Springer, Cham. https://doi.org/10.1007/978-3-319-10247-4
41. Hag, A.U., Zhang, G., Peng, H., Rahman, S.U.: Combining multiple feature-ranking techniques and clustering of variables for feature selection. IEEE Access **7**, 151482–151492 (2019). https://doi.org/10.1109/ACCESS.2019.2947701
42. Waylay: How to Choose a Rules Engines. Whitepaper (2019). https://www.waylay.io/publications/how-to-choose-a-rules-engine-seven-things-to-look-at-when-automating-for-iot. Accessed 20 Sept 2022
43. AWS: AWS IoT Core: Developer Guide. Amazon Web Services, Inc. (2022). https://docs.aws.amazon.com/iot/latest/developerguide/iot-dg.pdf#iot-rules. Accessed 20 Sept 2022
44. Seung-Kyu, Y., Sang-Young, C.: A light-weight rule engine for context-aware services. KIPS Trans. Softw. Data Eng. **5**(2), 59–68 (2016). https://doi.org/10.3745/ktsde.2016.5.2.59
45. Piegat, A.: Fuzzy Modelling and Control. Studies in Fuzziness and Soft Computing. Physica Heidelberg, New York (2001). https://doi.org/10.1007/978-3-7908-1824-6
46. Kargin, A., Petrenko, T.: Internet of things smart rules engine. In: 2018 IEEE International Scientific-Practical Conference on Problems of Infocommunications. Science and Technology (PIC S&T), pp. 639–644 (2018). https://doi.org/10.1109/infocommst.2018.8632027

47. Kargin, A., Petrenko, T.: Knowledge representation in smart rules engine. In: 3rd IEEE International Conference on Advanced Information and Communication Technologies (AICT 2019), pp. 231–236 (2019). https://doi.org/10.1109/aiact.2019.8847831
48. Kargin, A., Petrenko, T.A.: Abstraction as a way of uncertainty representation in Smart Rules Engine. In: 2019 XIth International Scientific and Practical Conference on Electronics and Information Technologies (ELIT), pp. 136–141 (2019). https://doi.org/10.1109/ELIT.2019.8892321
49. Clancey, W.J.: Heuristic classification. Artif. Intell. **27**(3), 289–350 (1985). https://doi.org/10.1016/0004-3702(85)90016-5
50. Chandrasekaran, B.: Generic tasks in knowledge-based reasoning: high-level building blocks for expert systems design. IEEE Expert **1**(3), 23–30 (1986). https://doi.org/10.1109/mex.1986.4306977
51. Jackson, P.: Introduction to Expert Systems, 3rd edn. Addison-Wesley, Boston (1998)
52. Pedrycz, W., Chen, S. (eds.): Granular Computing and Intelligent Systems. Design with Information Granules of Higher Order and Higher Type. Springer (2011). https://doi.org/10.1007/978-3-642-19820-5
53. Zadeh, L.A.: Generalized theory of uncertainty (GTU)—principal concept and ideas. Comput. Stat. Data Anal. **51**(1), 15–46 (2006). https://doi.org/10.1016/j.csda.2006.04.029
54. Sun, L., Xu, J., Wang, C., Xu, T., Ren, J.: Granular computing-based granular structure model and its application in knowledge retrieval. Inf. Technol. J. **11**(12), 1714–1721 (2012). https://doi.org/10.3923/itj.2012.1714.1721
55. Skowron, A., Jankowski, A., Dutta, S.: Interactive granular computing. Granular Comput. **1**(2), 95–113 (2016). https://doi.org/10.1007/s41066-015-0002-1
56. Liu, A., Gegov, A., Cocea, M.: Rule-based systems: a granular computing perspective. Granular Comput. **1**, 259–274 (2016). https://doi.org/10.1007/s41066-016-0021-6
57. Zadeh, L.A.: Computing with Words. Principal concepts and ideas. Studies in Fuzziness and Soft Computing, vol. 277. Springer, Berlin (2012). https://doi.org/10.1007/978-3-642-27473-2
58. Mendel, J.M., Wu, D.: Perceptual Computing: Adding People in Making Subjective Judgments. IEEE Press, Piscataway (2010). https://doi.org/10.1002/9780470599655
59. Zadeh, L.A.: Toward a theory of fuzzy information granulation and its centrality in human reasoning and fuzzy logic. FSs Syst. **90**(2), 111–127 (1997). https://doi.org/10.1016/s0165-0114(97)00077-8
60. Zadeh, L.A.: Toward a restriction-centered theory of truth and meaning (RCT). In: Magdalena, L., Verdegay, J., Esteva, F. (eds.) Enric Trillas: A Passion for FSs. A Collection of Recent Works on Fuzzy Logic. Studies in Fuzziness and Soft Computing, vol. 322, pp. 1–22. Springer, Cham (2015). https://doi.org/10.1007/978-3-319-16235-5_1
61. Gupta, P.K., Andreu-Perez, J.: A gentle introduction and survey on computing with words (CWW) methodologies. Neurocomputing **500**, 921–937 (2022). https://doi.org/10.1016/j.neucom.2022.05.097
62. Chen, J., Pham, T.T.: Introduction to FSs, Fuzzy Logic, and Fuzzy Control Systems, 1st edn. CRC Press, Boca Raton (2020). https://doi.org/10.1201/9781420039818
63. Castillo, O., Melin, P., Kacprzyk, J.: Design of Intelligent Systems Based on Fuzzy Logic, Neural Networks and Nature-Inspired Optimization. Studies in Computational Intelligence. Springer, Cham (2015). https://doi.org/10.1007/978-3-319-17747-2
64. Mendel, J.M.: Uncertain Rule-Based Fuzzy Systems: Introduction and New Directions, 2nd edn. Springer, Cham (2017). https://doi.org/10.1007/978-3-319-51370-6
65. Babuska, R., Mamdani, E.A.: Fuzzy control. Scholarpedia **3**(2), 2103 (2008). https://doi.org/10.4249/scholarpedia.2103
66. Prokopowicz, P., Czerniak, J., Mikolajewski, D., Apiecionek, L., Slezak, D.: Theory and Applications of Ordered Fuzzy Numbers: A Tribute to Professor Witold Kosinski, 1st edn. Studies in Fuzziness and Soft Computing. Springer, Cham (2017). https://doi.org/10.1007/978-3-319-59614-3
67. Sengupta, A., Pal, T.K.: Fuzzy preference ordering of intervals. In: Fuzzy Preference Ordering of Interval Numbers in Decision Problems. Chapter 4. Studies in Fuzziness and Soft Computing,

vol. 238, pp. 59–89. Springer, Berlin, Heidelberg (2009). https://doi.org/10.1007/978-3-540-89915-0_4

68. Stefanini, L., Sorini, L., Guerra, M.L.: Fuzzy numbers and fuzzy arithmetic. In: Pedrycz, W., Skowron, A., Kreinovich, V. (eds.) Handbook of Granular Computing, Chapter 12, pp. 249–283. Wiley (2008). https://doi.org/10.1002/9780470724163.ch12
69. Zhang, Z., Chen, S.M.: Group decision making with incomplete q-rung orthopair fuzzy preference relations. Inf. Sci. **535**, 376–396 (2021). https://doi.org/10.1016/j.ins.2020.10.015
70. Kargin, A., Petrenko, T.: Method of using data from intelligent machine short-term memory in fuzzy logic system. In: 2021 IEEE 7th World Forum on Internet of Things (WF-IoT) (2021). https://doi.org/10.1109/wf-iot51360.2021.9594918

Index

W. Pedrycz and S.-M. Chen (eds.), *Advancements in Knowledge Distillation: Towards New Horizons of Intelligent Systems*, Studies in Computational Intelligence 1100, https://doi.org/10.1007/978-3-031-32095-8

M

O

R

S

V